Smart Spaces and Places

Smart technologies have advanced rapidly throughout our society (e.g. smart energy, smart health, smart living, smart cities, smart environment, and smart society) and across geographic spaces and places. Behind these "smart" developments are a number of seminal drivers, such as social media (e.g. Twitter), sensors (drones, wearables), smartphone apps, and computing infrastructure (e.g. cloud computing). These developments have captured the enthusiasm of the public, while inevitably present unprecedented challenges and opportunities for the geographic research community. When meeting the smart challenges, are there emerging theories, methods, and observations that reveal new spatial phenomena, produce new knowledge, and foster new policies?

Smart Spaces and Places addresses questions such as how to make spaces and places "smart", how the "smartness" affects the way we think spaces and places, and what role geographies play in knowledge production and decision-making in a "smart" era. The collection of 21 chapters offers stimulating discussion over the meaning of spaces, places, and smartness; scientific insights into smartness; social-political views of smartness; and policy implications of smartness.

The chapters in this book were originally published as a special issue of *Annals of the American Association of Geographers*.

Ling Bian is Professor in the Department of Geography at the University at Buffalo, State University of New York, USA. Her current research interests include the conceptual frameworks to represent spatially dynamic phenomena, individual-based and spatially explicit epidemiological modeling, and network analytics.

Smart Spaces and Places

Edited by
Ling Bian

Routledge
Taylor & Francis Group

LONDON AND NEW YORK

First published 2021
by Routledge
2 Park Square, Milton Park, Abingdon, Oxon, OX14 4RN

and by Routledge
605 Third Avenue, New York, NY 10158

Routledge is an imprint of the Taylor & Francis Group, an informa business

Introduction, Chapters 1–17 and 19–21 © 2021 American Association of Geographers
Chapter 18 © 2020 Darcy Parks and Anna Wallsten. Originally published as Open Access.

British Library Cataloguing-in-Publication Data
A catalogue record for this book is available from the British Library

ISBN13: 978-0-367-70354-7 (hbk)
ISBN13: 978-0-367-70357-8 (pbk)
ISBN13: 978-1-003-14586-8 (ebk)

Typeset in Goudy Oldstyle Std
by codeMantra

Publisher's Note
The publisher accepts responsibility for any inconsistencies that may have arisen during the conversion of this book from journal articles to book chapters, namely the inclusion of journal terminology.

Disclaimer
Every effort has been made to contact copyright holders for their permission to reprint material in this book. The publishers would be grateful to hear from any copyright holder who is not here acknowledged and will undertake to rectify any errors or omissions in future editions of this book.

Contents

Citation Information

The chapters in this book were originally published in the *Annals of the American Association of Geographers*, volume 110, issue 2 (March 2020). When citing this material, please use the original page numbering for each article, as follows:

Contributors

Luis F. Alvarez León is Assistant Professor of Geography at Dartmouth College, Hanover, USA. He is a political economic geographer with substantive interests in geospatial data, media, and technologies. His research integrates the geographic, political, and regulatory dimensions of the informational and digital economy.

Abdullatif Alyaqout is PhD student in the Department of Geography at Texas State University, San Marcos, USA. His research interests include Cartography, Geovisualization, and applications of GIScience in Volunteered Geographic Information for natural hazards assessment and monitoring and citizen-government interaction studies.

Clinton Andrews is Professor of Urban Planning and Policy Development at Rutgers, The State University of New Jersey, New Brunswick, USA. His research addresses behavioral, policy, and planning questions related to how we use the built environment.

J. Colette Berbesque is Reader in the Department of Anthropology at the Centre for Research in Evolutionary and Inter-disciplinary Anthropology at Roehampton University, London, UK. Her interests are in human ecology; the evolution of the hominin diet; and the evolution of cooperation, prestige, and hierarchy.

Ling Bian is a Professor in the Department of Geography at the University at Buffalo, State University of New York, Amherst, NY 14261. Her research interests include GIScience, health behavior modeling, environmental modeling, and network analytics.

Angela Blatt is a Program Coordinator at the Initiative for Food and Agricultural Transformation at Ohio State University, Columbus OH 43210. Her interests include rural development, food security, farm animal welfare, and fair farming opportunities.

Yanwei Chai is Professor in the College of Urban and Environmental Sciences at Peking University, Beijing, P. R. China. His main research interests include behavioral geography and urban geography.

Kang Chen is an Assistant Professor in the Department of Electrical and Computer Engineering at Southern Illinois University, Carbondale, IL 62901. His research interests include emerging wireless networks, software-defined networking, and cloud computing.

Ximeng Cheng received his PhD in Cartology and GIScience from the School of Earth and Space Sciences at Peking University, Beijing, P. R. China. His research interests are in spatiotemporal data mining, social sensing, and deep learning.

T. Edwin Chow is Professor in the Department of Geography at Texas State University, San Marcos, USA. His research interests entail geocomputation, human dynamics, hazards, and environmental modeling.

Jeremy W. Crampton is Professor of Urban Data Analysis at Newcastle University, Newcastle Upon Tyne, UK. His research interests are in surveillance, spatial big data, and algorithmic decision-making.

Craig Dalton is Assistant Professor in the Department of Global Studies and Geography at Hofstra University, USA. His research interests include countermapping for social justice as well as critical data studies of personal location data and way-finding.

Vincent J. Del Casino Jr. is Provost and Senior Vice President for Academic Affairs and Professor of Urban and Regional Planning at Office of Provost, San José State University. His research interests include social and cultural, health geography, geographic thought and history, sexuality studies, and human-nonhuman relations.

Emma Fraser is Lecturer in media and cultural studies in the Department of Sociology at Lancaster University, UK. Her research covers urban decay and regeneration, philosophies of space and place, digital geographies, and meaning production in games and visual media.

Steve Graham is Professor of Cities and Society at Newcastle University, Newcastle Upon Tyne, UK. His interests are in vertical aspects of cities and urban life; links between cities, technology, and infrastructure; urban aspects of surveillance; and the mediation of urban life by digital technologies.

Kris Hartley is Assistant Professor in the Department of Asian and Policy Studies at The Education University of Hong Kong, P. R. China. His research interests include urban policy and development, technology and smart cities, environmental policy and water management, collaborative governance, and state–society relations.

Allan D. Hollander is a Geographic Information Systems Programmer in the Department of Environmental Science & Policy, University of California, Davis, CA, 95616. His research interests center on the use of information systems for conservation and environmental management, including developing systems and standards for sharing environmental information through making use of digital cataloguing systems, metadata standards, and linked open data technologies.

Kara C. Hoover is Professor in the Department of Anthropology at the University of Alaska Fairbanks, USA. Her interests are in the human sense of smell and the factors that shaped its evolutionary tuning as well as modern distribution of variation in human populations. She is also interested in links between smell and food preference and subsistence and evolutionary mismatches in modern environments.

Casey Hoy is the Faculty Director of the Initiative for Food and AgriCultural Transformation, and leads the Agroecosystems Management Program at Ohio State University, Wooster, USA. He has a background in systems analysis in applied ecology, and his interests focus on advancements in agroecosystem health for sustainable communities.

Zhou Huang is Associate Professor in the Institute of Remote Sensing and GIS at the School of Earth and Space Sciences at Peking University, Beijing, P. R. China. His research interests are geospatial data mining and high-performance geocomputing.

Patrick R. Huber is Project Scientist and colead of the Food Systems Lab in the Department of Environmental Science and Policy at the University of California, Davis, USA. He is a geographer with interests in regional systematic conservation planning, focusing on California.

Ayaz Hyder is Assistant Professor at the Division of Environmental Health Science in the College of Public Health and Translational Data Analytics Institute at Ohio State University, Columbus, USA. He is a systems modeler whose research goal is to understand the role of multiple determinants of human health operating at multiple levels of organization.

Kevin Kane is Program Manager at the Southern California Association of Governments, Los Angeles, USA. His research interests include migration, demographic forecasting, urban land use change, regional income dynamics, and smart cities.

Scott B. Kelley is Assistant Professor in the Department of Geography at the University of Nevada, Reno, USA. His research interests are in the adoption and use of emerging transportation technologies and services and their supporting infrastructure and their impact on urban areas.

Glen David Kuecker is Professor in the Department of History at DePauw University, Greencastle, USA. His research interests include the problem of how humanity will weather the perfect storm of multiple, large-scale, global, and concomitant crises, including climate change, thermodynamics (energy), food insecurity, demographic transformations (population growth and aging, and rapid urbanization), environmental and ecological degradation, and economic stress.

Mei-Po Kwan is Choh-Ming Li Professor of Geography and Resource Management, and Director of the Institute of Space and Earth Information Science at the Chinese University of Hong Kong, Shatin, Hong Kong. Her research interests include environmental health; sustainable cities; human mobility; urban, transport, and social issues in cities and GIScience.

Bradley W. Lane is Associate Professor in the School of Public Affairs and Administration at the University of Kansas, Lawrence, USA. His research interests center on the effect of the adoption and diffusion of changing technologies on the sustainability of transportation.

Matthew C. Lange is President and CEO of the International Center for Food Ontology Operability Data and Semantics (IC-FOODS.org), Davis, California. He is a food and health informatician with expertise in ontology development for the food system.

Rui Li is Associate Professor in the Department of Geography and Planning at University of Albany, State University of New York, USA. His research interests include way-finding behavior and the design of geospatial systems that support spatial orientation and learning of a new environment.

Ruopu Li is Associate Professor in the Geography and Environmental Resources program at Southern Illinois University, Carbondale, USA. His research interests include agricultural land use, water resources planning and management, coupled nature and human systems, and smart rurality.

Yu Liu is Professor of GIScience in the Institute of Remote Sensing and Geographical Information Systems at the School of Earth and Space Sciences at Peking University, Beijing, P. R. China. His research concentration is in GIScience and big geodata.

Yongmei Lu is Professor and Chair of the Department of Geography at Texas State University, San Marcos, USA. Her research interests include GIScience, and the application of GIS and spatial analysis and modeling in health research and crime studies.

Casey R. Lynch is Assistant Professor in the Department of Geography at the University of Nevada Reno, USA. His research interests include the politics and ethics of techno-capitalism and emerging digital technologies, imaginaries of urban futures, geographic thought, and critical social theory.

Jing Ma is Associate Professor of Human Geography in the Faculty of Geographical Science at Beijing Normal University, P. R. China. Her main research interests include environmental justice, health inequality, activity-travel behavior, and subjective well-being.

Sophia Maalsen is Senior Lecturer in the School of Architecture, Design and Planning at The University of Sydney, Australia. Her research interests include the intersection of the digital and material across cities, housing, and the everyday.

Michele Masucci is Professor in the Geography and Urban Studies Department, Director of the Information Technology and Society Research Group, and Vice President for Research at Temple University, Philadelphia, USA. Her research interests include innovation ecosystems, community geographic information systems, digital inclusion, and broadening participation in STEM fields of study.

Eric Nielsen is Graduate Research Assistant and master's student in the Department of Geography at the University of Nevada, Reno, USA. His research interests include where transportation projects are being placed in relation to communities and how those projects influence economic development.

David O'Sullivan is Professor of Geography and Geospatial Science in the School of Geography, Environment and Earth Science at Victoria University of Wellington, New Zealand. His research interests are in the appropriate application of novel geographical methods of complex geographical questions, particularly in urban settings, and also in the social implications of emerging geospatial technologies.

Darcy Parks is a Postdoctoral Researcher at Linköping University, Sweden. He has a PhD in Technology and Social Change from the same university. His research is based on socio-technical perspectives, and his research interests include cities, digitalization, science and technology studies, and sustainability transitions.

Will B. Payne is a Assistant Professor in the Bloustein School of Planning and Public Policy, Rutgers, the State University of New Jersey. His research uses quantitative and qualitative methods to study the relationship between geospatial technologies and urban inequality, examining how changing technical capabilities, labor relations, and competitive pressures in the location-based services (LBS) industry interact with processes of racialized and class-based segregation in American cities.

Hamil Pearsall is Associate Professor in the Geography and Urban Studies Department at Temple University, Philadelphia, USA. Her research interests include urban sustainability, environmental justice, and the science of broadening participation.

Sung-Yueh Perng is Associate Professor in the Institute of Science, Technology, and Society, National Yang-Ming University. His research focuses on the incorporation of digital and data-driven innovation into urban everyday life and governance.

Ate Poorthuis is Assistant Professor of Big Data and Human-Environment Systems at the Department of Earth and Environmental Sciences, KU Leuven, Belgium. His research explores the possibilities and limitations of big data, through quantitative analysis and visualization, to better understand urban processes.

James F. Quinn is Professor Emeritus in the Department of Environmental Science & Policy at the University of California, Davis, USA. He is cofounder of the Information Center for the Environment and has worked with multiple public agencies and conservation organizations to develop information systems applied to public environmental policy and ecological research.

Courtney M. Riggle is Sustainability Specialist and colead of the Food System Lab in the Department of Environmental Science and Policy at the University of California, Davis, USA. Her expertise is in public policy, stakeholder engagement, and coordination of broad-based research coalitions.

Stéphane Roche is Professor at the Department of Geomatic sciences, Laval University. His expertise lies in GIScience, cartography, and land use decision-making.

Jovanna Rosen is Assistant Professor in the Department of Public Policy and Administration at Rutgers University-Camden, USA. Her research interests include urban inequality, community development, and environmental justice.

Shih-Lung Shaw is Chancellor's Professor and Alvin and Sally Beaman Professor of Geography at the University of Tennessee, Knoxville, USA. His research interests include transportation geography, transportation planning and modeling, space–time analytics of human dynamics, space–time GIS, and GIS for transportation.

Renee Sieber is Associate Professor in the Department of Geography at McGill University, Montreal, Canada. She is best known for her research on public participation GIS/participatory GIS.

Harrison Smith is Lecturer in Digital Media and Society at the University of Sheffield's Department of Sociological Studies, UK. His interests include surveillance, data analytics industries, the geoweb, mobile digital culture, and digital marketing.

Benjamin W. Stanley is Research Analyst in the University–City Exchange at Arizona State University, Tempe, USA. His research interests revolve around modern urban sustainable development policy and comparative urban studies of contemporary and ancient cities.

Jonathan Stiles is Postdoctoral Researcher in the Department of Geography and the STEAM Factory at The Ohio State University, Columbus, USA. His research interests include the ways in which information technologies influence travel through changes to work practices.

Scotty Strachan is Director of Cyberinfrastructure in the Office of Information Technology at the University of Nevada, Reno, USA. His research interests include the design and implementation of digital edge networks across urban and rural geographies and expanding the concept of connected communities to include wide-area ecosystem services via the Internet of wild things (IoWT).

Daniel Sui is Vice President for The Office for Research and Innovation, Virginia Polytechnic Institute and State University. His research interests include environmental implications of the emerging sharing economy, development of smart cities, location-based social media, open/alternative GIS, and legal/ethical issues of using geospatial technology in society.

Yinhua Tao is PhD student in the Department of Urbanism, Delft University of Technology. His research interests include behavioral geography and health research.

Jim Thatcher is Associate Professor of Urban Studies at the University of Washington Tacoma, USA, and an affiliate with the graduate school of Geography at the University of Washington, USA. His work examines the recursive relations among extremely large geospatial data sets, the creation and analysis of those data sets, and society, with a focus on how data have come to create, shape, and sustain modern urban environments.

Thomas P. Tomich is Distinguished Professor of Sustainability Science and Policy in the Department of Environmental Science and Policy and founder of the Food Systems Lab at the University of California, Davis, USA, where he teaches in the Sustainable Agriculture and Food Systems program. His research spans sustainability science, integrated ecosystem assessment, and sustainability of food systems. Specific interests include agriculture and farming systems, economic development, food policy, and natural resource management.

Anna Wallsten is Researcher at the Swedish National Road and Transport Research Institute (VTI), Map Unit, Stockholm, Sweden. She has an interdisciplinary background and holds a Doctoral degree in Technology and Social Change. Her research focuses on issues within the field of sustainable transitions, in particular citizen engagements, digitalization, and social justice.

Le Wang is Professor in the Department of Geography, University at Buffalo, The State University of New York, Amherst, USA. His research interests include remote sensing, GIScience, forest characterization, environment modeling, land cover and land-use change, urban population estimation, invasive species modeling, and spatiotemporal analysis and modeling.

Shengyin Wang is PhD student at Peking University, Beijing, P. R. China. Her research interests concentrate in GIScience and place formalization.

Yaoli Wang is Postdoctoral Researcher at Peking University, Beijing, P. R. China. Her expertise is in transportation network analysis and optimization.

Alan Wiig is Assistant Professor of Urban Planning and Community Development in the School for the Environment at the University of Massachusetts, Boston, USA. His research examines global infrastructure; smart urbanization; and the form, function, and politics of economic development.

Clancy Wilmott was Vice Chancellor's Postdoctoral Research Fellow at RMIT Melbourne at the time this article was drafted. Wilmott is now with the Department of Geography at the University of California, Berkeley, USA. Her research interests include mobile mapping, digital cartographies, and postcolonial urbanisms.

Di Wu is a PhD student in the Environmental Resources and Policy program at Southern Illinois University, Carbondale, Il (??). Her research interests include spatial data mining, and remote sensing for water resources.

Jin Xing is Lecturer in Geospatial Analysis in the School of Engineering at Newcastle University, Newcastle Upon Tyne, UK. His research interests include smart cities, GIScience, remote sensing, and cyberGIS.

Chao Ye received his M.S. degree from the Institute of Remote Sensing and Geographic Information System at the School of Earth and Space Sciences at Peking University, Beijing, P. R. China. His research interest lies in place semantics and natural language processing.

Yihong Yuan is Associate Professor in the Department of Geography at Texas State University, San Marcos, USA. Her research interests include spatio-temporal data mining, big geo-data analytics, and human mobility modeling.

Fan Zhang received his PhD from The Chinese University of Hong Kong in 2017. His research interests are in GIScience, spatial data mining, and urban studies.

Di Zhu is Assistant Professor in the Department of Geography, Environment and Society, University of Minnesota, Twin Cities. His research interests include geospatial modeling, applied artificial intelligence, and urban social sensing.

Matthew Zook is University Research Professor of Digital and Economic Geography in the Department of Geography at the University of Kentucky, Lexington, USA. His research focuses on the production, practices, and uses of big geodata and the ways in which algorithms, space, and place interact in finance as well as every day, lived geographies.

Shengyuan Zou is PhD candidate in the Department of Geography at the University at Buffalo, The State University of New York, Amherst, USA. His research interests include urban remote sensing, image processing, and built environment modeling.

Introduction: Smart Spaces and Places

Ling Bian

Introduction

Smart technologies have advanced rapidly throughout our society (e.g. smart energy, smart health, smart living, smart cities, smart environment, and smart society) and across geographic spaces and places. These "smart" developments are embodied by a number of concurrent seminal drivers, such as novel data derived from social media (e.g. Twitter), sensors (e.g. drones, smartphones), spurting apps, and a new generation of computing infrastructure (e.g. cloud computing). These developments have captured the enthusiasm of a range of research communities, while inevitably presenting unprecedented challenges and opportunities for geography as a discipline.

Are there emerging theories, conceptual frameworks, methodologies, and observations that can harness mass data, reveal new spatial phenomena, produce new knowledge, and foster new policies? Ultimately, the question is whether a new paradigm can arise from smart challenges that would move the discipline forward.

To this end, the 2020 Special Issue of the *Annals of the Association of American Geographers*, the basis of this book, called for theoretical, methodological, and empirical contributions to address questions such as how to make spaces and places "smart"; how the "smartness" affects the way we perceive, analyze, and visualize spaces and places; and what role geographies play in knowledge production and decision-making in such a "smart" era. We sought for a broad range of views in understanding spaces, places, and smartness, including but not limited to social, cultural, political, ethical, legal, economic, behavioral, ontological, and cognitive perspectives. The geography community enthusiastically responded to the call. The final selection process was based on a number of considerations: (1) diversity in theme, perspective, approach, and regional focus; (2) adherence to the theme of Smart Spaces and Places; and (3) original contribution and potential impact to the discipline. Due to page limitations of the book, we were not able to include many promising submissions.

The collection of 21 chapters included in this book offers stimulating discussion about spaces, places, and smartness. They roughly belong to four topical categories: (1) spaces, places, and smartness; (2) analytical smartness; (3) critical smartness; and (4) smart sustainability and policy. Such a division is solely for organizational purposes and does not reflect the intellectual flow and overlap cross-categories.

Spaces, Places, and Smartness (six chapters)

A critical examination of spaces and places is crucial to how we face the smart challenges. Shaw and Sui (Chapter 1) argue for a four space-place dual framework: absolute space (location), relative space (locale), relational space (place identity and dynamics), and mental space (sense of place). Independently, Poorthuis and Zook (Chapter 2) also argue for the relational space, and to some extent the mental (cognitive) space, in their call for synthesizing quantitative methodology and critical geographical theory. Although from somewhat different perspectives, both chapters urge moving beyond the conventional absolute conceptualization of spaces and places defined by x and y coordinates. Mostly importantly, the authors sense an urgency for theoretical developments, a visionary call when the discipline approaches the smart challenges.

The next pair of chapters reveal how smart technologies cause changes in spatial perception and behaviors. Crampton et al. (Chapter 3) reveal changes in response to the rapid introduction of smart surveillance technology into public spaces. The changes are explicit in both attitudes toward and experiences in the new, smart spaces. Similarly, Stiles, and Andrews (Chapter 4) disclose how smart technologies (wireless networks in physical spaces and mobile collaboration software in cyberspaces) affect behaviors in the form of spatial fragmentation of work activities, especially among knowledge workers.

As spaces increasingly come to be described as smart, Lynch and Del Casino (Chapter 5) unpack smartness from the notion of intelligence. Drawing on the distinction between cognition and consciousness, the authors call for multiple forms of intelligence in the context of understanding the "humanness" of intelligence as geographers approach the theorization of smart spaces.

Equally interesting, while unique, is the chapter by Payne and O'Sullivan (Chapter 6) that offers a historical account of the development of city dictionary Web portals, an early attempt of location-based services and Web maps. The reconstruction of events tracks the political economic settings in the 1990s that fostered the city dictionary Web portals, a precursor of many popular smart technologies today.

Analytical Smartness (six chapters)

Data and their associated smart sensors have stimulated the exploitation of emerging methods to harness mass data and distill essential information about spatial phenomena not previously known to us. Three chapters by Xing, Sieber, and Roche (Chapter 7), Zhu et al. (Chapter 8), and Li (Chapter 9) embrace the articulation of relational space, cognitive space, and relative space in line with those in Shaw and Sui (Chapter 1) and in Poorthuis and Zook (Chapter 2), although independently arriving at the same perception. That is, spaces are topologically structured networks of places and flows. The chapter by Xing, Sieber, and Roche proposes to connect places with sensor networks by a dynamic tessellation to achieve a relational urban space. The chapter by Zhu et al. leverages spatial relations between places to infer unknown properties of places in a relational urban space. The last chapter of the trio by Li proposes a design to allow app users to navigate the cognitive space well beyond the locale (relative space) shown on the small screen of smartphones.

The next pair of chapters exploit a range of smart sensors to investigate spatial phenomena and their underlying mechanisms. Ma et al. (Chapter 10) use real-time data derived from portable sensors and smartphones to reveal spatial and temporal variations of individual-level pollution exposure and their associated microenvironments. Zou and Wang (Chapter 11) use very-high-resolution remote sensing images to detect abandoned houses. The information gained would inform city revitalization efforts. These aforementioned novel data and their associated sensors capture social processes and experiences that would have not been known to us otherwise.

While the novel data derived from social media have captured the enthusiasm of researchers, Yuan et al. (Chapter 12) alert the research community about biases in these data. The biases could considerably affect investigations into smart spaces and places from sociodemographic, spatiotemporal, and semantic perspectives. The warning should be extended to all types of smart data derived from social media or sensors.

Critical Smartness (five chapters)

In line with the critical view of smart spaces and places articulated in Lynch and Del Casino (Chapter 5), five chapters put the smart urban governance to critical contestation. Collectively, the authors argue that smart technologies have reinforced a dominant ideology that shapes the production of spaces and places. This ideology has acquired renewed potency with neoliberalized urbanism, where smart technologies lead to technocratic urban governance, tokenization of citizens, and reproduction of capitalist.

Each chapter argues this central theme from their specific perspectives. Masucci, Pearsall, and Wiig (Chapter 13) examine youth perspectives regarding the impact of smart technologies on urban transformation and the sense of ownership of their acquired skills, especially whether smart technologies reproduce urban inequalities. Using examples drawn from various data, Dalton et al. (Chapter 14) articulate whether the valorization of data and its analysis naturalizes constructions of space, place, and individual that elide the political and surveillant forms of technocratic governance. Alvarez León and Rosen (Chapter 15) explore the notion that technological urban governance is tailored for capital accumulation in the context of digital and surveillance capitalism as illustrated by recent events in a techno-economy-focused city. Perng and Maalsen (Chapter 16) call for inclusive and participatory ways of shaping urban futures. These practices focus on capabilities to devise diverse sociotechnical arrangements and power relations to dissent from technocratic visions and practices. Lastly, the chapter by Kuecker and Hartley (Chapter 17) analyzes the evolution of a testbed city to demonstrate the distinction between organically evolving entities and products of totalizing technocratic norms.

Smart Sustainability and Policy (four chapters)

Smart technology penetrates all aspects of our life and society, including sustainable energy, food, transportation, and digital infrastructure systems. Implementing smart systems could bring a host of benefits, but also encounter complex ramifications to the development of effective policies. Parks and Wallsten (Chapter 18) analyze localized smart energy places where new technologies are needed to make them an effective niche in the large-scale smart energy systems. In so doing, the authors call for policies that can provide protective spaces for these places to alleviate pressures from the established marketing, social, and technical regimes. Hollander et al. (Chapter 19) propose a smart "foodshed" (a catchment of food sources for a region). Ontologies serve as the backbone to improve information flows in the food system and ultimately ameliorate inequitable access to healthy and safe food.

Kelley et al. (Chapter 20) identify characteristics that affect the adoption of smart transportation technologies. A typology drawn from a comprehensive review provides the basis to analyze whether and how various characteristics, such as geographic scope and extent, demographic characteristics, and travel behaviors, underlie the motivation for the adoption. Li, Chen, and Wu (Chapter 21) raise the issue of a "smart divide", a concern rising from smart infrastructures, especially in rural America. A number of technological and planning strategies are proposed to inform policy making as society continues to progress toward sustainable smart communities.

Cross-Cutting Themes

As expected, cross-cutting themes emerge across the topical categories. Most prominent among them are the relational and cognitive spaces. In addition to those chapters in the Spaces, Places, and Smartness topical category that focus on the conceptual depth of the theme (Shaw and Sui; Poorthuis and Zook), a number of chapters in the other three categories ground their analytical smartness (Xing, Sieber, and Roche; Zhu et al.; Li), critical smartness (Dalton et al.), and policies (Hollander et al.) firmly in the theme, explicitly or implicitly. This theme of relational and cognitive spaces reflects a current, actively pursued interest in seeking fundamental theories to represent spaces and places well beyond the traditional, x- and y-coordinate defined absolute space.

Equally prominent is the on-going critical discussion of smart urban governance, including all the Critical Smartness chapters. Other themes that crosscut multiple topical categories include networks as a means to actualize the relational space (Xing, Sieber, and Roche; Zhu et al.; Hollander et al.), biases and uncertainties in data (Yuan et al.; Ma et al.; Zou and Wang), and lastly, the adoption of emerging machine learning approaches (Zhu et al.; Zou and Wang).

The collection of chapters in this book showcases the breadth and depth of research in spaces, places, and smartness. It is far from being comprehensive, but serves the intention of this volume to stimulate further intellectual exchanges. Although it may take time for the discipline to fully grasp whether new geographic paradigms have arisen when society moves into the smart era, new challenges will always advance the discipline forward.

Acknowledgments

We express our sincere gratefulness to all individuals who made this book possible. First, we deeply appreciate the authors for their high-quality contributions. We thank editorial board members of the *Annals of the American Association of Geographers* and a large number of reviewers who provided insightful and timely comments. Publication of this book would be impossible without the superb support of Jennifer Cassidento, the Managing Editor of the *Annals*; Dr. Stephen Hanna, the AAG Cartographic Editor; and Lea Cutler, the Production Editor at Taylor & Francis. A special appreciation goes to Drs. Mei-Po Kwan and Sarah Elwood for their enormous help and valuable advice in many aspects of this collection.

PART I
Spaces, Places, and Smartness

Understanding the New Human Dynamics in Smart Spaces and Places: Toward a Splatial Framework

Shih-Lung Shaw and Daniel Sui

The smart technologies led by advances in artificial intelligence, machine learning, and the emerging data science in recent years are transforming many facets of society in profound ways. One of these affected areas is the experience of human dynamics in general and human mobility in particular with the growing maturity of smart technologies. The goal of this article is to critically examine the concepts of space and place in geography in general and in geographic information science (GIScience) in particular so that intelligent geographic information systems incorporating concepts of smart space and smart place can be developed to support human dynamics research. We argue that the current discussions on smart technologies are conceptually constrained due to their confinement to absolute space and physical place. By engaging research on smart technologies with geography and GIScience, we seek to move beyond the crude, and often simplistic, conceptualizations of space and place by synthesizing the multiple dimensions of both space and place. By doing so, we can better understand human dynamics through a synergistic perspective of both space and place. The space–place (splatial) framework proposed in this article will enable us to creatively study the human dynamics in the age of smart technologies. Our approach will not only allow us to better understand human dynamics but also advance and enrich our theoretical and methodological frameworks for studying smart technologies and the profound social impacts from a geographic perspective. Challenges for the implementation of the proposed framework are discussed and directions for future research are highlighted. *Key Words: GIScience, human dynamics, place, space, splatial framework.*

晚近由人工智慧、机器学习和逐渐兴起的数据科学所推进的智能科技，正在以深刻的方式改变社会的诸多面向。其中一个受影响的面向，便是随着智能科技逐渐成熟的人类普遍动态经验，特别是人类移动。本文旨在批判性地检视普遍在地理学中、特别是地理信息科学 (GIScience) 的空间与地方概念，因而能够发展纳入智慧空间与智慧地方概念的智慧地理信息系统，以支持人类动态的研究。我们主张，当前对于智能技术的讨论，由于仅限于绝对空间和物理地方，因此在概念上受限。我们以地理学和地理信息科学进行智能技术的研究，通过综合空间与地方的多重面向，寻求超越粗糙且经常是简化的空间与地方之概念化。藉由这麽做，我们能够通过空间与地方的协力视角，更佳地理解人类动态。本文所提出的空间地方 (splatial) 架构，能够让我们有创意地在智能科技的时代研究人类动态。我们的方法不仅能够让我们更佳地理解人类动态，并可从地理学的视角推进并丰富我们研究智能科技及其深刻的社会影响之理论与方法架构。我们并探讨执行此一推荐架构的挑战，并强调未来的研究方向。关键词：地理信息科学，人类动态，地方，空间，空间地方 (splatial) 架构。

Las tecnologías inteligentes, impulsadas por los avances de años recientes en inteligencia artificial, aprendizaje con máquinas y la aparición de la ciencia de datos, están transformando muchas facetas de la sociedad de modo profundo. Una de las áreas afectadas es la experiencia de la dinámica humana, en general, y la movilidad humana, en particular, con la creciente madurez de las tecnologías inteligentes. El objetivo de este artículo es examinar críticamente los conceptos de espacio y lugar en geografía, en general, y la ciencia de la información geográfica (SIGciencia), en particular, de surte que los sistemas de información geográfica inteligentes, que incorporan conceptos de espacio inteligente y de lugar inteligente, puedan ser desarrollados para apoyar la dinámica de la investigación. Sostenemos que las discusiones actuales sobre tecnologías inteligentes están conceptualmente limitadas debido a su confinamiento dentro del espacio absoluto y el lugar físico. Al comprometernos en investigación de tecnologías inteligentes dentro de la geografía y SIGciencia, buscamos ir mucho más allá de las conceptualizaciones crudas y a menudo simplistas de espacio y lugar sintetizando las múltiples dimensiones tanto de espacio como de lugar. Al hacer esto, podemos entender mejor la dinámica humana a través de una perspectiva sinérgica de espacio y lugar. El marco espacio–lugar (splatial, esplugal) que se propone en este artículo nos capacitará para estudiar creativamente la dinámica humana en la era de las tecnologías inteligentes. Nuestro enfoque no solo nos permitirá entender mejor la dinámica humana, sino también desarrollar más y enriquecer nuestros marcos teóricos y metodológicos para estudiar tecnologías inteligentes y sus profundos impactos sociales, desde una perspectiva geográfica. Se discuten los retos asociados con la implementación del marco propuesto, lo mismo que se destacan las direcciones de futuras investigaciones. *Palabras clave: dinámica humana, espacio, lugar, marco esplugal, SIGciencia.*

The smart technologies led by advances in artificial intelligence, machine learning, and the emerging data science in recent years are transforming many facets of society in profound ways. One of these affected areas is the experience of human dynamics. The goal of this article is to critically examine the concepts of space and place in geography in general and in geographic information science (GIScience) in particular so that intelligent geographic information systems (GIS) incorporating concepts of smart space and smart place can be developed to support human dynamics research. *Human dynamics* in this article refers to all kinds of human activities and interactions in both physical and virtual spaces that constantly shape and are shaped by the physical, social, economic, cultural, and political environments (Shaw and Yu 2009; Shaw, Tsou, and Ye 2016; Shaw and Sui 2018). We choose human dynamics as the focus of this article for two reasons. First, we want to elevate humans as the focus for future GIS development as well as in geography in general. Although many geographers (especially human geographers) have considered humans as a focus in their studies, the development of GIS has traditionally neglected humans to a large degree. Humans often are represented in conventional GIS as an aggregate number associated with a polygon or a grid cell, points on a map, or paths reflecting their traces over time. Humans have not been considered in GIS and GIScience as autonomous, intelligent entities that possess not only locations and attributes but also behaviors, perceptions, feelings, and thoughts related to spaces and places. Second, we want to emphasize that human activities and interactions are dynamic in nature and they constantly evolve in place, across space, and over time. The new human dynamics are introducing major social, economic, political, and cultural changes and new challenges to human societies (West 2018). We need to better address the challenges brought up by the rapidly changing human dynamics. Smart technologies have enabled us to study the new human dynamics at much improved spatial and temporal granularity in a hybrid of physical and virtual spaces.

What Are Smart Spaces and Smart Places?

As an integral part of the fourth industrial revolution (Schwab 2017) that has been unfolding in front of us in recent years, smart technologies have advanced rapidly throughout society by introducing approaches such as autonomous vehicles, smart energy, smart health, smart living, smart cities, smart environment, smart society, and even smart planet. An implicit assumption underlying these smart systems is that these technologies can make human societies smart. Many projects of developing smart systems therefore have placed significant efforts and investments on technological infrastructures and technical solutions to become smart. A critical question, rarely asked or answered adequately, is how to measure the smartness or intelligence[1] of a system. There is no doubt that technologies play an important role in smart systems to facilitate faster and better communications, powerful data collection and data processing, and more convenient transactions and services. Yet, what is the purpose of developing these smart systems? We inevitably have to bring humans to the center of the smart systems because the ultimate goal of developing such systems is to improve human lives. Every smart system therefore should have a specific purpose of serving particular needs in human societies (e.g., reduce traffic congestion or have a sustainable world).

Egenhofer and Mark (1995) suggested that naive geography "captures and reflects the way people think and reason about geographic space and time" (4). They argued that future GIS can be built from the formal models of the commonsense geographic world, which can be used by average citizens in day-to-day tasks without receiving major training. In fact, they suggested that "naïve geography is also the basis for the design of intelligent GISs that will act and respond as a person would" (Egenhofer and Mark 1995, 2). In other words, making a system intuitive to human thinking and human behaviors is an important criterion of developing smart systems. Obviously, GIS today are far from being more intuitive to human thinking and human behaviors than two decades ago when Egenhofer and Mark published their paper. So what is qualified as a smart space? We suggest that a *smart space* is a space that is intuitive to human thinking and spatial cognition. Implementation of a smart space is aided by incorporating various technologies that can make the space easily comprehensible to humans. For example, Tobler's First Law of Geography states that "everything is related to everything else, but near things are more related than distant things" (Tobler 1970, 236). Tobler did not explicitly define the meanings of *near* and *distant* and in what kind of space, however. Although it is a statement with ambiguity, we nevertheless understand its premise and accept it as a common law regulating human activities. This kind of ambiguity exists in our everyday life and people can function fine with such ambiguities. Smart spaces need to be tolerant to such ambiguities rather than tying everything to specific coordinates in a space like the conventional GIS.

Regarding the definition of place, Tuan (1977) indicated that place is an area in a space to which humans have given meaning. Roche (2016) suggested that "a place is typically defined by a named event (or action, or dynamics, or sense) that takes place (or is associated) to a specific location at a specific time (punctual, recurrent, period)" (568). Place consequently is a social construction that is unique to particular people and varies over time (Goodchild and Li 2012; Roche 2016). Humans play an even more central role in defining place than in defining space. Places are difficult to work within GIS because they often have fuzzy or indeterminate boundaries such as downtown Manhattan or Chinatown in New York City (Goodchild and Li 2012; Chen and Shaw 2016). In the meantime, places must be identifiable objects to be handled in conventional GIS. The concept of place, which is an outcome of human dynamics, has not been seriously considered in GIS until the emergence of platial GIS in recent years. Increasingly available volunteered geographic information (VGI) and big geospatial data are often related to the concept of place (e.g., neighborhoods, points of interest) rather than the coordinates in space, which have motivated the GIScience communities to accommodate places in GIS and develop platial GIS (e.g., Elwood, Goodchild, and Sui 2013; Goodchild 2015; Miller and Goodchild 2015; Roche 2016).

In an urban context, recent research has argued that cities should be studied as networks of places and flows instead of a mosaic of various areal spaces (e.g., Castells 2010; Batty 2013, 2018). As stated by Jones (2009), "Place—the city, region or rural area—as a site of intersection between network topologies and territorial legacies ... it is a subtle folding together of the distant and the proximate, the virtual and the material, presence and absence, flow and stasis, into a single ontological plane upon which location—a place on the map—has come to be relationally and topologically defined" (487). The question "Where am I?" is being replaced by "Where am I in relation to everything else?" in a globalized network of places (Roche 2016). Whether or not a place is smart no longer can be decided by the location in a local space and local context. Instead, a smart place must be assessed locally as well as relationally in a networked space. This presents new meaning to Tobler's First Law of Geography because near and distant now can be both physical distance and topological distance. A physically distant place can be topologically near if it is well connected (e.g., multiple daily nonstop flights, high volumes of digital financial transactions) to other places. Roche (2016) pointed out, "Smart cities are then not only about data or representation, but also about sending and decoding senses of places and understanding complex relationships (correlation, cause–effects) between physically and digitally connected places" (570). These changes require new and innovative theories and methods in GIScience to develop smart platial GIS.

A Space–Place (Splatial) Framework for Understanding the New Human Dynamics

The conventional GIS conceptual framework is ill-equipped to capture this new reality where the physical and virtual, objective and subjective, territorial and topological worlds are increasingly coupled and entangled for most human activities from local

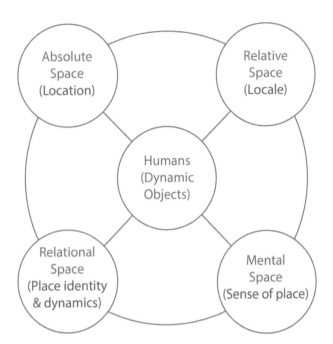

Figure 1. A space–place (splatial) GIScience framework for human dynamics research, an extension of the multispace framework from Shaw and Sui (2018).

to global scales. Shaw and Sui (2018) recently proposed a multispace GIScience framework for human dynamics research (see the red-color content in Figure 1). We argue that the current discussions on smart technologies are conceptually constrained due to their confinement to absolute space and physical place. By engaging research on smart technologies with geography and GIScience, we seek to move beyond the absolute conceptualization of space by expanding our multispace framework proposed earlier to a more inclusive space–place (splatial) framework in which we aim to encompass multiple dimensions of both space and place. By doing so, we can better understand human dynamics through a synergistic perspective of both space and place.

There have been many discussions about the concept of space in the literature (e.g., Nystuen 1963; Ullman 1974). Couclelis and Gale (1986) stated that "alternative conceptions of space abound in geography, and drawing a line between them is not always easy" (1). Curry (1996) further indicated that "the concept of space in geography, then, turns out to raise a set of difficult issues. … The geographer who wishes to adopt a view of space based on its own obvious merit finds all the ways blocked and is forced to adopt particular notions of space" (24). More than ever, we need not to be exclusive in our conceptualization of space. Instead, we need a more inclusive

framework that can better capture and elucidate the multiple dimensions of human dynamics in the age of smart technologies. Our splatial GIScience framework serves precisely that purpose and will make future GIS better align with naive geography to better support human dynamics research.

Curry (1996) summarized four main notions of space in Western thought:

> In Western thought there have been really only four main notions of space. … The first, codified by Aristotle, is static, hierarchical, and concrete. It gives greatest attention to a concept of place. The second, which we usually associate with Newton, imagines space as a kind of absolute grid, within which objects are located and events occur. The third, found in Leibniz's work, adopts the scientific outlook of Newton but argues that we need, as Aristotle does, to attend to the relationships among objects and events to the extent that we come to see space as fundamentally relational and defined entirely in terms of those relationships. The last, codified by Kant, turns the tables, where Aristotle and Newton had seen discussions of space as essentially about the world, he argued that we need to see space as a form imposed on the world by humans. (5)

Curry argued that the everyday practice of most contemporary geographers follows Aristotle's concept of space because they see the world as objects and events such as ethnic groups, women, vegetation, and economic activities that have their natural places, which is similar to Aristotle's argument of earth, air, fire, and water each having its own natural place. Although Curry's four main notions of space are not exactly the same as the four spaces (absolute, relative, relational, and mental) presented in our splatial GIScience framework, it is helpful to compare Curry's four notions of space with the absolute, relative, relational, and mental spaces in our framework for human dynamics research.

The *absolute space* in our splatial GIScience framework for human dynamics research is similar to Newton's concept of absolute space, which is infinite and immovable and can exist independent of anything external. Implementation of the absolute space concept is often based on Euclidean geometry and the Cartesian coordinate system. Most spatial analysis functions in GIS and many quantitative methods in geography are based on the concept of absolute space. Newton (1953), on the other hand, considered *relative space* as some movable dimension or measure of the absolute space. The relative space in our proposed splatial GIScience framework is consistent with this definition and allows the origin of a Cartesian coordinate system to move with a moving object. For example, a moving autonomous or driverless vehicle can detect the relative locations of objects in its surrounding environment with the vehicle itself as the origin point of a movable coordinate system. All autonomous vehicles are equipped with an array of sensors that can quickly and easily detect relative locations of objects around them at a high level of positional accuracy. If GIS can process location data measured in both absolute space and relative space efficiently, they can support autonomous vehicles without requiring a GIS database of recording everything (including dynamic moving objects such as pedestrians and other vehicles) at a very high positional accuracy (e.g., 3–5 cm), which is infeasible for the current GIS. This illustrates an example of a smarter GIS than the conventional GIS to meet a particular need of the modern society introduced by smart technologies.

In the splatial GIScience framework, we include *relational space* for how things are related to each other. The location of each object in a relational space is measured by its topological relationships to other objects. A relational space is like a topological space that is not affected by continuous deformations of shape or size of objects as far as their topological relationships (e.g., connectedness, continuity, compactness) are maintained. Social network graphs used in social network analysis are good examples of relational space because these graphs can be continuously transformed to other graphs while maintaining their topological relationships. Some publications use relative space and relational space interchangeably when they discuss Leibniz's concept of space. In our splatial GIScience framework, we clearly distinguish relative space and relational space to reflect different types of human activities and interactions (e.g., autonomous vehicles operation vs. social network analysis).

Mental space in the splatial GIScience framework is connected to the Kantian notion of space, which views the nature of space as a question about the nature of the observer who can have his or her own perceived space of things and events in the world (Curry 1996). Mental space covers feeling, emotion, perception, and cognition of the observer. For example, a neighborhood could be perceived by some people as safe, whereas other people might

perceive it as unsafe. In the current data science era, geographers and GIScientists wanting to extract semantics from various types of tracking data and other big geospatial data sets is a good example of going beyond absolute, relative, and relational space and moving into mental space to gain deeper insights of human behavior and human perception. In addition, mental space is a critical component in the splatial GIScience framework to move toward splatial GIS because humans give meaning to a place. Growing popularity of story maps, which tell a story with not only maps but also narrative texts, photos, videos, sounds, and so on, also indicates the limitations of conventional GIS and a need to express the feeling and perception of a storyteller that cannot be accomplished by the concepts of absolute, relative, and relational spaces alone.

The four spaces included in the splatial GIScience framework can be summarized as follows:

- *Absolute space* works with absolute locations in space and focuses on questions such as "Where are the different objects?"
- *Relative space* works with relative locations to a fixed or moving object and focuses on questions such as "What is around us?"
- *Relational space* works with relations to other objects and focuses on questions such as "What is related to us?"
- *Mental space* works with the cognitive and mental aspects of space and focuses on questions such as "What do people have in mind?"

Humans, who are dynamic objects themselves, are at the center of the splatial GIScience framework to replace locations as the central focus in conventional GIS. The circle that connects all four spaces in Figure 1 indicates that transformations or linkages can be performed among these four spaces such that they become an integrated, organic whole instead of four separate, independent spaces.

Until recently, GIS has been dominated by perspectives from absolute space using Cartesian coordinates according to Euclidean geometry. The massive amounts of volunteered geographic information (VGI) in general, and geotagged or location-based social media data in particular, seem to revive our approach to the world from the perspective of place, almost reaching the point of hyperlocalism dominated by "the tyranny of place" (Haklay 2010). The convergence of GIS and social media further prompts a new level of urgency for theoretical work to reconcile the world of space (traditional GIS) and the world of place (social media). How to formalize place in the GIS context is both interesting and challenging (Goodchild and Li 2012; Sui 2015). Our splatial framework could serve as a powerful bridge to link the two separate worlds of space and place.

By transcending the absolute conceptualization of space, the splatial framework just defined will expand and enrich our conceptualization of space, thus making space smarter. Furthermore, invoking the splatial framework is also consistent with the recent shift from space to place in GIS practices (Sui 2015). In contrast to space, *place* can be considered bottom-up and contingent, representing the perspectives and actions of people's daily routines. Geography practiced from the place tradition favors a slant or side (e.g., Google Street View or regular photos), qualitative, and multisensual perspective. According to Agnew (2011), the concept of place usually includes three pillars: location (as defined by latitude and longitude), locale (as defined by both physical/environmental and socioeconomic/cultural context), and a sense of place (as defined by human subjective perception or attachment to a particular location or locale). Massey (1991) further argued for a global and dynamic sense of place: (1) places do not have a single identity but are usually defined by multiple identities; (2) places are dynamic processes, not frozen in time; and (3) places are not enclosures with a clearly defined inside and outside.

Wainwright and Barnes (2009) discussed three views on the relationships between space and place in geography. The first view is *place trumps space*, which was advocated by humanistic geographers in the 1970s who argued that "place was good; it meant humanism, and opposed to a cold, heartless science represented by spatial science" (Wainwright and Barnes 2009, 968). Space is often considered as a background context for places under this view. The second view is *space trumps place*, which considers places as the sites, stops, pauses, or staging posts of actions, flows, and interconnections taking place in space. Space therefore plays a more critical role than places under this view. The third view is *relational space and place (splace)* which suggests that "space and place are relational. … It is the connections that are fundamental. There is no clear division between space and place because both are cut from the same cloth of multiplicitous relations" (Wainwright and Barnes 2009, 970). We are in favor of the third view and subsequently

extend our multispace framework to a splatial framework that integrates multiple aspects of both space and place to support human dynamics research.

Specifically, we argue that each of the four conceptualizations of space in our framework can be associated with a particular concept of place (see Figure 1). First, absolute space is associated with the concept of *location*. According to the Merriam-Webster dictionary, location is "a position or site occupied or available for occupancy or marked by some distinguishing feature" (see https://www.merriam-webster.com/dictionary/location). Location therefore suggests a specific position or site that can be conveniently represented by the coordinates based on the concepts of absolute space used in conventional GIS. Next, the concept of *locale*, which means "a place or locality especially when viewed in relation to a particular event or characteristic" (see https://www.merriam-webster.com/dictionary/locale), is closely related to relative space. In other words, our attention focuses mainly on the situation rather than the site of an object (e.g., a moving autonomous vehicle) that fits well with the concepts of relative space and locale. When we deal with data and analysis in relational space, the focus shifts to *place identity* and *dynamics* instead of location or locale. As relational space is based on topological relations rather than specific locations, place identity in a relational network (e.g., online social network) becomes critical for the identification of each entity as well as identifying their roles in a relational network. As Massey (1991) articulated for the increasingly global sense of place, these relational spaces as the massive social networks are also an integral part of place dynamics that constantly define and are also being defined by the identity of various places. Mental space in the splatial framework is associated with sense of place, which attempts to reflect what people have in mind about a location, a locale, or a place identity that are associated with absolute space, relative space, and relational space, respectively. Story maps that have gained popularity in recent years are one approach of sharing how people think about a place or a topic related to multiple places beyond the concepts of absolute space and conventional maps.

In a broader sense, our splatial framework has not only synthesized the alternative views of space as developed in Western thought (Curry 1996) but also integrated key elements of place as defined by Massey (1991) and Agnew (2011). Place has been off the intellectual radar screen of GIScientists, many of whom appear to treat place the same as space. Preliminary work has begun in the digital gazetteer literature to incorporate the concept of place. In a broader sense, the emerging critical and qualitative GIS literature has espoused a subtle shift of focus from space to place in GIS practices. If conventional GIS have been predominantly dealing with the spatial world, the emerging GIS practices have been increasingly focusing on the platial world using a plethora of smart technologies. As demonstrated by work on mapping networks (Abrams and Hall 2006) or deep maps (Bodenhamer 2013), multiple dimensions of human dynamics can be captured, analyzed, and mapped using these alternative conceptualizations of space and place as embodied in our splatial framework. Moving forward, we believe that our proposed splatial framework is more inclusive by incorporating the multiple dimensions of both space and place, thus providing us with a more robust conceptual tool to better understand the new human dynamics in the age of smart technologies.

Implementation Considerations of the Splatial Framework

There are several important considerations of implementing the splatial framework for human dynamics research illustrated in Figure 1. First of all, this new framework recognizes the success and usefulness of the conventional GIS using Euclidean geometry and a Cartesian coordination system based on the concept of absolute space. Functions supporting the conventional GIS will continue to be useful and further advance. One key consideration of implementing the splatial framework is to develop transformations or linkages between the absolute, relative, relational, and mental spaces. Relative space in the proposed framework is basically the same as the absolute space except that the origin point of its coordinate system moves with a particular object in relative space. Transformations between absolute space and relative space therefore are straightforward. As far as we know the absolute location of the origin point of a reference object in relative space, we can easily compute the absolute locations of other objects relative to the reference object. Conversely, we can easily compute the relative locations (e.g., direction and distance) of any objects based on the absolute coordinates of a given

object and its reference object. One potential challenge is the amount of computational power needed to process transformations for $N \times N$ pairs of objects, especially when N is a big number. Fortunately, most applications of relative space (e.g., autonomous vehicles or indoor navigation) only need to know the relative locations of a limited number of nearby objects around a give object. $N \times N$ transformations therefore can be reduced significantly to $n \times n$ transformations with a relatively small n that can be handled adequately by high-performance computers and parallel processing.

Relational space applications, on the other hand, could deal with absolute locations in Euclidean space as well as locations in non-Euclidean space. For applications that deal with specific locations in Euclidean space, they can be processed by conventional GIS with an additional function of tracking pairwise relationships between the objects. For relational space applications that do not require specific locations in Euclidean space (e.g., a schematic subway network map, Facebook friends network), a schematic map showing correct connections among the objects will suffice. This is similar to a topological network that can be generated by conventional GIS. Conventional GIS, however, start with specific coordinates to generate topological relationships. A schematic map, on the other hand, starts with topological relationships to generate a graph (e.g., a social network). This fundamental difference between conventional GIS and social network analysis leads to different requirements on analysis and visualization functions (Andris 2016). Regarding linkages between relational space and other spaces, one possible approach is to use unique object identification. For example, the location of Google could be represented by the coordinates of its headquarters in Mountain View, California, in absolute space, by the URLs of Google Web sites (e.g., www.google.com in the United States, www.google.ca in Canada, or www.google.co.uk in the United Kingdom) or by the Internet Protocol (IP) addresses of Google servers (e.g., 74.125.224.72). These various location references can all be linked to a unique object ID of Google. We then can choose the appropriate ones to use based on the specific applications and questions that we need to address.

Paralleling the four alternative dimensions of space, more work is needed to formalize the multiple dimensions of place, especially those related to locale, place identity and dynamics, and people's subjective sense of place. This will entail the incorporation of new sources of multiple geotagged data such as photos, videos, social media posts, and other digitized sources of information from traditional media, such as books, paper maps, and so on. Thus, full implementation of the splatial framework will be very data intensive and mandates fresh thinking in our ontological, epistemological, methodological, and ethical framework. Further, it will also necessitate the exploration and incorporation of new technologies that are on the horizon as the simultaneous demands for data of finer spatial and temporal granularity and the protection of privacy and confidentiality will be high. Existing technologies can only partially meet these conflicting demands. Moving forward, we believe that quantum computing, artificial intelligence (AI), and blockchain technologies (Zhang 2018) could be the game changer for the full implementation of our splatial framework, greatly facilitating the transition to smarter spaces and places with the computing power, intelligence analysis, and data security needed. To capture all four aspects of our splatial framework will require massive amount of data, which in many ways might be beyond the reach of existing technologies. Advances in AI, however, especially in affective computing (also known in the literature as emotional AI), sentiment analysis, and deep learning potentially can be applied to better capture mental space and people's sense of place (Picard [2000] 2018; Cambria et al. 2017; Lee 2018). For example, the Face++ system (see https://www.faceplusplus.com) that is currently being developed can not only perform simple face recognition, but also infer the person's emotional state according to the facial expression and analysis. We foresee that all of these advances in affective computing and sentiment analysis at the individual level can potentially improve our understanding of people's sense of place down the road at the collective level.

As we are about to start the third decade in the twenty-first century, more and more people will move into cities of different sizes globally. We foresee that smart cities, which are ultimately a synergy of both smart spaces and smart places, will play increasingly important roles in addressing the challenges of our growing cities related to the economy, society, and environment. The development of smart cities is currently at a crossroads due to the five competing frameworks (focusing on technology, human,

institution, energy, and data management, respectively; for details, see https://en.wikipedia.org/wiki/Smart_city) developed during the past two decades. Moving forward, we do not believe any single framework will serve the goals of the next phase of smart city development well. Instead, we need a new comprehensive strategy that is capable of synthesizing the core elements of all five existing frameworks for smart city development. The splatial framework as outlined in this article is uniquely suited to serve as the conceptual foundation to connect the dots as articulated by these five competing frameworks to move our cities toward the goal of economic efficiency, social equity, and environmental sustainability.

Conclusion

The accelerated pace of development of various smart technologies has not only done things for us but also to us, which has triggered an unprecedented change in human dynamics. In this article, we aim to argue that we should refocus our attention on the human dynamics by moving beyond the simplistic, absolute conceptualization of space. We present a splatial framework in which we have tried to synthesize the multiple dimensions of both space and place. Conceptually, the framework we are trying to articulate in this article is a new synthesis of previous discussions on space and place in the context of studying the new human dynamics, which is more robust and liberating because it captures the new human conditions in the age of smart technologies more inclusively and comprehensively than the existing framework. The current discussions on smart technologies are conceptually constrained due to their confinement to absolute space and physical place. By engaging research on smart technologies with geography and GIScience through our splatial framework, we seek to better understand the complexity of human dynamics through a synergistic perspective of both space and place, with people being placed at the center of GIS research. Exploring the alternative conceptualizations of space and place will enable us to creatively study the human dynamics in the age of smart technologies. Our approach will not only allow us to better understand human dynamics but also advance and enrich our theoretical and methodological frameworks in studying smart technologies and the profound social impacts from a geographic perspective.

We are fully cognizant of the technical challenges for the implementation of the proposed framework, but recent breakthroughs in quantum computing, AI, and blockchain technologies seem to be promising. More than ever, space and place are ontologically interconnected via a plethora of information, computing, and communication technologies (Graham 1998). In an age that increasingly demands convergence research, a new level of conceptual synthesis for human dynamics research based on GIScience is long overdue. We hope that the splatial framework discussed in this article is a major step forward toward unpacking the complexity of the new human dynamics as a result of the fourth industrial revolution.

Note

1. There have been many discussions of smart system versus intelligent system, and some people prefer intelligent system over smart system. We choose to use smart rather than intelligent in the remaining parts of this article to be consistent with the theme of smart spaces and places of this special issue.

References

Abrams, J., and P. Hall. 2006. *Else/where: Mapping new cartographies of networks and territories.* Minneapolis: University of Minnesota Design Institute/University of Minnesota Press.

Agnew, J. 2011. Space and place. In *The Sage handbook of geographical knowledge,* ed. J. Agnew and D. N. Livingston, 316–30. Los Angeles: Sage.

Andris, C. 2016. Integrating social network data into GISystems. *International Journal of Geographical Information Science* 30 (10):2009–31. doi: 10.1080/13658816.2016.1153103.

Batty, M. 2013. *The new science of cities.* Cambridge, MA: The MIT Press.

Batty, M. 2018. *Inventing future cities.* Cambridge, MA: The MIT Press.

Bodenhamer, D. J. 2013. An exploration of deep maps. Accessed July 21, 2014. http://thepoliscenter.iupui.edu/index.php/an-exploration-of-deep-maps.

Cambria, E., D. Das, S. Bandyopadhyay, and A. Feraco, eds. 2017. *A practical guide to sentiment analysis.* Berlin: Springer.

Castells, M. 2010. Globalisation, networking, urbanisation: Reflections on the spatial dynamics of the information age. *Urban Studies* 47 (13):2737–45. doi: 10.1177/0042098010377365.

Chen, J., and S.-L. Shaw. 2016. Representing the spatial extent of places based on Flickr photos with a representativeness-weighted kernel density estimation. In *Lecture notes in computer science (LNCS) 9927: GIScience 2016,* ed. J. Miller, D. O'Sullivan, and N.

Wiegand, 130–44. Cham, Switzerland: Springer International.

Couclelis, H., and N. Gale. 1986. Space and spaces. *Geografiska Annaler Series B: Human Geography* 68 (1):1–12. doi: 10.1080/04353684.1986.11879523.

Curry, M. R. 1996. On space and spatial practice in contemporary geography. In *Concepts in human geography,* ed. C. Earle, M. S. Kenzer, and K. Mathewson, 1–32. Lanham, MD: Rowman and Littlefield.

Egenhofer, M. J., and D. M. Mark. 1995. Naïve geography. Report 95-8, National Center for Geographic Information and Analysis, State University of New York at Buffalo, Buffalo, New York.

Elwood, S., M. F. Goodchild, and D. S. Sui. 2013. Prospects for VGI research and the emerging fourth paradigm. In *Crowdsourcing geographic knowledge,* ed. D. Sui, S. Elwood, and M. F. Goodchild, 361–75. Amsterdam: Springer.

Goodchild, M. F. 2015. Space, place and health. *Annals of GIS* 21 (2):97–100. doi: 10.1080/19475683.2015.1007895.

Goodchild, M. F., and L. Li. 2012. Formalizing space and place. In *Fonder les sciences du territoire,* ed. P. Beckouche, C. Grasland, F. Guérin-Pace, and J.-Y. Moisseron, 83–94. Paris: Editions Karthala.

Graham, S. 1998. The end of geography or the explosion of place? Conceptualizing space, place and information technology. *Progress in Human Geography* 22 (2):165–85. doi: 10.1191/030913298671334137.

Haklay, M. 2010. The tyranny of place and Openstreetmap. Accessed July 21, 2014. http://povesham.wordpress.com/2010/07/10/the-tyranny-of-place-and-openstreetmap/.

Jones, M. 2009. Phase space: Geography, relational thinking, and beyond. *Progress in Human Geography* 33 (4):487–506. doi: 10.1177/0309132508101599.

Lee, K. F. 2018. *AI superpowers: China, Silicon Valley, and the new world order.* New York: Houghton Mifflin Harcourt.

Massey, D. 1991. A global sense of place. *Marxism Today* June:24–29.

Miller, H. M., and M. F. Goodchild. 2015. Data-driven geography. *GeoJournal* 80 (4):449–61. doi: 10.1007/s10708-014-9602-6.

Newton, I. 1953. *Newton's philosophy of nature: Selections from his writing,* ed. H. S. Thayer. New York: The Hafner Library of Classics.

Nystuen, J. D. 1963. Identification of some fundamental spatial concepts. *Papers of Michigan Academy of Science, Arts, and Letters* 48:373–84.

Picard, R. W. [2000] 2018. *Affective computing.* Cambridge, MA: The MIT Press.

Roche, S. 2016. Geographic information science II: Less space, more places in smart cities. *Progress in Human Geography* 40 (4):565–73. doi: 10.1177/0309132515586296.

Schwab, K. 2017. *The fourth industrial revolution.* New York: Random House.

Shaw, S.-L., and D. Sui. 2018. GIScience for human dynamics research in a changing world. *Transactions in GIS* 22 (4):891–99. doi: 10.1111/tgis.12474.

Shaw, S.-L., M.-H. Tsou, and X. Ye. 2016. Editorial: Human dynamics in the mobile and big data era. *International Journal of Geographical Information Science* 30 (9):1687–93. doi: 10.1080/13658816.2016.1164317.

Shaw, S.-L., and H. Yu. 2009. A GIS-based time-geographic approach of studying individual activities and interactions in a hybrid physical–virtual space. *Journal of Transport Geography* 17 (2):141–49. doi: 10.1016/j.jtrangeo.2008.11.012.

Sui, D. Z. 2015. Emerging GIS themes and the six senses of the new mind: Is GIS becoming a liberation technology? *Annals of GIS* 21 (1):1–13. doi: 10.1080/19475683.2014.992958.

Tobler, W. 1970. A computer movie simulating urban growth in the Detroit region. *Economic Geography* 46 (Suppl.):234–40. doi: 10.2307/143141.

Tuan, Y. F. 1977. *Space and place: The perspective of experience*. Minneapolis: University of Minnesota Press.

Ullman, E. 1974. Space and/or time: Opportunity for substitution and prediction. *Transactions of the Institute of British Geographers* 63:125–39. doi: 10.2307/621536.

Wainwright, J., and T. J. Barnes. 2009. Nature, economy, and the space–place distinction. *Environment and Planning D: Society and Space* 27 (6):966–86. doi: 10.1068/d7707.

West, D. M. 2018. *The future of work: Robots, AI, and automation*. Washington, DC: Brookings Institution Press.

Zhang, S. 2018. Quantum computing, AI and blockchain: The future of IT. Talks @ Google. Accessed December 30, 2018. https://www.youtube.com/watch?v=MozDSajpLTY.

Being Smarter about Space: Drawing Lessons from Spatial Science

Ate Poorthuis ⓘ and Matthew Zook ⓘ

Smart technology—in its many facets—is often critiqued within geography in ways that parallel the critiques of quantitative geography in the 1960s and GIScience in the 1990s. In this way, both the development of "smart" technology itself and its criticisms are the latest chapter in a long-standing disciplinary debate around quantification and technology. We reevaluate this history and argue that quantitative methodology and its theoretical critiques are not as incompatible as often claimed. To illustrate how we might address this apparent tension between theory and quantitative methods, we review how both approaches conceptualize one of geography's core concepts—space—and highlight opportunities for symbiosis. Although smart technologies can further orthodox positivist approaches, we argue that the actual practice is more nuanced and not necessarily absolute or totalizing. For example, recent computational work builds on critical geographic theories to analyze and visualize topological and relational spaces, relevant to topics such as gentrification and segregation. The result is not a geography in which smart technology and algorithms remove the need for human input but rather a rejoinder in line with the recent resurgence of a critical quantitative geography. In short, the result is a geography where social theory and the human intellect play a key role in guiding computational approaches to analyze the largest, most versatile, most and relevant data sets on social space that we have ever had. *Key Words: geocomputation, GIScience, smart technology, space, spatial science.*

地质学领域经常从不同的方面批评智能技术，一如20世纪60年代对定量地理学的批判和90年代对地理科学的批判。但在这场围绕量化和技术的漫长学术辩论中，无论是"智能"技术本身的发展还是对其进行的批评，都属于新时期的内容。我们在重新梳理这段历史后发现，定量方法论及其理论批判并不像人们以为的那样水火不容。为了说明我们如何解决理论和定量方法之间的明显矛盾，我们回顾了这两种方法如何让一个核心地理学概念真正概念化的方式，即"空间"，强调了两种方法共生的机会。尽管智能技术可以进一步推动正统的实证主义方法，但我们认为，实际操作存在很大的细微差异性，不一定绝对或笼统化。例如，在重要的地理理论基础上开展最先进的计算工作，分析并可视化拓扑空间和关系空间，这些空间与主题相关，诸如贵族化和种族隔离。这样做的结果不会让地理学变成不再需要人类输入的智能技术和算法，而是与最近重新兴起的关键定量地理学相呼应。简而言之，以这种方式构建的地理学中，社会理论和人类智力会发挥关键作用，指导计算方法分析我们所拥有的最庞大、最通用、最相关的社会空间数据集。*关键词: 地理计算、地理科学、智能技术、空间、空间科学。*

La tecnología inteligente—en sus variadas facetas—a menudo es criticada en geografía a la usanza de las críticas formuladas a la geografía cuantitativa en los años 1960, y a la SIGciencia en los 1990. Con esa perspectiva, tanto el desarrollo de la propia tecnología "inteligente," como sus críticas, aparecen como el propio capítulo final en un debate disciplinario de vieja data acerca de la cuantificación y la tecnología. Nosotros reevaluamos esta historia y argüimos que la metodología cuantitativa y sus críticas teóricas no son tan incompatibles como a menudo se sostiene. Para ilustrar el modo como podríamos abocar esta aparente tensión entre teoría y métodos cuantitativos, hacemos una revisión sobre cómo los dos enfoques conceptualizan uno de los conceptos medulares de la geografía—el espacio—y destacamos las posibles oportunidades de simbiosis. Aunque las tecnologías inteligentes pueden promover más los enfoques positivistas ortodoxos, sostenemos que la práctica real es más matizada y no necesariamente absoluta o totalizante. Por ejemplo, un reciente trabajo computacional edifica sobre teorías geográficas críticas para analizar y visualizar espacios topológicos y relacionales, que son relevantes para tratar tópicos como la gentrificación y la segregación. El resultado no es una geografía en la que la tecnología inteligente y los algoritmos remueven la necesidad del aporte humano, sino mejor una réplica en línea que el reciente resurgimiento de una geografía cuantitativa crítica. En pocas palabras, el resultado es una geografía donde la teoría social y el intelecto humano juegan un papel crucial para guiar los enfoques computacionales para analizar los más grandes, versátiles y muy relevantes conjuntos de datos del espacio social que hayamos tenido jamás. *Palabras clave: ciencia espacial, espacio, geocomputación, SIGciencia, tecnología inteligente.*

It has been a decade since Anderson (2008) published his polemic *Wired* article on big data and the subsequent "end of theory." In the intervening years, our discipline has written extensively on the theoretical, methodological, and practical implications of a broad range of new technologies (using interconnected monikers varying from geoweb and volunteered geographic information to big data and, in this issue, labeling such technology as "smart"), perhaps even heralding in a "digital turn" within the discipline (Ash, Kitchin, and Leszczynski 2018). Much attention has been paid to the sociospatial implications of digital technology on our everyday lives (e.g., Kitchin and Dodge 2011) and, in the context of *smart* specifically, the effect on urban governance (Shelton, Zook, and Wiig 2015; Leszczynski 2016). Building on this foundation, we focus instead on the methodological implications of smart technology for geography. This, too, has received ample attention recently and is sometimes labeled a new "quantitative revolution" (Wyly 2014), one that is not necessarily welcomed by all. These quantitative approaches, reinvigorated by access to new computing technology and data sets, have been critiqued from (social) theory in similar ways as GIScience in the 1990s and the earlier onset of spatial science in 1960s. Thus, both the current smart technology itself and the critiques leveled against it are but the latest chapter in a long-standing disciplinary debate around quantification and technology.

Using one of geography's core concepts—space—and drawing from the spatial science turn of the 1960s, we highlight the significant potential for synthesis and symbiosis between quantitative methodology and geographical theory. In short, we argue that although new technologies might further an orthodox positivist ontology and epistemology, they also can enable the analysis of the rich, intricate fabric of space in myriad ways that are quantitative but are not necessarily absolute or totalizing. In particular, we seek to connect the ways in which the quantitative movement of the 1960s theorized space—including acknowledging the limitations of a Euclidian approach (Bunge 1966; Harvey 1969)—with the social theories used in theoretical geography today.

Geography as a discipline is at a unique moment in its relationship with smart technologies. We use the smart label here as an umbrella term to indicate the onset of new computational technologies enabling ever larger, faster, more exhaustive and varied data sets about social life to be created and collected (big data; cf. Kitchin and McArdle 2016) as well as the associated technologies (e.g., cloud computing, machine learning, etc.) allowing for the analysis of such data (Shelton, Zook, and Wiig 2015; Leszczynski 2016). If geography chooses to engage with smart technologies, it is essential that it is not constrained to just extending existing geographic information systems (GIS), city dashboards, or sophisticated governance control rooms. Nor should we simply let software giants like ESRI and IBM push geographic software development. Although creating useful tools, they are ultimately responding to market demands, not the pursuit of science. Moreover, we should not take a backseat to other disciplines entering the smart arena: Physics, computer science, and complexity science all have important contributions to make but are coming from significantly different perspectives than geography. Rather, we should learn from our experience with the spatial science turn to leverage the onset of smart technologies into opportunities to build a tighter integration between geographic theory and quantitative methodology.

Evolving Theories of Space

Traditionally, space was treated as a container or geometric system holding the object(s) under study and allowed users to lay an abstract plane over the Earth's surface. In most cases this abstract plane is simply a two-dimensional Euclidean (or, after the seventeenth century, Cartesian) space with x- and y-axes useful for describing location. Examples range from Ptolemaios (AD 90–168) pinpointing all then-known places within his atlas, *Geographia*, or the regional geography of the prewar era (e.g., Hartshorne 1939) that tasked itself with locating cities, countries, and people. Importantly, objects within such a space are generally considered to be completely independent of each other. Their only relationship is defined through measuring their distance to other objects.

In the second half of the twentieth century, geographers—and social scientists in general—critiqued this conceptualization of space as hampering the actual understanding of social (and spatial) processes. For example, Merleau-Ponty's ([1945] 2012) phenomenological analysis of space (forbearing the subdiscipline of humanistic geography) argues for a primordial *situated* space that is neither object nor subject or a space where people and things are no longer independent of each other. Likewise, Lefebvre's well-known argument completely breaks away from the notion of space as a container, asserting explicitly that space is socially produced and non-Euclidean (Lefebvre 1991). Moreover, space as a product is never final but rather produced, consumed, and reproduced in a never-ending and iterative process. As such, space cannot simply be thought of as something finished—a static object of study. To understand space, we must study the process of space.

In this way, geography has moved far from the Euclidean conceptualization of space as a simple container and instead engages with space that is produced and reproduced in a continuous fashion. This production of space is messy—consisting of many interconnected and contradictory layers of everyday social life—such as emblemized by Massey's (1994) walk through Kilburn in which she argues that places do not just have one singular identity, or one sense of place. More recently, geography has seen a renewed interest in a relational understanding of space (Boggs and Rantisi 2003; Jones 2009) that foregrounds actors (or actants) within space and how their relationships continuously make and remake space (Brenner, Madden, and Wachsmuth 2011).

Continued Reliance on Euclidean Spatial Ontologies

This review highlights the complexity of the social theories of space used by geographers and yet we must also note the continued reliance on a Euclidean definition of space within quantitative research. Making this dependence even more remarkable are the convincing critiques of its problematic nature. First, space as an absolute container enables a reductionist, totalizing perspective on society. Harvey (1996) was very clear about the

consequences of this perspective: "to produce one dominant cartographic image out of all this multiplicity is a power-laden act of domination" (284). In other words, a Euclidean perspective on space is just one of many views on space. A single, true representation of space does not exist, and proceeding otherwise neglects the heterogeneity of social life. Second, space as an absolute container can enable a static, immutable perspective on society. The Euclidean container cannot change itself—it is preexisting after all—and thus might direct our attention to static patterns and less so to social processes. Third and finally, space as an absolute container focuses on the location of independent objects and ignores the relative and relational connections between people, institutions, and other actors that together make up social space.

This contrast begs the question of why there is such a disconnect between theoretical and quantitative work on space within geography. Why are we still content with the spatial constraints of traditional desktop GIS software rather than developing Lefebvrian GIS or Masseyian mapping software? In sum, why are we still actively using a quantitative methodology that is inherently and ontologically incompatible with how we (theoretically) think about society and space? To be clear, we are not calling for an abolishment of Euclidean space altogether, because its conceptualization is useful for many applications. Instead, our point is that it is but one of many possible conceptualizations, and our reliance on it shapes our understanding of the world.

Although there are many historical factors beyond our simple summary here (cf. Schuurman 2006), we use this understanding to document a moment during the 1950s and 1960s when competing conceptualizations were in play, including the surprisingly sophisticated use of relational spatial ontologies in early work by spatial scientists. We suggest that, ultimately, the emergence of the personal computer and desktop GIS in the 1980s helped solidify a standard for quantitative analysis of space that leaves less room for alternative conceptualizations. We further contend that this history can provide useful guidance in how the discipline approaches smart technologies today. This is similar to the impetus behind the field of geocomputation—as Gahegan (1999) put it: "GIS saw to it that geographers became the slaves of the computer, having to adopt the impoverished representational and analysis capabilities that GIS provided" (203). It is our contention that a similar impoverishment of space need not, and should not, occur in the present day.

Developing Spatial Science

Amidst an academic landscape in which geography was increasingly threatened with departments closed down, a reorientation toward spatial science began in the 1950s. As a spatial science, the goal was not to describe things within space but rather to find general laws and theories explaining space. With this, space shifted from just a container or handy filing system to become the very nexus of geographic research. It is sometimes argued that this "revolution" reinforced the use of an absolute or naive spatial ontology within geography (e.g., Cresswell 2013). The centrality of space to the new spatial science, however, meant that many of its proponents and the contemporarily cited works of the decade grappled explicitly with the limitations of Euclidean space. For example, in his standard work on spatial science, Harvey (1969) wrote:

> Given the philosophy of absolute space, the metric in that space must remain isotropic and constant. [...] Relations between objects on the earth's surface, the extent of areal units, and so on, could be measured by the direct extension of Euclidean concepts of space and distance to the surface of a sphere. [...] This view is no longer generally acceptable. Thus Watson (1955) has pointed out that distance can and must be measured in terms of cost, time, social interaction and so on, if we were to gain any deep insight in to the forces moulding geographic patterns. [...] The metric is thus determined by activity and by the influence of objects. Such a concept of distance is purely relative. (210–11)

Importantly, Harvey was not isolated in this spatial thinking. Alternative, relational conceptualizations of space (and the accompanying operationalization through mathematical topology and graph theory) were *en vogue* well before Harvey's observation. We highlight two particularly noteworthy strains of this relational thinking within spatial science. The first removes some of the limitations of Euclidean space by looking at non-Euclidean measures of distance (e.g., perception, cost, etc.). The second strain is more explicit and forgoes two-dimensional space altogether in favor of graph theory and network analysis.

Nonabsolute Conceptualizations of Space

Some of the earliest traces of geographers using a nonabsolute space are the so-called potential maps of the 1950s. Here, geographers (and physicists turned geographers; see Barnes and Wilson 2014) borrowed heavily from physics and used the concept of potential to study spatial phenomena such as population distribution and transportation accessibility (Stewart and Warntz 1958). In a potential model, the value of a point is determined by its connections—based on variables such as accessibility or economic pricing—to all other points. For example, Harris (1954) used it to calculate the market potential in the United States, making it an implicitly relational, not territorial, concept.

The ideas within potential models were extended to map transformation (Hägerstrand 1957; Tobler 1963), used not to project the three-dimensional earth on a two-dimensional plane but rather to transform Euclidean space to a more phenomenon-appropriate form. Hägerstrand (1957) used an innovative azimuthal projection centered on Asby, Sweden, that is scaled logarithmically to show that physical distance is not the most important determinant in understanding migration (see Figure 1A). Tobler (1963) extended this innovation and argued that maps can be transformed based on other variables as well (similar to cartograms) and can even be used to (quantitatively) compare and contrast these different spaces. Building from Christaller's theoretical model of central places (which assumes an underlying isotropic space), Bunge (1966) and Getis (1963) demonstrated that it was possible to transform space based on population and income levels to better "fit" the real-world empirical application (see Figure 1B).

In methodologically similar approaches, other spatial scientists studied cognitive spaces to better understand how space is actually perceived. In the same way as Hägerstrand transformed physical distance, the perception of space and distance can be drawn, compared, and analyzed—eventually kickstarting an entire subdiscipline of behavioral geography (Gould and White 1974; Golledge 1977). One technique used in spatial science to deal with these new spaces is multidimensional scaling, in which objects are defined by the strength of their relationship to other objects rather than treated as independent. In cognitive mapping, this can be used to show cognitive distortion (Golledge, Rivizzigno, and Spector 1976) but can be equally useful in mapping

A

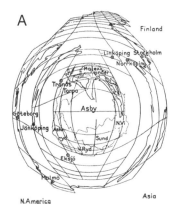

B

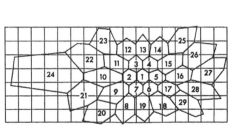

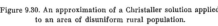

Figure 9.30. An approximation of a Christaller solution applied to an area of disuniform rural population.

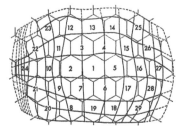

Figure 9.31. Map transformed into uniform density.
Note: The reader might benefit from mental comparison of the two *base* maps in Figures 9.30 and 9.31, i.e., a comparison of the two coordinate systems.

Figure 1. Examples of nonabsolute space in geography from the 1950 and 1960s. (A) Hägerstrand's (1957) use of an azimuthal projection centered on Asby that is scaled logarithmically to illustrate that physical distance is not the most important determinant in understanding migration. (B) Bunge's (1966) transformation of Christaller's hexagonal space based on the underlying population.

other spaces, ranging from time–space (Marchand 1973) to cost space (Forer 1978) and even "marriage" space (Kendall 1971). In short, far from only using Euclidean spatial ontologies, this period represents an intense engagement with spatial theories, including relational space. As Gatrell (1981) argued, "The proliferation of different concepts of space has placed new demands [...] for techniques to give tangible expression to such concepts. The realization that a relative, rather than absolute, view of space may offer a more appropriate context within which to map geographical objects [...]" (151).

Experimenting with Graph Theory

The most explicit use of relational space, however, can perhaps be found in the adoption of graph theory in which space does not preexist at all but is defined through its constituent parts. These parts consist of nodes (also called vertices) and the edges (or ties or links) between them. Both vertices and edges can have certain characteristics indicating, for example, the strength of a tie or the size of a node. Together, these nodes and their ties form a network, which is why the term *network analysis* is often used to refer to this type of analysis.

Arguably the first explicit use of graph theory in geography was Garrison's (1960) study of the new U.S. interstate highway system. Garrison searched for a way to analyze these new highways as a system and tentatively identified graph theory as a solution. Nystuen and Dacey (1961) took Garrison's first pass a step further and applied graph theory to the problem of regionalization. Interested in the concept of

the nodal region—a dominant central place or nodal point with surrounding hinterland—they argued that graph theory was well suited to express and analyze this hierarchy of cities or nodes. In an empirical study that could be labeled novel even today, they used the number of direct long-distance telephone messages between cities in and around Washington State to build a weighted network. In so doing, and without explicitly expressing this, they projected a two-mode (person–city) network to a one-mode (city–city) network and used weighted degree centrality to detect the nodal structure within this network—forty years before the oft-cited Barrat et al. (2004) study formalized weighted degree centrality in the same way. A final note should be made of Hemmens's (1966) dissertation on "the spatial structure of urban activities" in which he used trip data to determine linkages between different zones in Buffalo, New York. He showed how zones cluster together in distinct groups that are—very importantly—not necessary spatially contiguous, a conclusion that would have been impossible to derive from the common, topographical perspective on spatial clusters.

After the initial wave of work from U.S. geographers, British geographers took over the baton with Haggett and Chorley (1969) publishing a comprehensive standard work, *Network Analysis in Human Geography*. It spends more than 300 pages discussing topological structures and graph theory and their application to geographic problems. Even in the 1970s, network analysis still formed an important part of British geography, with Taylor's (1977) *Quantitative Methods in Geography*—an introduction to spatial analysis for undergraduate students—discussing graph theory in its very first pages.

Despite this early work, by and large, graph theory and topological space dropped off the quantitative geography radar during the 1980s, not to reappear with any strength until the twenty-first century. Perhaps coincidentally, this decline paralleled the appearance of the personal computer in academic research and the standardization of tools—most notably ESRI's launch of the ARC/INFO GIS in the early 1980s. Although these technologies heightened the speed and ease of quantitative analysis, they also constrained the possibility frontier. After all, using off-the-shelf software means accepting the techniques, methods, and ontologies preprogrammed in the software and potentially forgoing other approaches (cf. Rey 2009). A simple example of the power of the affordances and standards offered by software is the de facto standard spatial data format—the shapefile—which, crucially for this article, has no awareness of topology. Instead, it uses a topographical ontology, where each object is defined by its exact location in a Euclidean container. Thus, the narrowing of quantitative research within geography might be seen as an outcome of software sorting (cf. Graham 2005) in which the code and algorithms embedded in our tools shape our methods and ontologies.

Assembling a "Smart" Symbiosis for Geography Today

This short exposition of different spatial ontologies in spatial science illustrates the historical interest in and potential for non-Euclidean spaces within quantitative geography. As geography grapples with how best to incorporate smart technologies in our research praxis, this history and the ways in which some spatial ontologies were elevated and others dismissed are particularly germane. One could imagine adoption of new technologies leading to an extension of conventional GIS software or being applied to urban dashboards and governance. As such, smart technology and data might enhance these systems by virtue of its data covering a larger part of the population and social life or even enable the adoption of participatory frameworks as is often discussed or foreshadowed in the discourse around smart cities (Stratigea, Papadopoulou, and Panagiotopoulou 2015), but this would not entail a more radical

reimagining of such software and methods. In our assessment, this would be a disappointing outcome.

Instead, we posit that these new technologies could usefully catalyze a more comprehensive reanchoring of computational geography to the theoretical foundations of our discipline, in line with the calls for critical quantitative geography made in the past decade (Kwan and Schwanen 2009; Wyly 2011). In the recent discourse around critical GIS (O'Sullivan, Bergmann, and Thatcher 2018; Thatcher, Bergmann, and O'Sullivan 2018) we find several contributions arguing for a reimagining of GIS (Gahegan 2018) and an opening up to different spatial ontologies (Bergmann and O'Sullivan 2018). Likewise, the field of geocomputation (Cheng, Haworth, and Manley 2012) brings together geography and computer science and, in Gahegan's (1999) reading, is specifically meant to alleviate the "shortcomings" of GIS. Arribas-Bel and Reades (2018) similarly posited geocomputation as a separate tradition focused on what the intersection of geography and computing makes possible. In short, if we are to enter a smart era of geography, we would do well to heed these calls in earnest and use the unique opportunity offered by new technologies and data sources to radically reimagine how we think geographically through computation (methods), (new) data, and software. Luckily, against this backdrop of new technologies, there are several examples of recent work that reimagines quantitative geography across these three dimensions.

Reimagining Methods, Data, and Software

In terms of methods, similar to the experiments with different types of distance and relations and multidimensional scaling, geographers have started to use self-organizing maps (SOMs). SOMs are a type of neural network and thus need to be rained. This training is computationally intensive, so this (similar to other machine and deep learning methods) is a methodological approach enabled by the affordances of new, smart technology. Spielman and Thill (2008) used SOMs to understand the complex attribute space of urban neighborhoods in New York (see Figure 2A). The same technique can be expanded to understand spatial processes such as neighborhood change (Delmelle 2017). Similarly, other authors have explicitly experimented with relational spaces in quantitative work. For example,

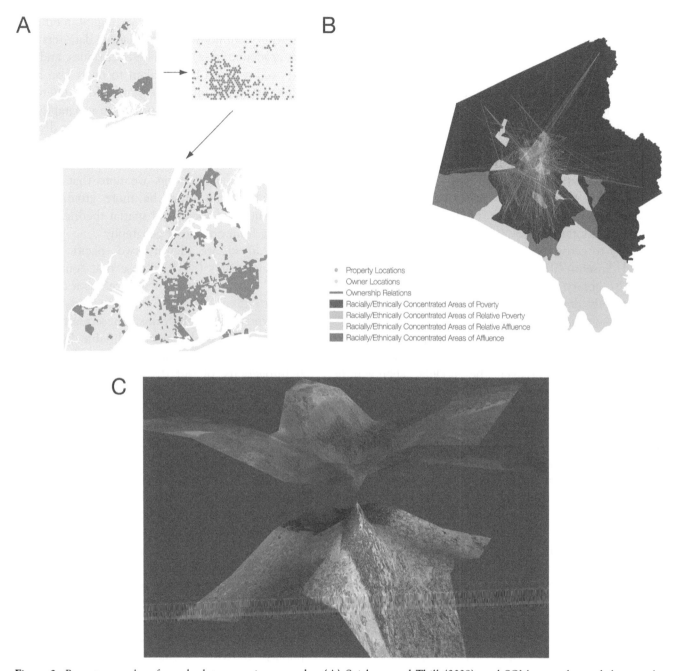

Figure 2. Recent examples of nonabsolute space in geography. (A) Spielman and Thill (2008) used SOMs to understand the complex attribute space of urban neighborhoods. Here they show how SOMs can be used to select neighborhoods that are similar to the input selection (upper left corner). (B) Shelton (2018) used a relational perspective to show that property owners of rental housing in concentrated areas of poverty often live outside those areas themselves. (C) Lally and Bergmann (forthcoming) use their software '*enfolding: a prototype Geographical Imagination System (GIS)*' to create a "wormhole" that connects Creech Air Force Base in Nevada, from which drones are operated, with the location of a U.S. drone strike in Marjah, Pakistan. SOM = self-organizing map.

Shelton (2018) used a relational perspective on poverty and constructed networks of relations between rental properties in concentrated areas of poverty and their owners, who are often located in other neighborhoods—showing that poverty itself is networked process (Figure 2B; see also Elwood, Lawson, and Sheppard 2017).

New technologies, such as smartphones, and new social practices that come along with these technologies produce new types of finely grained data. These data sets range from social media data and cell phone records to sensor data and smart card transactions and help to further enable the aforementioned reimagination. For example, Shelton and

Poorthuis (2019) used social media data to redraw Atlanta's neighborhood boundaries based on the spatial networks of its inhabitants and contrasted this with the official administrative and territorial boundaries. Boeing (2018) used OpenStreetMap data to perform a network analysis of street networks in every U.S. city at different levels of scale, and Wachsmuth and Weisler (2018) used data derived from Airbnb to analyze the effect of short-term rentals on gentrification in New York.

Finally, we have seen a major resurgence of open-source geographic software, ranging from the desktop GIS platform QGIS (QGIS Development Team 2019) to ecosystems around the pySAL library (Rey and Anselin 2010) for Python and the sf (Pebesma 2018) library for R—allowing geographers to more creatively develop and use research tools. Many of these software packages contain relational concepts in addition to Euclidean or topographical tools, opening up possibilities for adoption of diverse theories of space. Moreover, the modular approach of open-source software provides a pathway for reconnecting geography with network analysis, because network analysis libraries for the Python and R ecosystems exist parallel to their spatial counterparts. Ironically, network analysis has paid considerable attention to the spatial layout and visualization of network graphs in recent years—so far without input from geography. As an example, building on top of the projection software proj.4, Bergmann and O'Sullivan (2017) explored how Tobler's ideas for applying map projections to different spaces can be implemented in modern geographic software. Although their suggested software hyperproj is still only conceptual, an actual implementation is not far-fetched (see also Figure 2C).

Reproducible research and the open sharing of code and data, if not yet commonplace, are also becoming more prevalent (Brunsdon 2016), and such approaches are increasingly an alternative way of working next to, or in addition to, more conventional desktop GIS, as evidenced in Arribas-Bel and Reades's (2018) call for a geographic data science.

Building a Critical and Smart Symbiosis

Smart technologies and their implications have received well-deserved critical evaluation. Wyly (2014) feared the cooptation by a "neo-liberal digital capitalism" and Leszczynski (2018) highlighted epistemological limitations of digital methods that are now commonplace when smart data sets are used in academic research. These critiques and worries are well-founded but if taken to heart also highlight a unique opportunity: We can build a different, new vision on the use of technology in geography, paying particular attention to avoid some of the disciplinary mistakes of the last thirty years. Indeed, from our own research collaborations across fields of computer science and engineering, we note that quantitative geographers can often be more grounded in and appreciative of social and spatial theories than those coming from outside of geography.

To do so, we identify two specific points that echo Gahegan's (2018) phrase that "our GIS is too small." First, on a technical level, as a discipline we need to take charge of building and maintaining our own software platforms. These platforms should be open, accessible, and modifiable by the entire academic community and reflect the diversity and heterogeneity of our discipline. Similarly, as data are becoming increasingly commodified, we need to guard against walled data gardens that provide access only to selected researchers. A more widespread adoption of the principles behind reproducible research (Brunsdon 2016) in our research, publishing, and teaching would be an important cornerstone in achieving this.

Second, on a more substantive level, the computational approaches we use should allow us to think creatively about geographic processes. We can draw inspiration from all corners of our heterogenous field and its history and build on top of and together with geographical theory without constraining ourselves by what is possible with off-the-shelf software. This is especially important for space—one of our core disciplinary concepts. Computational approaches must enable researchers to freely reason with and about spatial processes. If we are to enable an engaged pluralism (Sheppard and Plummer 2007) in geography, our systems should be designed to allow for contrasting spatial ontologies rather than constrain users to a single perspective on the world.

The history of spatial science shows us that this connection between geographic theory and quantitative methods is indeed possible. We now have the technology and data sets to enable this connection in new, computational ways. To be clear, this is not a geography in which smart technology and computation completely remove the need for human input.

Instead, this is a geography where social theory and the human intellect, in conjunction with computational technologies, play a key role in analyzing the largest, most versatile, and relevant data set on social space that we have ever had.

ORCID

Ate Poorthuis ⓘ http://orcid.org/0000-0002-3808-7493
Matthew Zook ⓘ http://orcid.org/0000-0002-6034-3262

References

Anderson, C. 2008. The end of theory: The data deluge makes the scientific method obsolete. *WIRED.* Accessed October 20, 2019. http://archive.wired.com/science/discoveries/magazine/16-07/pb_theory.

Arribas-Bel, D., and J. Reades. 2018. Geography and computers: Past, present, and future. *Geography Compass* 12 (10):e12403. doi: 10.1111/gec3.12403.

Ash, J., R. Kitchin, and A. Leszczynski. 2018. Digital turn, digital geographies? *Progress in Human Geography* 42 (1):25–43. doi: 10.1177/0309132516664800.

Barnes, T. J., and M. W. Wilson. 2014. Big data, social physics, and spatial analysis: The early years. *Big Data & Society* 1 (1). doi: 10.1177/2053951714535365.

Barrat, A., M. Barthelemy, R. Pastor-Satorras, and A. Vespignani. 2004. The architecture of complex weighted networks. *Proceedings of the National Academy of Sciences* 101 (11):3747–52. doi: 10.1073/pnas.0400087101.

Bergmann, L. R., and D. O'Sullivan. 2017. Computing with many spaces: Generalizing projections for the digital geohumanities and GIScience. Paper presented at the 1st ACM SIGSPATIAL Workshop on Geospatial Humanities, GeoHumanities'17, New York. doi: 10.1145/3149858.3149866.

Bergmann, L. R., and D. O'Sullivan. 2018. Reimagining GIScience for relational spaces. *The Canadian Geographer / Le Géographe Canadien* 62 (1):7–14. doi: 10.1111/cag.12405.

Boeing, G. 2018. A multi-scale analysis of 27,000 urban street networks: Every U.S. city, town, urbanized area, and Zillow neighborhood. *Environment and Planning B: Urban Analytics and City Science.* Advance online publication. doi: 10.1177/2399808318784595.

Boggs, J. S., and N. M. Rantisi. 2003. The "relational turn" in economic geography. *Journal of Economic Geography* 3 (2):109–16. doi: 10.1093/jeg/3.2.109.

Brenner, N., D. J. Madden, and D. Wachsmuth. 2011. Assemblage urbanism and the challenges of critical urban theory. *City* 15 (2):225–40. doi: 10.1080/13604813.2011.568717.

Brunsdon, C. 2016. Quantitative methods I: Reproducible research and quantitative geography. *Progress in Human Geography* 40 (5):687–96. doi: 10.1177/0309132515599625.

Bunge, W. 1966. *Theoretical geography.* Lund, Sweden: Gleerup.

Cheng, T., J. Haworth, and E. Manley. 2012. Advances in geocomputation (1996–2011). *Computers, Environment and Urban Systems* 36 (6):481–87. doi: 10.1016/j.compenvurbsys.2012.10.002.

Cresswell, T. 2013. *Geographic thought: A critical introduction.* West Sussex, UK: Wiley-Blackwell.

Delmelle, E. C. 2017. Differentiating pathways of neighborhood change in 50 U.S. metropolitan areas. *Environment and Planning A* 49 (10):2402–24. doi: 10.1177/0308518X17722564.

Elwood, S., V. Lawson, and E. Sheppard. 2017. Geographical relational poverty studies. *Progress in Human Geography* 41 (6):745–65. doi: 10.1177/0309132516659706.

Forer, P. 1978. A place for plastic space? *Progress in Geography* 2 (2):230–67. doi: 10.1177/030913257800200203.

Gahegan, M. 1999. Guest editorial: What is geocomputation? *Transactions in GIS* 3 (3):203–6. doi: 10.1111/1467-9671.00017.

Gahegan, M. 2018. Our GIS is too small. *The Canadian Geographer / Le Géographe Canadien* 62 (1):15–26. doi: 10.1111/cag.12434.

Garrison, W. L. 1960. Connectivity of the interstate highway system. *Papers in Regional Science* 6 (1):121–37. doi: 10.1111/j.1435-5597.1960.tb01707.x.

Gatrell, A. C. 1981. Multidimensional scaling. In *Quantitative geography: A British view,* ed. N. Wrigley and R. J. Bennett, 151–63. Abingdon, UK: Routledge & Kegan Paul.

Getis, A. 1963. The determination of the location of retail activities with the use of a map transformation. *Economic Geography* 39 (1):14–22. doi: 10.2307/142492.

Golledge, R. G. 1977. Multidimensional analysis in the study of environmental behavior and environmental design. In *Human behavior and environment,* ed. I. Altman and J. F. Wohlwil, 1–42. Boston: Springer.

Golledge, R. G., V. L. Rivizzigno, and A. Spector. 1976. Learning about a city: Analysis by multidimensional scaling. In *Spatial choice and spatial behavior,* ed. R. G. Golledge and G. Rushton, 95–116. Columbus: Ohio State University Press.

Gould, P., and R. White. 1974. *Mental maps.* New York: Penguin.

Graham, S. D. N. 2005. Software-sorted geographies. *Progress in Human Geography* 29 (5):562–80. doi: 10.1191/0309132505ph568oa.

Hägerstrand, T. 1957. Migration and area. In *Migration in Sweden: A symposium,* 27–158. Lund Studies in Geography, Series B. Lund, Sweden: Gleerup.

Haggett, P., and R. J. Chorley. 1969. *Network analysis in human geography.* London: Edward Arnold.

Harris, C. D. 1954. The market as a factor in the localization of industry in the United States. *Annals of the Association of American Geographers* 44 (4):315–48. doi: 10.1080/00045605409352140.

Hartshorne, R. 1939. The nature of geography: A critical survey of current thought in the light of the past. *Annals of the Association of American Geographers* 29 (3):173–412. doi: 10.2307/2561063.

Harvey, D. 1969. *Explanation in geography.* London: Edward Arnold.

Harvey, D. 1996. *Justice, nature and the geography of difference.* Hoboken, NJ: Wiley-Blackwell.

Hemmens, G. C. 1966. *An analysis of urban travel and the spatial structure of urban activities.* PhD diss., Massachusetts Institute of Technology. Accessed October 20, 2019. http://dspace.mit.edu/handle/1721.1/75560.

Jones, M. 2009. Phase space: Geography, relational thinking, and beyond. *Progress in Human Geography* 33 (4):487–506. doi: 10.1177/0309132508101599.

Kendall, D. G. 1971. Maps from marriages: An application of non-metric multi-dimensional scaling to parish register data. In *Mathematics in the archaeological and historical sciences,* ed. P. R. Hodson, D. G. Kendall, and P. Tăutu, 303–18. Edinburgh, UK: Edinburgh University Press.

Kitchin, R. M., and M. Dodge. 2011. *Code/space: Software and everyday life.* Cambridge, MA: MIT Press.

Kitchin, R. M., and G. McArdle. 2016. What makes big data, big data? Exploring the ontological characteristics of 26 datasets. *Big Data & Society* 3 (1):1–10. doi: 10.1177/2053951716631130.

Kwan, M.-P., and T. Schwanen. 2009. Critical quantitative geographies. *Environment and Planning A: Economy and Space* 41 (2):261–64. doi: 10.1068/a41350.

Lally, N., and L. Bergmann. Forthcoming. Enfolding: A geographical imagination system (GIS). In *Into the void,* ed. A. Secor and P. Kingsbury. Lincoln: University of Nebraska Press.

Lefebvre, H. 1991. *The production of space.* Hoboken, NJ: Wiley-Blackwell.

Leszczynski, A. 2016. Speculative futures: Cities, data, and governance beyond smart urbanism. *Environment and Planning A: Economy and Space* 48 (9):1691–1708. doi: 10.1177/0308518X16651445.

Leszczynski, A. 2018. Digital methods I: Wicked tensions. *Progress in Human Geography* 42 (3):473–81. doi: 10.1177/0309132517711779.

Marchand, B. 1973. Deformation of a transportation surface. *Annals of the Association of American Geographers* 63 (4):507–21. doi:10.2307/2562050. doi: 10.1111/j.1467-8306.1973.tb00944.x.

Massey, D. B. 1994. *Space, place, and gender.* Minneapolis: University of Minnesota Press.

Merleau-Ponty, M. [1945] 2012. *Phenomenology of perception,* trans. D. A. Landes. London and New York: Routledge.

Nystuen, J. D., and M. F. Dacey. 1961. A graph theory interpretation of nodal regions. *Papers of the Regional Science Association* 7 (1):29–42. doi: 10.1007/BF01969070.

O'Sullivan, D., L. Bergmann, and J. E. Thatcher. 2018. Spatiality, maps, and mathematics in critical human geography: Toward a repetition with difference. *The Professional Geographer* 70 (1):129–39. doi: 10.1080/00330124.2017.1326081.

Pebesma, E. 2018. Simple features for R: Standardized support for spatial vector data. *The R Journal* 10 (1):439–46. doi: 10.32614/RJ-2018-009.

QGIS Development Team. 2019. QGIS geographic information system: Open Source Geospatial Foundation Project. Accessed October 20, 2019. http://qgis.osgeo.org/.

Rey, S. J. 2009. Show me the code: Spatial analysis and open source. *Journal of Geographical Systems* 11 (2):191–207. doi: 10.1007/s10109-009-0086-8.

Rey, S. J., and L. Anselin. 2010. PySAL: A Python library of spatial analytical methods. In *Handbook of applied spatial analysis: Software tools, methods and applications,* ed. M. M. Fischer and A. Getis, 175–93. Berlin: Springer.

Schuurman, N. 2006. Formalization matters: Critical GIS and ontology research. *Annals of the Association of American Geographers* 96 (4):726–39. doi: 10.1111/j.1467-8306.2006.00513.x.

Shelton, T. 2018. Rethinking the RECAP: Mapping the relational geographies of concentrated poverty and affluence in Lexington, Kentucky. *Urban Geography* 39 (7):1070–91. doi: 10.1080/02723638.2018.1433927.

Shelton, T., and A. Poorthuis. 2019. The nature of neighborhoods: Using Big Data to rethink the geographies of Atlanta's neighborhood planning unit system. *Annals of the American Association of Geographers* 109 (5):1341–61. doi: 10.1080/24694452.2019.1571895.

Shelton, T., M. A. Zook, and A. Wiig. 2015. The actually existing smart city. *Cambridge Journal of Regions, Economy and Society* 8 (1):13–25. doi: 10.1093/cjres/rsu026.

Sheppard, E., and P. Plummer. 2007. Toward engaged pluralism in geographical debate. *Environment and Planning A: Economy and Space* 39 (11):2545–48. doi: 10.1068/a40205.

Spielman, S. E., and J.-C. Thill. 2008. Social area analysis, data mining, and GIS. *Computers, Environment and Urban Systems* 32 (2):110–22. doi: 10.1016/j.compenvurbsys.2007.11.004.

Stewart, J. Q., and W. Warntz. 1958. Macrogeography and social science. *Geographical Review* 48 (2):167–84. doi: 10.2307/212129.

Stratigea, A., C. Papadopoulou, and M. Panagiotopoulou. 2015. Tools and technologies for planning the development of smart cities. *Journal of Urban Technology* 22 (2):43–62. doi: 10.1080/10630732.2015.1018725.

Taylor, P. J. 1977. *Quantitative methods in geography: An introduction to spatial analysis.* Boston: Houghton Mifflin.

Thatcher, J. E., L. Bergmann, and D. O'Sullivan. 2018. Speculative and constructively critical GIS. *The Canadian Geographer / Le Géographe Canadien* 62 (1):4–6. doi: 10.1111/cag.12441.

Tobler, W. R. 1963. Geographic area and map projections. *Geographical Review* 53 (1):59–78. doi: 10.2307/212809.

Wachsmuth, D., and A. Weisler. 2018. Airbnb and the rent gap: Gentrification through the sharing economy. *Environment and Planning A: Economy and Space* 50 (6):1147–70. doi: 10.1177/0308518X18778038.

Wyly, E. 2011. Positively radical. *International Journal of Urban and Regional Research* 35:889–912. doi: 10.1111/j.1468-2427.2011.01047.x.

Wyly, E. 2014. The new quantitative revolution. *Dialogues in Human Geography* 4 (1):26–38. doi: 10.1177/2043820614525732.

Smart Festivals? Security and Freedom for Well-Being in Urban Smart Spaces

Jeremy W. Crampton, ⓘ Kara C. Hoover, ⓘ Harrison Smith, ⓘ Steve Graham, and J. Colette Berbesque

In this article we use the natural lab of music festivals to examine behavioral change in response to the rapid introduction of smart surveillance technology into formerly unpoliced spaces. Festivals are liminal spaces, free from the governance of everyday social norms and regulations, permitting participants to assert a desired self. Due to a number of recent festival deaths, drug confiscations, pickpockets, and a terroristic mass shooting, festivals have quickly introduced smart security measures such as drones and facial recognition technologies. Such a rapid introduction contrasts with urban spaces where surveillance is introduced gradually and unnoticeably. In this article we use some findings from an online survey of festivalgoers to reveal explicit attitudes and experiences of surveillance. We found that surveillance is often discomforting because it changes experience of place, it diminishes feelings of safety, and bottom-up measures (health tents, being in contact with friends) are preferred to top-down surveillance. We also found marked variation between men, women, and nonbinary people's feelings toward surveillance. Men were much less affected by surveillance. Women have very mixed views on surveillance; they simultaneously have greater safety concerns (especially sexual assault in public) and are keener on surveillance than men but also feel that it is ineffective in preventing assault (but might be useful in providing evidence subsequently). Our findings have significant ramifications for the efficacy of a one-size-fits-all solution of increased surveillance and security in smart places and cities and point to the need for more bottom-up safety measures. *Key Words: anxiety, festivals, smart city, surveillance, well-being.*

在本文中，我们将音乐节视为一个"天然实验室"，探讨将智能监控科技快速引入未监控区域后，人们对之做出的行为变化。节日活动属于模糊临界区域，脱离日常社会规范和规则管理，让参与者释放自我。由于最近节日期间发生多起死亡、毒品、盗窃和大规模恐怖主义枪击事件，越来越多的节日活动开始启用无人机和面部识别等智能安全措施。如此快速的引入速度，与城市的监控网络构建截然不同，后者会采用逐步推进而又低调的方式进行。在本文中，我们通过对一项在线节日参与者的调研结果，说明人们对监视的明确态度和体验。我们发现，监控会降低人们的场所体验，常常令人感到不安和缺乏安全感。此外，采取自下而上的措施（急救站，与朋友保持联系）要比自上而下的监控更为可取。我们还发现男性、女性和单独出行人士对监控的感受截然不同。男性受监视的影响要小得多。女性对监控的观点尤为复杂：他们一方面更关注安全问题（特别是在公共场合的性侵犯），因此更倾向于使用监控，但同时也觉得监控对预防暴力袭击作用不大（但可能对事后举证有用）。很多人认为，在各场所和城市使用强化的"一刀切"式智能监控和安全解决方案，会有很大的效果，但我们的研究结果与之相悖。结果还指出，为了实现理想效果，采取更多自下而上的安全措施。关键词：*焦虑、节日、智能城市、监控、健康。*

En este artículo usamos el laboratorio natural de los festivales de música para examinar el cambio conductual presentado en respuesta a la rápida introducción de tecnología de vigilancia inteligente en espacios que antes carecían de control policial. Los festivales son espacios liminales, libres de la gobernanza de las normas y regulaciones sociales cotidianas, que permiten a los participantes reafirmar su propia personalidad. Pero, debido a un número reciente de muertes en los festivales, a las confiscaciones de drogas, robos de billeteras y tiroteos terroristas indiscriminados, los festivales introdujeron de manera expedita medidas de seguridad inteligente, tales como drones y tecnologías de reconocimiento facial. Tan rápida introducción contrasta con los espacios urbanos donde la vigilancia se introduce de manera gradual e inadvertida. En este artículo usamos ciertos hallazgos de una exploración realizada en red con participantes de los festivales para sacar a flote actitudes y experiencias explícitas de vigilancia. Encontramos que con frecuencia la vigilancia genera

Color versions of one or more figures in the article can be found online at www.tandfonline.com/raag

incomodidad porque cambia la experiencia de lugar, disminuye la sensación de seguridad y porque las medidas ascendentes (carpas saludables, estar en contacto con amigos) son preferidas a la vigilancia que viene de arriba a abajo. Encontramos también una marcada variación de los sentimientos de hombres y mujeres hacia la vigilancia. A los hombres les afecta mucho menos la vigilancia. Las mujeres tienen percepciones muy mezcladas sobre la vigilancia; simultáneamente, ellas tienen una mayor preocupación por sí mismas (especialmente en relación con el asalto sexual en público), y son más fanáticas sobre la vigilancia que los hombres, aunque también sienten que tiene poco efecto en la prevención del asalto (pero que subsiguientemente podría ser útil en proporcionar evidencia). Nuestros descubrimientos tienen ramificaciones significativas sobre la eficacia de una solución estándar de una vigilancia y seguridad más intensas en lugares y ciudades inteligentes y son puntuales sobre la necesidad de adoptar más medidas de prevención desde la base. *Palabras clave: ansiedad, bienestar, ciudades inteligentes, festivales, vigilancia.*

Smart city innovations promise much in the way of improved efficiency, reliability, and real-time optimization of resources (Townsend 2013). City governments will spend an estimated \$135 billion on smart city innovations by 2021 (International Data Corporation 2018). An essential selling point of the smart city is that increased surveillance of citizens and the environment (including the workplace and smart homes) will promote well-being, a sense of security, real-time governance, and citizen-centricity (Cardullo and Kitchin 2018). To this end, a smart city resident might be subject to surveillance via closed-circuit television (CCTV), automatic facial recognition at smart borders (e.g., Ireland–Northern Ireland, U.S.–Mexico), and corporate collection of personal data such as geolocational tracking.

Often the introduction of these security surveillance measures in smart places and cities is incremental and might pass unnoticed. It is therefore extremely hard to gauge what people feel in response to surveillance due to habituation. To address this question, we examine what effects sudden, increased surveillance measures have had on public places previously exempt from observation. Music festivals have recently had to grapple with a number of high-profile incidents (e.g., the 2017 Las Vegas shooting, a drug overdose death in 2014 at Coachella, and widespread sexual assaults and harassment) that resulted in the rapid introduction of smart surveillance technology into formerly less policed spaces. We used a survey targeting tool to ask participants to comment on security measures and personal safety. Unsurprising, females and nonbinary individuals had more safety concerns than males. More important, unwanted security was a common safety concern across all demographics. One of our strongest findings was that surveillance changes the experience of a place and contributes to an anxious sense of being at risk from authorities. We report these survey results and

consider how our findings are applicable to smart city planning. In particular, we argue that a one-size-fits-all approach to smart places and cities of increased surveillance can be counterproductive and that bottom-up measures are more desirable.

Surveillance Anxiety

Definition and Impacts

Surveillance anxiety is acute persistent worry and stress experienced by individuals and groups as a result of known or suspected surveillance. A critical component of surveillance anxiety is that surveilled subjects experience worries about how they are being (mis)judged and that judgment will result in exploitation, punishment, or social disparagement (Crawford 2014; Leszczynski 2015). Surveillance anxiety emerges when subjects are aware of surveillance, even in the absence of surveillance.

The history of the social impact of surveillance is substantial. In the eighteenth century, the social reformer Jeremy Bentham proposed that the space of institutions such as penitentiaries could be organized around an all-seeing or panoptic gaze, which itself could not be discerned (Bentham 1995). According to Foucault, such a spatial arrangement was instrumental in instituting a disciplinary society of normalized subjects (Foucault 1977). Smart cities might also be increasingly subject to real-time nudging through surveillance (Kitchin 2019). There is relatively little work, however, on how surveillance feels subjectively or what behaviors it affects. As surveillance reaches into the interiority of the subject via biometrics and psychographic profiling, this interiority has become a site of political contestation (Ball 2009). Data from a recent field study suggest that if neighborhood residents might initially desire increased CCTV surveillance of their homes by a security company, they eventually reported little benefit from it and experienced increased worry about their privacy. As one resident stated:

> When I see the CCTV, I still don't feel safe, in fact sometimes it makes me feel more anxious. Because I think I'm in a bad area, I get into a panic sometimes … it raises emotions of feeling frightened and intimidated. (Minton 2013, 12)

Likewise, a major survey of crime fear levels in Glasgow found majority support for new CCTV installation in residential areas but no decrease in fear levels after installation (Ditton 2000). Indeed, other studies have found that better street lighting is more cost effective than CCTV (Lawson, Rogerson, and Barnacle 2018). Visible security (chain link fences, warning notices, "smart water," gated access, CCTV) has been found to make people feel less safe (Gill, Bryan, and Allen 2007).

One aspect of modern capitalism is the increasing pressure felt by individuals to conduct themselves appropriately across different spaces. In his later work, Foucault complemented his analysis of the disciplinary society with one of governmentality, where expectations are internalized as forms of self-conduct (Foucault et al. 1991). Examples today include responsibilized behaviors such as required daily steps or target heart rates in health apps that are rewarded with lower insurance premiums or workplace wellness accreditation (Zuboff 2019). As Deleuze pointed out in his work on Foucault, it is not a question of hegemonically imposing these behaviors on passive subjects but of assigning the responsibility for them to the subject, who then also takes on additional burdens of performance anxiety (Deleuze 1992). Indeed, studies show that overfrequent health checks (a seemingly benign form of surveillance) can exacerbate feelings of guilt and worry (Stol, Schermer, and Asscher 2017).

It has also been argued that modern capitalism itself exacts a mental health cost to living under conditions of heightened surveillance (McGowan 2016). Here, private subjects fantasize that they could, if only surveillance were not so intrusive, truly be themselves. The lack of opportunity to do so, however, causes distress: Capitalism promises a sanctuary from surveillance in the private sphere, but such spaces continually recede—a "cruel optimism" (Berlant 2011). Finally, the privatization of formerly public spaces under neoliberalism (and the security surveillance entailed) has potentially transformative impacts on public space (Minton 2018).

Disproportionality

Surveillance effects will likely be experienced disproportionately, because people of color, people in poverty, LGBTQ + people, and those in urban areas experience higher levels of surveillance. Case studies of digital surveillance of the poor reveal that they are subject to a "digital poorhouse" of rules and checks (Eubanks 2018). People of color are disproportionately represented in police databases and, consequently, algorithms built to assess reoffense likelihood can discriminate against them (Angwin et al. 2016). An analysis of more than 60 million stop and searches across twenty states revealed that black drivers were stopped more often than whites and that Hispanic drivers were stopped less often (Pierson et al. 2017). In the United Kingdom, government figures indicate that in 2017–2018, black people were subject to stop and search at ten times the rate of white people (HM Government 2019).

The geography of surveillance is also uneven, being more preponderant in urban than rural areas, and because urban areas are typically more diverse, this will result in higher surveillance of minorities. Urban music such as grime often protests excessive surveillance in everyday lived spaces. East London's grime scene arose due to a variety of sociocultural factors, mainly socioeconomic marginalization and the poor quality of life in council estates. The press of gentrification on these areas means that spaces where grime arose are subject to more surveillance (Hancox 2018) by the London Metropolitan Police (Fatsis 2018). The lyrics "It was only yesterday, there was less bobbies on the beat" (Dizzie Rascal's "Sittin Here [Boy In Da Corner]") and "Out there tryna survive on the streets/Tryin' not to get killed by the police" (Skepta's "Man [Konnichiwa]") exemplify this surveillant pressure.

Such disparities in surveillance will inequitably affect well-being. Well-being is increasingly promoted as a better index "beyond the GNP" of national and global health (Organization for Economic Cooperation and Development 2017). Although holistic approaches that integrate psychological health into measures of well-being are becoming more common, there is little standardization across mental health services to collect these data at a time when mental health disorders are rising in Western industrialized nations (Ohrnberger, Fichera, and Sutton 2017). Further, ample data have demonstrated that psychosocial stresses are embodied, and there is demographic disparity in exposure and vulnerability to stresses. Anxiety is manifest biologically as a deregulated hypothalamic–pituitary–adrenal (HPA) axis (Schulkin, Gold, and McEwen 1998) that controls the stress response. Deregulation of HPA results in elevated cortisol levels—either chronic or several acute episodes, both of which contribute to cumulative stress burden—that will damage soft tissues and impede immune functioning with a long-term outcome of higher morbidity, higher mortality, and reduced life span. The end result is a weakened immune response (greater morbidity) and shortened life span (increased mortality; Goosby, Cheadle, and Mitchell 2018). Stress burden is socially patterned, with minorities bearing greater loads than whites and females bearing greater loads than males.

Why Festivals?

Festivals were chosen as a natural field lab for studying the affective relationship between citizens and surveillance. In referring to the festival as a natural field lab, we point to the real-world and authentic nature of the festival. The field setting provides a representative environment (List 2007) and a natural, rather than self-selected or convenience, sample of research (Arnett 2008; Henrich, Heine, and Norenzayan 2010). Thus, given the rapidly changing surveillance environment at festivals, we propose that they can be used as key risk indicators for slower, more long-term changes in surveillance in urban areas. Our fundamental question is whether surveillance measures have been noticed by attendees and how this affects their festival-going experience.

The modern music festival is perhaps most associated in the public mind with Woodstock (first held in 1969) and Glastonbury (first held 1970). More recently, the Burning Man festival (first held in 1986 in San Francisco and in the desert since 1991) has attracted participants from around the world. Burning Man is especially notable for attracting a wide diversity of "free spirits" originally loosely affiliated with the San Francisco Cacophony Society, an inspiration behind *Fight Club*, as well as urban exploration and flash mobs. There are now dozens of Burning Man events in the United States, Europe, South Africa, and Australasia.

Festivals are a particularly important type of event because they promise to provide a "transformative" experience (as Burning Man advertises itself) or

liminal space. In festivals, attendees have traditionally been able to act out, disrupt day–night sleeping schedules, indulge in illicit substances, or experience new sexual liaisons unmonitored by authority. As a liminal space, they offer a temporary escape from the constraints of everyday life (Kim and Jamal 2007). The festival experience (commercially aimed at a particular demographic market) is one that is cocreated by organizers and festivalgoers, which allows for the creation of an authentic sociospatial experience (Szmigin et al. 2017). In other words, the vibe or experience of place at a festival is a key attractor.

Over the past few years, however, festivals have experienced a sharp increase in assaults, pick-pockets, drug overdoses, and gun violence. A June 2018 Yougov survey in the United Kingdom revealed that 43 percent of women have experienced unwanted sexual behavior at festivals (Snapes 2018). Introduction of surveillance measures has been dramatic. By contrast, in the smart city, the gradual introduction and escalation of surveillance tools is experienced over longer periods of time. In a festival setting, the comparative experience between two festival events where drones and extra security personnel have been introduced is striking (Winton 2018) and we can better capture the primary affective reaction to changes in surveillance.

In 2018, Coachella implemented aerial surveillance of the festival perimeter by drone in the wake of the mass shooting at the Las Vegas Route 91 music festival, in which fifty-one people were killed and more than 800 were injured. In 2015, Leicester (UK) police scanned the 90,000 attendees at the Download Festival and checked them against criminal databases with automated facial recognition technology (BBC 2015). Festivals and police have also experimented with using facial recognition technology for security and access control. The technology was used at the Notting Hill Carnival in 2016–2017, but after controversy over its efficacy it was replaced in 2018 by "super recognisers" and "knife arches" (walk-through metal detectors). Ticketmaster is partnering with biometric identification firm Blink Identity to embed biometric data in tickets to reduce ticket scalping (Holbrook 2018). Consumer products such as LynQ allow users to locate others through real-time distance and direction tracking via a wearable device that can be useful for finding a friend in large crowds or where cell-phone coverage is limited or nonexistent (e.g., at Burning Man).

These speculative surveillance futures suggest several key patterns that might define how festival landscapes are experienced and governed and, by extension, the governance of smart spaces. We emphasize four developments. First, the intensification and normalization of pervasive and ubiquitous forms of surveillance and data analytics. Soon, biometric observation might be impossible to escape when accessing public spaces—whether by digital face or body movement (gestures, gait analysis, posture) capture, body tagging by wristband, chip implantation, or geolocational tracking. Second, the intensification is compounded by security and marketing logics that mutually reinforce a broader strategy of crowd control along political and economic lines. Third, leading-edge biometric technologies promise to infer not just identity (who someone is) but characteristics of people's nature such as sexual orientation (Wang and Kosinski 2018) and criminality (Wu and Zhang 2017). Finally, these logics of control fundamentally depend on surveillance strategies that invest in the power of the corporeal body to serve as both the object for governance and as subject for enabling new forms of audience engagement and affective atmospheres. This is particularly evident in the case of wearable technologies that can be exploited to encourage audience engagement with branded spaces for social media marketing.

Festival Survey

We conducted an online pilot survey designed to capture affective responses to the introduction of surveillance to festival environments. The survey was hosted on Survey Monkey. All data are fully anonymized and cannot be linked to an individual. Sample size per question varies depending on participant choice, with a total $N = 201$. Because it was a pilot survey, we sought to capture the views of festivalgoers rather than of youth in general and

Table 1. Demographic data

	18–24	25–34	35–44	45–54	55–64	65+	Total
Female	28	19	13	9	2	1	72
Male	25	33	25	22	13	1	119
Nonbinary	4	2	2	1	—	—	9
Total	57	54	40	32	15	2	200[a]

Notes: N = 201.
[a]Totals are less than 201 where respondents omitted question or selected "prefer not to say."

emphasize that although results are indicative rather than definitive, they do provide actionable empirical data on surveillance attitudes. The full survey instrument and results are presented elsewhere (Hoover et al. 2019), but here we present the responses to three key questions that focus on adult (older than eighteen) festivalgoer attitudes about safety and security (Table 1; note that *nonbinary* is an umbrella term for intersex, transgender, and nonbinary.)

Festival Personal Safety Concerns

Survey completion required that respondents have attended at least one festival. Participants were asked to select but not rank their personal safety concerns from a list and, if they had ones not listed, to describe them. Figure 1 shows the list of choices and the percentage of the sample choosing each item. The highest concern is crowd violence (71 percent on average), followed by unwanted security (47 percent).

Concerns over sexual harassment and sexual assault are split by gender, with the majority of women and nonbinaries identifying sexual harassment as a key safety concern. All women, but only 54 percent of men, identified potential crowd violence as a safety concern (Figure 2). Males reported very little fear over sexual harassment or assault (13 percent compared to 51 percent for females and 56 percent for nonbinary). One male wrote, "Im [sic] not personally worried about sexual assault. I am nervous for my female friends." Twenty-eight participants wrote in additional safety concerns (fifteen males, twelve females, and one nonbinary). Putting aside idiosyncratic concerns ("being hit by flying bottles of urine"), theft was the biggest write-in item ($n = 8$), with more males expressing this concern ($n = 6$). Only two males were concerned over security, compared to four females who specifically mentioned fear of power abuse. More females than males expressed concerns about substance abuse ruining the festival and equal numbers were concerned about crowd crushing.

Safety Measures Taken by Festivalgoers or Implemented at Festivals

What is it that makes festivals feel safer? Here we were interested in what role security surveillance plays compared to other measures (Figure 3). We found that "going with friends" and the "availability of health tents" were rated most highly in all

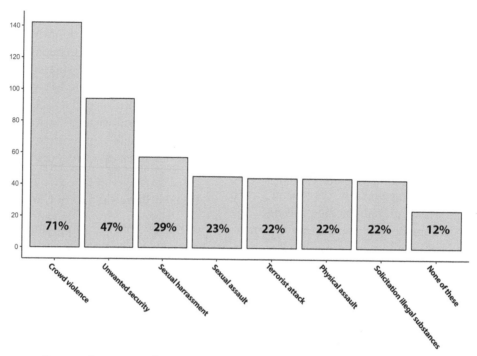

Figure 1. Safety concerns: Count and percentage choosing each option (participants could select more than one option). *Note:* N = 201.

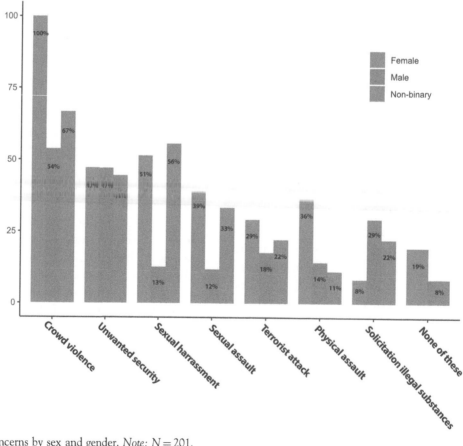

Figure 2. Safety concerns by sex and gender. *Note:* N = 201.

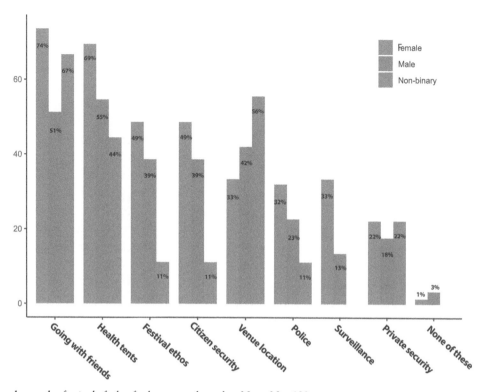

Figure 3. Measures that make festivals feel safer by sex and gender. *Note:* N = 199.

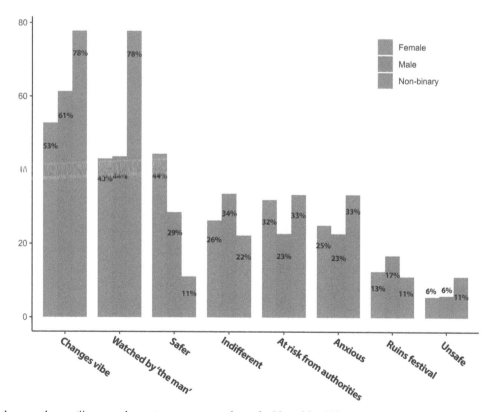

Figure 4. Attitude toward surveillance and security measures at festivals. *Note:* N = 200.

demographic categories. Additional surveillance was selected by 0 percent of nonbinary, 13 percent of men, and 33 percent of women. Women chose going with friends first and health tents second. Nonbinary chose going with friends first and venue location second. More men chose health tents first and going with friends second. Women felt consistently safer with any measure compared to men and nonbinary, with the exception of venue location. Generally, we found that security in the form of surveillance, police, or private security was the least popular choice for effective safety measures. Measures that helped participants feel safe were crowd-sourced (citizen security, health tents), having a support network, and the festival itself (venue and ethos). Three female and five male respondents wrote in additional measures but these were variants on listed choices (e.g., "medical facilities on site," festival tradition, metal detectors). One felt divided because having police to report sexual harassment to was preferred but the participant did not feel that the police could do anything after the fact, which recalls the statistic given earlier indicating that sexual harassment cases at festivals are rarely reported. Another person specifically stated that private security forces were dangerous: "They seem to feel entitled and that they just have a status/a power over you."

Attitude toward Surveillance and Security Measures at Festivals

We also specifically asked participants for their feelings about surveillance. Nonbinary participants continue to exhibit a mistrust of authority by rating the presence of surveillance and security measures as negative (Figure 4). The strongest feeling across categories was that surveillance "changed the vibe" and felt like being "watched by the man," a finding that we interpret to mean that their experience of place was affected by surveillance. We again observe gender differences in that more women felt it made them safer (44 percent) compared to men and nonbinaries (29 percent and 11 percent, respectively). Female and nonbinary participants were split on the issue, with some saying that they felt they were at risk from authorities and others saying that it made them feel safer. Most suggestively, substantial numbers (23–33 percent) of people reported being made more anxious about surveillance.

Discussion

Our survey results suggest a complex and mixed attitude about security. Contrary to our initial expectations from the literature, festivalgoers do not rank the liminality of a festival or escapism as primary reasons for attending. More significant for our work, there is a negative attitude about security surveillance. Responses across questions indicate a strong preference for community-based or bottom-up safety measures—being with a group, having a health tent, and citizen security. Women, in particular, voiced concerns that they might be harmed by figures of authority, but at the same time they expressed some element of feeling safer with security present—this contrast might reflect the particular concerns of women over sexual infractions. Finally, we found that between one in four (men and women) and one in three (nonbinary) respondents reported feeling explicitly anxious about surveillance. Although there is still much work to be done on elucidating when and whether surveillance anxiety is a persistent factor of the smart city in the age of the Internet of Things, our pilot survey does provide indicators for preliminary conclusions that can inform planning for the smart city. We argue, therefore, for the following four positions in implementing security surveillance in smart places and cities.

Surveillance Is Not a Universal Public Good

First, surveillance is not a "good" that universally increases well-being. Notable proportions of our respondents said that surveillance made them anxious, changed the experience of place, and made them feel subject to authoritarian judgment. This is quite remarkable, given the youth of our respondents (median age younger than thirty-five). This means that respondents came of age during the advent of social media and the rollout of governmental security measures (11 September 2001 occurred seventeen years prior to our survey) and might be expected to be inured to having open, digitally documented lives. Yet respondents not only disliked surveillance but felt that policing, private security, and surveillance made them feel less safe.

Disproportionality across Gender

Second, we must do more to account for the fact that security and surveillance are experienced disproportionately. Our evidence in particular suggests strong gender divisions. Perhaps the most suggestive here are the different concerns about being "watched by the man," as our survey put it. Whereas nearly 80 percent of those identifying as nonbinary identified this authoritative surveillance as a concern, fewer than half of men and women did. Another difference, not exclusive to festival spaces, is fear of sexual harassment. Only about one in ten men feared this, whereas nearly half of women and one in three nonbinary respondents did. Yet women remain ambivalent about the efficacy of surveillance in preventing attacks. We can speculate that if the smart city credibly planned measures to prevent sexual harassment in public, it would be much more likely trusted.

Peer and Bottom-Up Solutions Are Preferred

Third, measures that are within the control of the community such as going with friends, health tents, and festival ethos were all deemed to be the best way to enhance well-being, especially for women. This finding is in stark contrast to outsider-imposed security (surveillance or policing). Not a single nonbinary respondent identified surveillance as a contributor to safety. Given that the biggest concern overall is "crowd violence," an internally generated problem, survey participants look to peers and community as preferable remediations. Although more work needs to be done, and our survey was a pilot with relatively small numbers, we can tentatively conclude that planning and designing the smart city will be more successful if it is cocreated with residents rather than top-down.

Security Surveillance and the Production of Violence

Finally, we point to our finding that crowd violence is the biggest single concern (and the only concern felt by a majority of men). The proliferation of unrest and protest in cities dating to the 1960s and the Watts Riots, often in association with police use of deadly force, is patently a massive concern. We argue from our results and other evidence that overuse of security surveillance is both unwanted and materially produces less safe communities. Our findings support other data on excessive and often racialized policing of some communities. Of the

1,000 people killed in the United States by police in each of the last several years, African Americans and Hispanics are disproportionately affected (Mapping Police Violence 2018). A parallel concern in the United Kingdom is inequitable policing, such as stop and search procedures, where blacks are affected at ten times the rates of white people (HM Government 2019). The Macpherson Report of 1999 into a racially motivated murder admitted for the first time that UK policing suffered from "institutional racism" and new laws were enacted (the Race Relations Amendment Act) that make the police more accountable.

Our article draws on survey results about surveillance and safety to contribute to the ongoing smart city question of how and whether surveillance can make them safer. We raise key concerns about the normalization of smart logics and technologies in urban governance through the surveilled citizen. Utopian discourses of smart cities might act to reproduce existing inequalities, digital divides, and power structures, as well as to register increased burdens on well-being, both physically and mentally. We still know too little about the effects of surveillance over the long term, who is more likely to experience it, under what conditions, and how it changes people's experiences of place. Given the foundational basis of surveillance in enabling the smart city to function, a cooperative rather than top-down approach is clearly warranted. Our findings indicate that a one-size-fits-all approach that valorizes more surveillance in the smart city will likely be counterproductive.

Acknowledgments

We acknowledge the helpful suggestions of the anonymous reviewers and the editorial guidance of Dr. Ling Bian. Thank you to the respondents to our survey and those who helped to publicize it.

ORCID

Jeremy W. Crampton ⓘ http://orcid.org/0000-0001-5702-0430
Kara C. Hoover ⓘ https://orcid.org/0000-0001-6394-930X
Harrison Smith ⓘ http://orcid.org/0000-0001-8144-5754

References

Angwin, J., J. Larson, S. Mattu, and L. Kirchner. 2016. Machine bias. *ProPublica*, May 23. Accessed May 27, 2016. https://tinyurl.com/jzdpyz.

Arnett, J. J. 2008. The neglected 95%: Why American psychology needs to become less American. *American Psychologist* 63 (7):602–14. doi: 10.1037/0003-066X.63.7.602.

Author Redacted. 2018. An empirical study of suveillance anxiety. *SocArXiv*.

Ball, K. 2009. EXPOSURE. *Information, Communication & Society* 12:639–57. doi: 10.1080/13691180802270386.

BBC. 2015. Facewatch "thief recognition" CCTV on trial in UK stores. Accessed November 15, 2018. https://tinyurl.com/y4xexn9v.

Bentham, J. 1995. *The panopticon writings*. London: Verso.

Berlant, L. G. 2011. *Cruel optimism*. Durham, NC: Duke University Press.

Cardullo, P., and R. Kitchin. 2018. Being a "citizen" in the smart city: Up and down the scaffold of smart citizen participation in Dublin, Ireland. *GeoJournal* 84:1–13. doi: 10.1007/s10708-018-9845-8.

Crawford, K. 2014. The anxieties of big data. *The New Inquiry*, May 30: 28.

Deleuze, G. 1992. Postscript on the societies of control. *October* 59:3–7.

Ditton, J. 2000. Crime and the city. *British Journal of Criminology* 40 (4):692–709. doi: 10.1093/bjc/40.4.692.

Eubanks, V. 2018. *Automating inequality: How high-tech tools profile, police, and punish the poor*. New York: St. Martin's.

Fatsis, L. 2018. Grime: Criminal subculture or public counterculture? A critical investigation into the criminalization of black musical subcultures in the UK. *Crime, Media, Culture: An International Journal* 15(3). doi: 10.1177/1741659018784111.

Foucault, M. 1977. *Discipline and punish: The birth of the prison*. New York: Pantheon.

Foucault, M., G. Burchell, C. Gordon, and P. Miller. 1991. *The Foucault effect: Studies in governmentality: With two lectures by and an interview with Michel Foucault*. Chicago: University of Chicago Press.

Gill, M., J. Bryan, and J. Allen. 2007. Public perceptions of CCTV in residential areas: "It is not as good as we thought it would be." *International Criminal Justice Review* 17 (4):304–24. doi: 10.1177/1057567707311584.

Goosby, B. J., J. E. Cheadle, and C. Mitchell. 2018. Stress-related biosocial mechanisms of discrimination and African American health inequities. *Annual Review of Sociology* 44 (1):319–40. doi: 10.1146/annurev-soc-060116-053403.

Hancox, D. 2018. *Inner city pressure: The story of grime*. London: William Collons.

Henrich, J., S. J. Heine, and A. Norenzayan. 2010. Most people are not WEIRD. *Nature* 466 (7302):29. doi: 10.1038/466029a.

HM Government. 2019. *Stop and search: Ethnicity facts and figures*. London: Home Office. Accessed March 4, 2019. https://www.ethnicity-facts-figures.service.gov.uk/crime-justice-and-the-law/policing/stop-and-search/latest.

Holbrook, C. 2018. Concerts may soon be using facial recognition technology for tickets. *MixMag*, May 8. Accessed November 15, 2018. https://mixmag.net/read/concerts-may-soon-be-using-facial-recognition-technology-for-tickets-news.

Hoover, K. C., J. W. Crampton, H. Smith, and J. C. Berbesque. 2019. An empirical study of surveillance anxiety. *SocArXiv Papers*. doi: osf.io/osf.io/yx6nk.

International Data Corporation. 2018. *Worldwide semiannual smart cities spending guide*. Framington, MA: International Data Corporation. Accessed April 8, 2019. https://www.idc.com/getdoc.jsp?containerId=prUS43576718.

Kim, H., and T. Jamal. 2007. Touristic quest for existential authenticity. *Annals of Tourism Research* 34 (1):181–201. doi: 10.1016/j.annals.2006.07.009.

Kitchin, R. 2019. The timescape of smart cities. *Annals of the American Association of Geographers* 109 (3):775–90. doi: 10.1080/24694452.2018.1497475.

Lawson, T., R. Rogerson, and M. Barnacle. 2018. A comparison between the cost effectiveness of CCTV and improved street lighting as a means of crime reduction. *Computers, Environment and Urban Systems* 68:17–25. doi: 10.1016/j.compenvurbsys.2017.09.008.

Leszczynski, A. 2015. Spatial big data and anxieties of control. *Environment and Planning D: Society and Space* 33 (6):965–84. doi: 10.1177/0263775815595814.

List, J. A. 2007. *Field experiments: A bridge between lab and naturally occurring data*. Chicago: University of Chicago. Accessed April 8, 2019. https://www.nber.org/papers/w16062.

Mapping Police Violence. 2018. 2017 Police violence report. Accessed November 15, 2018. https://policeviolencereport.org/.

McGowan, T. 2016. *Capitalism and desire: The psychic cost of free markets*. New York: Columbia University Press.

Minton, A. 2013. *"Fortress Britain." High security, insecurity and the challenge of preventing harm*. London: New Economics Foundation.

Minton, A. 2018. The paradox of safety and fear: Security in public space. *Architectural Design* 88 (3):84–91.

Ohrnberger, J., E. Fichera, and M. Sutton. 2017. The relationship between physical and mental health: A mediation analysis. *Social Science & Medicine* 195:42–49. doi: 10.1016/j.socscimed.2017.11.008.

Organization for Economic Cooperation and Development. 2017. *How's life 2017: Measuring well being*. Paris: Organization for Economic Cooperation and Development.

Pierson, E., C. Simoiu, J. Overgoor, S. Corbett-Davis, V. Ramachandran, C. Phillips, and S. Goal. 2017. A large-scale analysis of racial disparities in police stops across the United States. Accessed November 15, 2018. https://5harad.com/papers/traffic-stops.pdf.

Schulkin, J., P. W. Gold, and B. S. McEwen. 1998. Induction of corticotropin-releasing hormone gene expression by glucocorticoids: Implication for understanding the states of fear and anxiety and allostatic load. *Psychoneuroendocrinology* 23 (3):219–43. doi: 10.1016/S0306-4530(97)00099-1.

Snapes, L. 2018. One in five at UK festivals sexually assaulted or harrassed—Survey. *The Guardian*, June 18. Accessed November 15, 2018. https://www.theguardian.com/music/2018/jun/18/one-in-five-at-uk-festivals-sexually-assaulted-or-harassed-survey.

Stol, Y. H., M. H. N. Schermer, and E. C. A. Asscher. 2017. Omnipresent health checks may result in over-responsibilization. *Public Health Ethics* 10:35–48.

Szmigin, I., A. Bengry-Howell, Y. Morey, C. Griffin, and S. Riley. 2017. Socio-spatial authenticity at co-created music festivals. *Annals of Tourism Research* 63:1–11. doi: 10.1016/j.annals.2016.12.007.

Townsend, A. M. 2013. *Smart cities: Big data, civic hackers, and the quest for a New Utopia.* New York: Norton.

Wang, Y., and M. Kosinski. 2018. Deep neural networks are more accurate than humans at detecting sexual orientation from facial images. *Journal of Personality and Social Psychology* 114 (2):246–57.

Winton, R. 2018. Coachella: Police using army of drones to boost security in wake of Las Vegas massacre. *LA Times*, April 12. Accessed November 15, 2018. http://www.latimes.com/local/lanow/la-me-coachella-security-drones-cops-20180412-story.html.

Wu, X., and X. Zhang. 2017. Automated inference on criminality using face images. Accessed August 15, 2018. http://arxiv.org/abs/1611.04135.

Zuboff, S. 2019. *The age of surveillance capitalism: The fight for a human future at the new frontier of power.* New York: Public Affairs Books.

Powers of Division: "Smart" Spaces as Controlling Workplace Activity Fragmentation

Jonathan Stiles and Clinton Andrews

In this article we argue that "smart" work spaces are created through access to wireless networks and mobile cloud computing collaboration software. Yet the power relations embedded in these overlapping physical and cyberspaces function to control the spatial and temporal fragmentation of related work activities. Through a review of the literature, we first describe how scholars have considered the relationship between space, technology, and the workplace and how conceptions of work activity fragmentation developed over time in relation to computing technology. We provide data from the American Time Use Survey to show the extent of work activity fragmentation among workers in the United States, finding that spatial fragmentation is more prominent among knowledge workers in large metropolitan areas. Drawing on original interview and observation data in the New York metropolitan area, we then describe both the cyberspaces of work enabled by mobile cloud collaboration software and the physical spaces of work opened up by the availability of wireless networks in diverse locations. Finally, we consider how the exercise of power and control in these overlapping physical and cyberspaces can either enable or prevent the fragmentation of work activities at multiple points. *Key Words: activity fragmentation, cyberspace, information and communication technology, power relations, smart space.*

本文认为，可通过访问无线网络和移动云计算协作软件，创建"智能"工作空间。但在这些重叠的实际空间和网络空间中，其内在的权力关系会控制相关工作活动的时空分散性。我们首先回顾了相关文献内容，说明了其他学者如何考虑空间、技术和工作场所之间的关系，讨论了工作活动分散概念这一概念，如何随着时间开发与计算机技术同步发展。我们根据《美国人时间使用调查》提供的数据，列出了美国人工作活动的分散程度。结果显示，空间分散性最突出的群体，是大都市地区中的知识型工作人员。我们还利用对纽约都会区的原始采访和观察数据，对两个不同的空间进行了描述：由移动云协作软件支持的工作网络空间，在不同地点提供无线网而实现的开放式实际工作空间。文章最后，我们思考了一个问题：在这些重叠的实际空间和网络空间中，行使权力和控制权如何实现(或妨碍) 多点分散式工作模式。关键词：活动分散、网络空间、信息和通信技术、权力关系，智能空间。*关键词：活动分散、网络空间、信息和通信技术、权力关系，智能空间。*

En este artículo sostenemos que los espacios de trabajo "inteligentes" se crean mediante el acceso a redes inalámbricas y la colaboración del software de computación móvil en la nube. Sin embargo, las relaciones de poder incrustadas en estos espacios cíber y físico trasladados funcionan para controlar la fragmentación espacial y temporal de actividades de trabajo relacionadas. A través de una revisión de la literatura, describimos primero el modo como los eruditos han considerado la relación entre espacio, tecnología y lugar de trabajo, y cómo las concepciones de la fragmentacion de la actividad laboral se desarrollaron a través del tiempo en relación con la tecnología de la computación. Suministramos datos del Estudio Americano sobre el Uso del Tiempo para mostrar el alcance de la fragmentación de la actividad laboral entre los trabajadores de los Estados Unidos, hallando que la fragmentación espacial es más pronunciada entre los trabajadores del conocimiento en las áreas metropolitanas grandes. A partir de entrevistas originales y datos de observación en el área metropolitana de Nueva York, describimos después los ciberespacios de trabajo activados en diversas localizaciones con software de colaboración de la nube móvil y los espacios físicos de trabajo abiertos por la disponibilidad de redes inalámbricas. Por último, consideramos cómo el ejercicio de poder y control de estos espacios físicos y ciber que se traslapan pueden bien habilitar o prevenir la fragmentación de las actividades de trabajo en múltiples puntos. *Palabras clave: ciberespacio, espacios inteligentes, fragmentación de la actividad, relaciones de poder, tecnología de la información y la comunicación.*

Reductions in distance frictions and increases in information flows facilitate activity fragmentation for both work and play. When activities fragment, they are subdivided and dispersed in both time and space. This allows greater freedom to choose when and where to do what, with attendant benefits in convenience and efficiency. It could also make it more difficult to maintain a separation between work and home life and allow distraction to diminish the quality of work. Examples of activity fragmentation are easy to locate in daily lives. A couple continues a morning conversation about what to cook for dinner by texting from their workplaces throughout the day. A shopper browses appliances online while commuting but purchases later from home. Workers from different office locations and from home collaboratively produce a project report. Activity fragmentation researchers have investigated the role of mobile devices in enabling fragmentation for individual workers. This article's contribution is that it considers the "smart" work spaces that are created through embedded wireless networks and access to mobile cloud computing collaboration software. We argue that for knowledge workers, physical and cyberspaces of work can overlap and function to control the fragmentation of their work activities.

This article proceeds as follows. First, through a review of literature, we describe how scholars have considered the relationship between space, technology, and the workplace and offer our definition of a smart work space. Second, we look at how conceptions of work activity fragmentation developed over time in relation to computing technology and use data from the American Time Use Survey to show the extent of work activity fragmentation in the United States. Third, we draw on original interview and observation data in the New York metropolitan area to describe both cyberspaces of work enabled by online collaboration software and physical spaces of work opened up by the availability of wireless network access. We provide a typology of these spaces and their territorial controls. Finally, we consider how the exercise of power and control in these overlapping physical and cyberspaces can either enable or prevent the fragmentation of work activities at multiple points.

Approaching Smart Spaces for Work

For human beings, space is a social phenomenon. Social relationships—in particular power relations—give form to the shared societal spaces we inhabit and also shape our individual experiences of space and our efforts to represent particular spaces as we seek to manage them (Lefebvre 1991). The order and materiality of these spaces depend on the purposes of the power relations through which they are produced. For example, Foucault (2012) documented the ways in which efforts to discipline the human

body in service of military, political, and economic needs of society manifested in spaces of schools, prisons, and factories. Yet space is also nonexclusive and unfixed, which means that spaces overlap and comingle as they change and move through time (Thrift 2006). Even as a body is confined by one space, the possibility exists for the inhabiting of multiple other spaces. These changes derive from an unending process of social construction that supports the coexistence of a multiplicity of social processes within the diversity of spaces (Massey 2005).

If space is produced through power relations, then spaces of work depend on the ways in which organizations seek to control employees toward achieving organizational goals. Edwards (1982) defined three forms of organizational control: Simple control consists of bosses exercising power directly, technical control relies on physical machinery to direct workers, and bureaucratic control depends on a hierarchy of roles. Since the 1990s, a "projectification" of work grew in some industries, as teams were given autonomy (Midler 1995). Such teams function through a fourth form of organizational control—concertive control—in which teams self-manage by enforcing practices through peer review (Barker 1993). These forms of organizational control can also be framed by how apparent they are: Simple and technical controls are more obtrusive, whereas bureaucratic and concertive controls are more unobtrusive (Tompkins and Cheney 1985).

The workplace represents a practice of territoriality, as organizational controls are extended to manage access to an area, which is only allowed—or required—at certain times and for some people (Sack 1986). The description of the space of a factory given by Foucault speaks to both bureaucratic and technical forms of control, as the disciplining of the human body demanded by the machinery of the modern factory defined the geographic distribution of workers along with their ranks and positions (Foucault 2012). In office spaces, employees are similarly directed yet might also act territorially in marking or defending particular spaces as their own (Brown 2009). What distinguishes "place" as a particular form of space is the embedding of values by social actors (Gieryn 2000). For a workplace, such values are derived from the organization. Organizations establish shared values among workers, in part as a means of unobtrusive control (Tompkins and Cheney 1985). Organizational values contribute to the shaping of worker identities, which establishes and fulfills roles within bureaucratic hierarchies and on project teams and guides the standards to which organization members hold themselves and others (Alvesson and Willmott 2002).

This role of technologies in mediating organizational control and in the production of space broadly has been seen in both positive and negative lights. For example, Perrow (1991) described the bureaucracy derived from the machinery of the factory as "an unobtrusive control device of unprecedented power" (743) that supersedes nonorganizational influences on employees such as community and family. Perrow's and Foucault's visions of the technology of the factory tend toward both the dystopian and the deterministic. Yet empirical research supports a similar view of the intersection of information technology, control, and space. Sewell (1998) found that project teams subjected to information technology–enabled surveillance of work simultaneously experience top-down bureaucratic control and horizontal concertive control, which together support a centralization of power. Levy (2015) similarly found that new electronic surveillance of truck drivers served to centralize control in the firm, in part by extending influence into their personal lives. More utopian visions see information technology as having a liberating effect. The "death of distance" of Cairncross (1995) viewed the "communications revolution" as largely positive for organizations and individuals. Additionally, Castells (2015) saw it as empowering social movements seeking societal change. In some organizational work contexts, team members might offer a "concertive resistance" to centralized power (Zanin and Bisel 2019).

With the expansion of network infrastructure, workspace can emerge within shared public space. Public spaces are those that are considered to be accessible, yet in practice, accessibility is contested and in flux (Smith and Low 2013). In New York City, the many small privately owned public spaces as a result of zoning incentives tend to be securitized, and their public use is subject to rules (Németh 2009). Nonetheless, these spaces contain a diversity of uses (Huang and Franck 2018). Public spaces embedded with wireless Internet networks are blends of physical and digital space—what Forlano (2009) described as "codescapes"—and used by freelancers as work locations. Yet Varnelis and Friedberg (2008) pointed out that spaces can function at

odds: Although customers are present in the physical space of a corporate café, they are more engaged in some cyberspace. Alternately they function together as hybrid spaces (de Souza e Silva 2006). For mobile knowledge workers, the embedded networks in diverse public spaces create the possibility to conduct work in new locations, yet also represent infrastructural challenges that must be overcome because access might be restricted or equipment might be lacking (Erickson and Jarrahi 2016)

If space is a social phenomenon, then it follows that a "smart space" possesses this essential quality, in additional to another. In this article, we consider "smartness" to be the integration of information technology in a capacity that creates or mediates the social interactions—and in particular power relations—that produce space. Yet we do not take a stance on the direction of its effect as positive or negative, liberating or controlling, but rather see these effects as rooted in particular context. In physical space, smartness can be derived from smart objects that are aware of the environment around them and can interpret and act on sensory data (Kortuem et al. 2010). Yet it also is supplied by infrastructures that enable connectivity such as wireless networks. In cyberspaces, smartness might be algorithms designed to promote organizational knowledge (Liebowitz 2001). Yet it also is supplied by software that makes information and communication available to users. Rather than existing distinct from one another, smart physical spaces and cyberspaces can be overlaid and coexisting.

Work Activity Fragmentation

Conceptual understandings of the effects of information and communication technology (ICT) on activities have evolved toward an understanding of its capability to fragment activities into smaller units that occur across different spaces or points in time. This concept has especially been applied to work activities. The notion of telecommuting emerged in the 1970s as a means to alleviate problems related to automobile commuting. This earliest conception of telework envisioned clusters of computer terminals in an employer-owned suburban center that communicated with a firm's central computer system (Nilles et al. 1976). For individual workers, this imagining represented a wholesale spatial and temporal shift of their work activities, with workers attending the nearby alternate office location for a full workday. Other framings of this period similarly see telecommunications as substituting for an activity that otherwise necessitated travel to an office (Coss and Lennox 1975; Kraemer and King 1982).

Researchers subsequently developed more complex understandings of the relationship of telecommunications with activity patterns (Salomon 1985; Mokhtarian 1990; Couclelis 1996). Under the growing dominance of mobile computing and the Internet, Couclelis (2009) offered the concept of activity fragmentation, arguing that activities spread across time and space in ways they could not before ICTs and that this change is embedded in perceptions of action. Dijst (2004) considered activity fragmentation, such as work activities, as a possibility but one constrained by the needs of daily life such as parenting. Schwanen and Kwan (2008) examined the effects of ICTs on such space–time constraints in enabling activity fragmentation, considering three types of constraints: capability (getting stuff done), coupling (social interactions), and authority (observance of laws and rules), finding that ICT can relax constraints but that those tied to formal settings or necessitating face-to-face contact persist.

Findings concerning the costs and benefits of fragmentation have been mixed. A meta-analysis considered thirty-five empirical studies finding that nearly all showed that telecommuting reduced travel (Andreev, Salomon, and Pliskin 2010). Evaluations of productivity, however, found that although telecommuting can increase productivity, in wider application it has not greatly affected it. In an analysis of pilot telework programs among Fortune 100 companies, Olson (1989) found that managers preferred to have workers in the office, and productivity was not changed among home workers compared to in-office workers. Fragmentation also depends on participant satisfaction. A study of attitudes toward telecommuting in 1988 found that both employees and managers worried about the effect of working at home on career development (Duxbury, Higgins, and Irving 1987). Professional isolation was found to be problematic because professional development activities are often spontaneous, such as the learning that takes place outside a cubicle (Cooper and Kurland 2002).

Scholars have further defined and operationalized activity fragmentation. Hubers, Schwanen, and Dijst (2008) measured activities through dimensions of component episodes, including number of episodes, sizes, and particular configuration. Cao (2012) considered the mechanisms through which e-commerce affects travel, including fragmentation, such as trips to try out items in-store before buying online or in-store pick up. Lenz and Nobis (2007) operationalized the concept through considerations of both spatial and temporal fragmentation and compared fragmentation among workers with differing levels of mobile phone and computer usage. Finally, Alexander, Ettema, and Dijst (2010) defined fragmentation as "the decomposition of work into multiple segments of subtasks that can be performed in different times and/or locations" (55). They observed different forms of fragmentation and different tendencies based on ICT usage, finding handheld devices associated with temporal fragmentation and laptops associated with spatial fragmentation.

Work Activity Fragmentation in the United States

For some Americans, work is fragmented across different spaces and times. To view this magnitude, we use pooled 2003 to 2017 cross-sectional data from the American Time Use Survey, which asks a sample of Americans to record activities and locations throughout a diary day. Time use analysis of work has explored a decline in overnight work and coordination of work hours for working couples (Hamermesh 2003). It has also been combined with analysis of work locations in the Canadian context (Lachapelle, Tanguay, and Neumark-Gaudet 2018). We compare activity fragmentation for all U.S. workers against fragmentation for just knowledge workers, defined as those in management, business, computer, engineering, community, legal, education, arts, health care practice, sales, and office occupations. Additionally, to see the extent to which fragmentation is a global city phenomenon, we look at fragmentation for just those knowledge workers living in large urban areas, measured as metropolitan statistical areas (MSAs) with a population greater than 2.5 million people.

In terms of spatial work fragmentation, knowledge workers are more likely to work from home or from multiple types of locations for their primary job in a day (Table 1). Over 16 percent of U.S. knowledge workers indicated that they conducted work from multiple locations in a day, with over 11 percent reporting conducting work from both home and the

Table 1. Work activity fragmentation for main job: U.S. workers on workday, 2003–2017

	All workers (n = 51,326) (%)	Knowledge workers (n = 34,917) (%)	Knowledge workers in large MSAs (n = 11,790) (%)
Spatial: Daily work locations			
Workplace only	77.4	71.2	68.7
Home only	9.5	12.4	14.0
Workplace and home only	9.0	11.6	12.6
Workplace and one other only (café, public, etc.)	2.0	2.2	2.0
Home and one other only (café, public, etc.)	1.0	1.2	1.2
More than two types of location	1.2	1.4	1.5
Temporal: Daily episodes of work			
One episode	19.3	19.9	19.5
Two episodes	41.5	43.2	45.0
Three episodes	21.3	20.4	20.2
Four or more episodes	18.0	16.5	15.3

Notes: Knowledge occupations exclude protective service, food preparation, personal care, agricultural, production, construction, maintenance, and transportation occupations. Large MSAs are designated as MSAs greater than 2.5 million persons. MSA = metropolitan statistical area.
Source: American Time Use Survey.

workplace in the same day. Additionally, nearly 5 percent of U.S. knowledge workers indicated that they worked from a location beyond home and work, such as a café, vehicle, airport, or public space. Among those knowledge workers in large MSAs, more than 30 percent reported working from a location or locations other than a workplace only, with over 17 percent working from multiple locations. Although knowledge workers in the United States, especially those in large cities, are more likely to experience spatial fragmentation, this relationship does not hold for temporal fragmentation. Nearly 40 percent of all workers reported more than two episodes of work for their main job in a day. The incidence of temporal fragmentation beyond two episodes slightly decreases, however, for knowledge workers and for knowledge workers in large MSAs (Table 1).

Territoriality and Work Activity Fragmentation in Overlapping Space

We argue that smart work spaces enact power relations through the control of activities, supported by ICT. These power relations can be understood through a concept of territoriality. The architect John Habraken defined territoriality as "we are in control of a space if we can decide who comes in and who goes out" (Andrews 2004, 8). We extend this to consider not only control over physical access for one's body but also control over access to the materials and collaborators through which activities

could be accomplished. Given the potential for spaces to be overlapping and changing through time, the distinctions and interactions between physical spaces and cyberspaces is an important consideration, and each will be explored in turn based on empirical findings. First, based on interviews with knowledge workers in the New York metropolitan area, we provide examples of smart cyberspaces of work enabled through mobile cloud computing infrastructure. Second, based on observational research in New York City, we describe the diverse physical spaces of work that are used by knowledge workers in a global city.

Smart Cyberspaces through Team Collaboration Software

Cyberspaces enabled through Internet networks are best seen as multiple particular spaces based on social relations, rather than as a singular generalized cyberspace (Graham 2013). Employees inhabit particular cyberspaces that are formed through virtual social interactions with colleagues, work materials, and managers. To gain insight into cyberspaces of work, we draw on interview data collected from a nonprobability sample of thirty-one knowledge workers in the New York metropolitan area who reported teleworking at least occasionally. Interviewees had higher levels of education and income and were more likely to live in an urban environment than the overall metropolitan workforce. Recruitment was done through online and flyer advertisements and the networks of one author. Twenty-nine of the interviews were conducted using Skype audio, and the others were conducted in person. Findings show how knowledge workers used mobile cloud computing–based software to communicate with team members and to access work materials. In a mobile cloud software model, rather than being wholly executed on a local device, software is delivered "as a service" through network connections (Mell and Grance 2011). In practice, cloud software provides the same data and services to users across multiple devices, including mobile devices such as smartphones.

More than three quarters of interviewees discussed using mobile cloud collaboration software beyond e-mail to communicate with colleagues or clients and to access work information. These applications included messaging software such as Slack, shared database software such as Salesforce, video or audio conferencing software such as Skype, document-sharing software such as Google Docs, and specialized project management software. Slack, a team messaging platform, connects users to multiple channels within an organization built around skills or shared projects and makes information and communications available on both mobile phones and personal computers. Slack also uses machine learning to identify and suggest experts within an organization on a particular subject (Woyke 2018). The relationship management platform Salesforce similarly makes data available through the Internet across multiple devices and uses machine learning to identify sales opportunities as well as automate interactions with customers (Salesforce 2018). By injecting machine intelligence into virtual interactions, these smart elements of cyberspaces give an advantage over face-to-face alternatives.

In using such software to collaborate, teams of knowledge workers inhabit particular cyberspaces that support their work activities. In the words of an interviewee regarding the messaging platform Slack, "Our team has our own channel within Slack so we can talk about our own stuff." One interviewee referred to Slack as a place, stating, "That's pretty much where everyone is working all day." Employees can participate in multiple cyberspaces. One interviewee drew a distinction between the informality of her team's messaging software in which "we sometimes send funny messages" and the formality of a note-sharing platform that was used internally for more official

reporting to management. These cyberspaces can also support the fragmenting of activities, because "you don't need to be physically at your computer to use Slack and communicate with everyone else." Another interviewee reported, "The project management software has a mobile platform so [I] can keep track of projects and tasks remotely via smartphone."

Territoriality also emerged as a theme in interviews, with some interviewees reporting being required to use virtual private networks (VPNs) to access work materials and services. VPNs allow secure access to a private network, such as a corporate network, through the larger public Internet. Some interviewees also discussed using two-factor authentication to work remotely on a laptop, which requires them to have a smartphone installed with "an app that authenticates you." Access to some services was also restricted to work devices, with one interviewee reporting, "I'm only allowed to use [Salesforce] on my work phone, because of security reasons." Cyberspaces also showed elements of surveillance by management. One interviewee reported being surprised when she "figured out recently that [my] head boss can read [my] Slack messages." Another employee whose work consisted of managing information in a database reported that her employer's Web site tracked her number of hours worked, which she would review with a manager. One freelance worker, however, practiced a form of protective self-surveillance in using a time-tracking app, which she described as standard for local freelancers.

Smart Physical Spaces through Wireless Network Connectivity

Such cyberspaces of work must be accessed through a network connection that is grounded in physical space, such as a WiFi hotspot or node in a mobile network. Therefore, considerations of territoriality for the physical spaces of work include both physical access to the space and access to any network connections available within the space. Indeed, the spatial fragmentation of work depends on access to such network connections. To gain insight into territoriality for a diversity of physical work spaces beyond offices and homes, we conducted observations at fifteen publicly accessible New York City public spaces including Brookfield Place, the David Rubenstein Atrium at Lincoln Center, Bryant

Table 2. Typology of physical work spaces by territorial control and network access

Category	Work location	Territorial control	Network access
Employer-managed spaces	Private office with exclusive workspaces	Employer User should be employee/guest	Authentication protected
	Private office with shared workspaces (i.e., hoteling)	Employer User should be employee/guest	Authentication protected
Third-party-managed spaces	Coworking space	Space owner User should be member	Authentication protected
	Retail space or coffee shop	Space owner User should be customer	Free if available but might require registration
	Privately owned public space	Space owner User should follow rules	Free if available but might require registration
	Publicly owned public space	Space manager User should follow rules	Free if available but might require registration
	Hotel	Space owner User should be guest	Might be free, require registration, or paywall
	Airport, plane, or train	Space manager User should be traveler	Might be free, require registration, or paywall
Employee-managed spaces	Home	Householder User should be resident or guest	Authentication protected
	Own vehicle	Vehicle owner User should be occupant	None; depends on mobile networks

Park, and multiple coffee shops where workers congregated. Our observation protocol included observing layout, security, the workers and nonworkers using the space, and its network connection protocols.

Different configurations of territoriality exist for different physical spaces. In Table 2 we offer a typology of work locations that designates who controls access to a space and its networks, as well as a description of how workers access networks based on our observations. In employer-managed spaces, forms of organizational control create power asymmetries and determine access to the assets that permit engaging in work activities. If the employment relationship is severed altogether, the employee loses the right to access the workspace and its assets both in physical space and in cyberspace. For third-party-managed public spaces, determinants of territoriality are more nebulous. Government often loosely regulates public spaces, either directly or in partnerships; hence, public use is permitted and direct exclusion is unusual. Furthermore, some commercial private spaces that function as public spaces, such as coffee shops, also see much public use, but exclusion exists based on the ability to pay. Public space might also have rules that are established—known or unknown—the violation of which can provide grounds for expulsion. Based on our observation

research we consider four brief cases of third-party-managed spaces. Systematic observations of such spaces are emerging in the literature (Griffiths and Gilly 2012; Huang and Franck 2018).

Working at Brookfield Place in Lower Manhattan. Brookfield Place in Lower Manhattan's Battery Park City combines a market, food court, retail mall, and an indoor public space called the "winter garden" at the center of four towers of office space. On a Tuesday morning its central public space was buzzing with foot traffic including office workers, residents of nearby housing, and tourists from the adjacent World Trade Center. During an hour-long observation, nine people were working on laptops in the public space, and ten more were working on laptops in the food court. Most workers appeared affiliated with nearby offices, with some wearing the access cards needed to bypass the gates that protect upstairs offices. Most were working alone, some while dining, although there were two meeting pairs with laptops as well. The Brookfield Place WiFi network is free to use but requires an e-mail registration and also encourages signing up for the Brookfield Place newsletter, which advertises upcoming events. Security in the public space was noticeable, with at least two guards actively roaming the central public space, but no attempts to limit access were observed. In 2011, however, a group of Occupy protesters were removed from the space, and Brookfield tightened its rules to prohibit lying down and constructing dwellings (Tower 2014).

Working at the David Rubenstein Atrium at Lincoln Center.   The David Rubenstein Atrium is operated by the Lincoln Center for the Performing Arts and combines a ticket office, a café, and ample public seating. During an early afternoon weekday observation, the space was highly staffed with café and ticket office employees, as well as a security guard and information kiosk employee. The space also was busy with a diverse mix of users, including five people working from laptops, several tables of socializing café customers, tourists waiting for ticket purchases, and people resting in the space, some possibly homeless. This diversity of uses observed reinforces recent findings from Huang and Franck (2018). In contrast to Brookfield Place, here the security guard was not actively patrolling and seemed more relaxed. The WiFi is free yet requires registration with a full name and e-mail address and asks users to complete a survey and view upcoming Lincoln Center events. The vendor that provides the WiFi registration service promises to allow site managers to collect email addresses of visitors for marketing purposes. Several of the workers with laptops were present for the entire duration of the hour-long observation.

Coffee Shop–Based Working in Brooklyn.   At a large coffee shop in North Brooklyn in the late afternoon on a Tuesday there were sixteen people working on laptops, most in the large interior but a few outside on the sunny patio. Workers far outnumbered the handful here not working. Most workers were alone but one group of young women was collaborating on a project. The WiFi network here is free and the login information is prominently posted; however, a sign at the counter reads, "No laptops after 8PM." A similar scene played out at a South Brooklyn coffee shop, which also operates as a bar in the evenings, where there were thirteen people working on laptops at communal tables on a Monday afternoon. This space is not quiet like some cafés with a large proportion of workers, because there were also some groups socializing, as well as an event in an upstairs venue. Access to the WiFi network is free but is password protected. Even though the information is posted, one shop employee noticed a wandering customer and said, "I've come to recognize that face—looking for WiFi—it's a very

distinct face." A Starbucks in a neighborhood distant from Manhattan also had workers present but fewer of them and mixed among students and shoppers. During an hour-long observation, four tables were occupied by workers, one of whom was alternating between working on a laptop and making sales calls. The WiFi at Starbucks is free but requires a one-time e-mail signup. Finally, at another Brooklyn coffee shop, a newly instituted closing time of 5 p.m. led to nine customers—mostly workers with laptops—scrambling to finish and pack up when the barista announced, "We are closing in ten minutes."

Working in Bryant Park in Midtown Manhattan. Bryant Park, a small green space surrounded by commercial office buildings on three sides, is known for its plentiful wooden chairs that can be moved about the space. The day of observation was cloudy, yet the park was busy with a mix of lunch-goers and tourists enjoying the park's attractions such as a café, dining outlets, a reading library, and a carousel. Yet workers were also present. Altogether fifteen people were counted working on laptops in the park; the largest number was clustered on the park's southern border, across from a coworking space. The Bryant Park Public WiFi network available in the park is free. Yet other networks are also available, including the coworking network requiring a membership, which extended well into the park, and an AT&T network, which requires being a customer.

Powers of Division: The Control of Smart Spaces

Our interviews with New York City–area knowledge workers show that the cyberspaces of work they inhabit are supported by collaborative technologies based on mobile cloud computing infrastructure, yet these spaces are subject to organizational control through territoriality and surveillance. Our observations of New York City public spaces show that third-party-managed spaces beyond the workplace and home are being used for work activities yet are also subject to limitations on physical and network access. We contend that control over the territory of overlapping smart physical work spaces and smart cyber work spaces represents a power to divide activities across time and space. This fragmenting of activities can be enabled or prevented at multiple points within these overlapping physical and cyberspaces. This includes controlling physical access to a space, controlling access to wireless networks, controlling access to private networks through a VPN, and controlling access to work data and collaboration software. It is a power that is largely held by employers, yet third-party managers play an increasing role, and employees must contend with a relative lack of control. Economic inequalities that are evident elsewhere in society replicate themselves in smart spaces, both physically, when the user sometimes must pay for access, and virtually, when the user sometimes must either pass through a paywall or use a mobile phone with a data plan.

Smart spaces are not created accidentally or by evolutionary processes. Rather, these spaces are planned for and made actors in the solving of organizational problems related to activity fragmentation. We consider it appropriate, then, to problematize smartness as an intentional planning act, asking questions about who has the power to control outcomes. Empirical work certainly needs to continue characterizing activity fragmentation in terms of the number, distribution, and configuration of fragments in space and time (Alexander, Ettema, and Dijst 2010). It also needs to characterize the associated power relationships in emerging smart spaces. A recent study of the efforts of mobile knowledge workers to overcome a lack of access to the infrastructure needed to conduct their work is an example of such a focus (Erickson and Jarrahi 2016). Another example is a study that speaks to concertive control in terms of how organizational culture can support an extension of work space into the home and of home space into the workplace (Fleming and Spicer 2004). In terms of policy, the design of public spaces could strive for the minimum acceptable set of characteristics to support work activities, with consideration of seating, tables, shelter, noise, and public network access.

Finally, smart spaces are also an instantiation of the abstract intentions of the designers of their component technologies. Proprietary cross-platform productivity suites such as Salesforce and Slack are important, because they highlight corporate power and influence in the supply of smartness. The cloud is not really without boundaries, and individuals are pushed to live within one or another of these ecosystems for different aspects of their lives—and are confounded by poor interoperability if they try to cross boundaries or go against a designated platform. The research community is challenged by the relative newness of the mobile cloud computing platforms and network configurations discussed in this article. An additional challenge will be the incorporation of power and space—both physical and cyber—into understanding their effects on society.

References

Alexander, B., D. Ettema, and M. Dijst. 2010. Fragmentation of work activity as a multi-dimensional construct and its association with ICT, employment and sociodemographic characteristics. *Journal of Transport Geography* 18 (1):55–64. doi: 10.1016/j.jtrangeo.2009.05.010.

Alvesson, M., and H. Willmott. 2002. Identity regulation as organizational control: Producing the appropriate individual. *Journal of Management Studies* 39 (5):619–44. doi: 10.1111/1467-6486.00305.

Andreev, P., I. Salomon, and N. Pliskin. 2010. Review: State of teleactivities. *Transportation Research Part C: Emerging Technologies* 18 (1):3–20. doi: 10.1016/j.trc.2009.04.017.

Andrews, C. J. 2004. Security and the built environment: An interview with John Habraken. *IEEE Technology and Society Magazine* 23 (3):7–12. doi: 10.1109/MTAS.2004.1337874.

Barker, J. R. 1993. Tightening the iron cage: Concertive control in self-managing teams. *Administrative Science Quarterly* 38 (3):408–37.

Brown, G. 2009. Claiming a corner at work: Measuring employee territoriality in their workspaces. *Journal of Environmental Psychology* 29 (1):44–52. doi: 10.1016/j.jenvp.2008.05.004.

Cairncross, F. 1995. Telecommunications: The death of distance. *The Economist* 30:5–28.

Cao, X. J. 2012. The relationships between e-shopping and store shopping in the shopping process of search goods. *Transportation Research Part A: Policy and Practice* 46 (7):993–1002. doi: 10.1016/j.tra.2012.04.007.

Castells, M. 2015. *Networks of outrage and hope: Social movements in the Internet age*. Hoboken, NJ: Wiley.

Cooper, C. D., and N. B. Kurland. 2002. Telecommuting, professional isolation, and employee development in public and private organization. *Journal of Organizational Behavior* 23 (4):511–32. doi: 10.1002/job.145.

Coss, F., and D. D. Lennox 1975. Teletransportation—An answer to 21st-century problems. *Management Review* 64 (1):54–56.

Couclelis, H. 1996. The death of distance. *Environment and Planning B: Planning and Design* 23 (4):387–89. doi: 10.1068/b230387.

Couclelis, H. 2009. Rethinking time geography in the information age. *Environment and Planning A: Economy and Space* 41 (7):1556–75. doi: 10.1068/a4151.

De Souza e Silva, A. 2006. From cyber to hybrid: Mobile technologies as interfaces of hybrid spaces. *Space and*

Culture 9 (3):261–78. doi: 10.1177/1206331206 289022.

Dijst, M. 2004. ICTs and accessibility: An action space perspective on the impact of new information and communication technologies. In *Transport developments and innovations in an evolving world*, ed. M. Beuthe, V. Himanen, A. Reggiani, and L. Zamparini, 27–46. Berlin: Springer.

Duxbury, L., C. Higgins, and R. Irving. 1987. Attitudes of managers and employees to telecommuting. *INFOR: Information Systems and Operational Research* 25:273–85. doi: 10.1080/03155986.1987.11732043.

Edwards, R. 1982. *Contested terrain: The transformation of the workplace in the twentieth century*. New York: Basic.

Erickson, I., and M. H. Jarrahi. 2016. Infrastructuring and the challenge of dynamic seams in mobile knowledge work. In *Proceedings of the 19th ACM Conference on Computer-Supported Cooperative Work & Social Computing*, 1323–36. New York: ACM.

Fleming, P., and A. Spicer. 2004. "You can checkout anytime, but you can never leave": Spatial boundaries in a high commitment organization. *Human Relations* 57 (1):75–94. doi: 10.1177/0018726704042715.

Forlano, L. 2009. WiFi geographies: When code meets place. *The Information Society* 25 (5):344–52. doi: 10. 1080/01972240903213076.

Foucault, M. 2012. *Discipline and punish: The birth of the prison*. New York: Vintage.

Gieryn, T. F. 2000. A space for place in sociology. *Annual Review of Sociology* 26 (1):463–96. doi: 10.1146/ annurev.soc.26.1.463.

Graham, M. 2013. Geography/Internet: Ethereal alternate dimensions of cyberspace or grounded augmented realities? *The Geographical Journal* 179 (2):177–82. doi: 10.1111/geoj.12009.

Griffiths, M. A., and M. C. Gilly. 2012. Dibs! Customer territorial behaviors. *Journal of Service Research* 15 (2): 131–49. doi: 10.1177/1094670511430530.

Hamermesh, D. S. 2003. Timing, togetherness and time windfalls. In *Family, household and work*, ed. K. F. Zimmermann and M. Vogler, 1–23. Berlin: Springer.

Huang, T., and K. A. Franck. 2018. Let's meet at Citicorp: Can privately owned public spaces be inclusive? *Journal of Urban Design* 23 (4):499–517. doi: 10.1080/13574809.2018.1429214.

Hubers, C., T. Schwanen, and M. Dijst. 2008. ICT and temporal fragmentation of activities: An analytical framework and initial empirical findings. *Tijdschrift voor Economische en Sociale Geografie* 99 (5):528–46. doi: 10.1111/j.1467-9663.2008.00490.x.

Kortuem, G., F. Kawsar, V. Sundramoorthy, and D. Fitton. 2010. Smart objects as building blocks for the Internet of things. *IEEE Internet Computing* 14 (1):44–51. doi: 10.1109/MIC.2009.143.

Kraemer, K. L., and J. L. King. 1982. Telecommunications/transportation substitution and energy conservation Part 2. *Telecommunications Policy* 6 (2):87–99. doi: 10.1016/0308-5961(82)90027-1.

Lachapelle, U., G. A. Tanguay, and L. Neumark-Gaudet. 2018. Telecommuting and sustainable travel: Reduction of overall travel time, increases in non-

motorised travel and congestion relief? *Urban Studies* 55 (10):2226–44. doi: 10.1177/0042098017708985.

Lefebvre, H. 1991. *The production of space*. Oxford, UK: Blackwell.

Lenz, B., and C. Nobis. 2007. The changing allocation of activities in space and time by the use of ICT— "Fragmentation" as a new concept and empirical results. *Transportation Research Part A: Policy and Practice* 41:190–204. doi: 10.1016/j.tra.2006.03.004.

Levy, K. E. 2015. The contexts of control: Information, power, and truck-driving work. *The Information Society* 31 (2):160–74. doi: 10.1080/01972243.2015.998105.

Liebowitz, J. 2001. Knowledge management and its link to artificial intelligence. *Expert Systems with Applications* 20 (1):1–6. doi: 10.1016/S0957-4174(00) 00044-0.

Massey, D. 2005. *For space*. Thousand Oaks, CA: Sage.

Mell, P., and T. Grance. 2011. *The NIST definition of cloud computing*. Washington, DC: U.S. Department of Commerce.

Midler, C. 1995. "Projectification" of the firm: The Renault case. *Scandinavian Journal of Management* 11 (4):363–75. doi: 10.1016/0956-5221(95)00035-T.

Mokhtarian, P. L. 1990. A typology of relationships between telecommunications and transportation. *Transportation Research Part A: General* 24 (3): 231–42. doi: 10.1016/0191-2607(90)90060-J.

Németh, J. 2009. Defining a public: The management of privately owned public space. *Urban Studies* 46 (11): 2463–90. doi: 10.1177/0042098009342903.

Nilles, J. M., R. Carlson, P. Gray, and H. Gerard. 1976. *Telecommunications-transportation trade-off: Options for tomorrow*. Hoboken, NJ: Wiley.

Olson, M. 1989. Work at home for computer professionals: Current attitudes and future prospects. *ACM Transactions on Information Systems* 7 (4):317–38. doi: 10.1145/76158.76891.

Perrow, C. 1991. A society of organizations. *Theory and Society* 20 (6):725–62. doi: 10.1007/BF00678095.

Sack, R. D. 1986. *Human territoriality: Its theory and history*. Cambridge, UK: Cambridge University Press. doi: 10.1086/ahr/94.1.103.

Salesforce. 2018. Salesforce Einstein. Accessed November 26, 2018. https://www.salesforce.com/products/einstein/ overview/.

Salomon, I. 1985. Telecommunications and travel: Substitution or modified mobility? *Journal of Transport Economics and Policy* 19:219–35.

Schwanen, T., and M. P. Kwan. 2008. The Internet, mobile phone and space-time constraints. *Geoforum* 39 (3):1362–77. doi: 10.1016/j.geoforum.2007.11.005.

Sewell, G. 1998. The discipline of teams: The control of team-based industrial work through electronic and peer surveillance. *Administrative Science Quarterly* 43 (2):397–428. doi: 10.2307/2393857.

Smith, N., and S. Low. 2013. Introduction: The imperative of public space. In *The politics of public space*, ed. N. Smith and S. Low, 7–22. London and New York: Routledge.

Thrift, N. 2006. Space. *Theory, Culture & Society* 23:139–46. doi: 10.1177/0263276406063780.

Tompkins, P. K., and G. Cheney. 1985. Communication and unobtrusive control in contemporary organizations. *Organizational Communication: Traditional Themes and New Directions* 13:179–210.

Tower, S. 2014. Not in my front yard: Freedom of speech and state action in New York City's privately owned public spaces. *Journal of Law and Policy* 22 (1):433–81.

Varnelis, K., and A. Friedberg. 2008. Place: The networking of public space. In *Networked publics*, ed. K. Varnelis, 15–42. Cambridge, MA: MIT Press.

Woyke, E. 2018. Slack hopes its AI will keep you from hating slack. *MIT Technology Review* 121 (2):13–14.

Zanin, A. C., and R. S. Bisel. 2019. Concertive resistance: How overlapping team identifications enable collective organizational resistance. *Culture and Organization* 1–19. Advance online publication. doi: 10.1080/14759551.2019.1566233.

Smart Spaces, Information Processing, and the Question of Intelligence

Casey R. Lynch ⓘD and Vincent J. Del Casino, Jr.

As spaces increasingly come to be described as "smart," "sentient," or "thinking," scholars remain in disagreement as to the nature of intelligence, knowledge, or the "human mind." This article opens the notion of intelligence to contestation, examining differing conceptions of intelligence and what they might mean for how geographers approach the theorization of "smart" spaces. Engaging debates on the distinction between cognition and consciousness, we argue for a view of intelligence as multiple, partial, and situated in and in-between spaces, bodies, objects, and technologies. This article calls on geographers to be attentive to the *multiple* forms of intelligence made possible by innovations in information processing and to the ways in which particular intelligences are prioritized—as others might be neglected or suppressed—through the production of smart spaces in the context of our rapidly changing understandings of the "humanness" of intelligence. *Key Words: cognition, consciousness, digital technology, intelligence, space.*

隨著空間逐漸被描繪為"智慧的"、"知覺的"或"思考的", 學者仍對智能、知識, 抑或"人類心靈"的本質有所爭議。本文開啟有關智能概念的爭論, 檢視不同的智能概念, 及其對地理學者如何處理"智慧"空間的理論化而言有何意義。我們涉入區分認知和意識的辯論, 主張將智能視為多重、部分、且處於空間、身體、物件和科技之中與之間。本文呼籲地理學者在我們對於智能的"人性"快速變化的理解脈絡中, 關照由信息處理中的創新所造就的多重智能形式, 以及特定的智能通過智慧空間的生產取得優先順位的方式——其他智能則可能受到忽略或壓迫。关键词: 认知, 意识, 数码科技, 智能, 空间。

Frente a la tendencia en aumento de describir los espacios como "inteligentes", "conscientes" o "pensantes", los estudiosos siguen en desacuerdo sobre la naturaleza de la inteligencia, el conocimiento, o la "mente humana". Este artículo pone la noción de inteligencia en discusión, examinando diferentes concepciones de la misma y lo que estos conceptos podrían significar en la manera como los geógrafos se aproximan a la teorización de los espacios "inteligentes". Adentrándonos en los debates sobre la distinción entre cognición y conciencia, abogamos por una visión de la inteligencia como múltiple, parcial y situada entre y dentro de espacios, cuerpos, objetos y tecnologías. Este artículo hace un llamado a los geógrafos para que estén atentos a las múltiples formas de inteligencia que se hacen posibles por las innovaciones en el procesamiento de información y a las maneras como inteligencias particulares se priorizan—del mismo modo como otras podrían desecharse o suprimirse—mediante la producción de espacios inteligentes en el contexto de nuestros rápidamente cambiantes entendimientos de la "humanidad" de la inteligencia. Palabras clave: cognición, conciencia, espacio, inteligencia, tecnología digital.

Recently, geographical inquiry into "smart" spaces, particularly but not exclusively cities, has expanded. Geographers have examined the political economy of the smart city as a development model and new urban planning paradigm (Batty 2013; Marvin and Luque-Ayala 2017), while asking questions about the logics of securitization and control on which smart models are based (Söderström, Paasche, and Klauser 2014; Pötzsch 2015; Leszczynski 2016; Amoore and Raley 2017). Authors have highlighted the fuzzy definition of *smart*, from approaches driven by technology to those stressing the "smartness" of coupled human–technical systems or the importance of human capital (e.g., smart people) to the information economy (Kitchin 2014). In all, the notion of smart spaces is generally described in its entanglements with shifting ideologies of planning and governance and the expanding domains of contemporary techno-capitalism. This approach is appropriate for understanding the spread and power of smart discourses and practices. Yet, geographers have engaged less with the concept of *intelligence*—a key topic of debate in psychology, neurobiology, philosophy, computer science, and digital humanities. We address this lacunae by resituating geographical discussions around smart spaces within these broader debates around intelligence.

We argue that expanded understandings of intelligence as multidimensional, variegated, and exceeding the human open up new ways to imagine so-called smart futures. We lay out a vocabulary for examining the various forms of intelligence that are proliferating into the spaces of everyday life and how these might be understood in relation to classically understood human forms of intelligence. We draw on the distinction between cognition and consciousness and their relationship—supported by Hayles (2017) and Shanahan (2010)—to interrogate both the evolving agency of technical systems and the complex sociotechnical milieu within which (post)human consciousness is entangled. We reframe smart spaces as cognitive—or as ongoing processes in the evolving spatialization of cognition—albeit in a way that is quite different from cognitive and behavioral geography (cf. Aitken and Bjorklund 1988). At the same time, we reassert a role for differentiated forms of (post)human consciousness, often ignored in literature that tends to focus on forms of technical agency with little attention given to the human (Rose 2017).

In what follows, we review the existing literature on smart spaces in geography. We argue that within this literature, terms like *smart*, *intelligent*, *sentient*, and *thinking* have been undertheorized—often inadvertently reinforcing reductive cognitivist notions of intelligence. This cognitivist approach limits geographers' capacities to think creatively about intelligence because it operates in a binary logic of conscious and unconscious, human and nonhuman. To challenge this logic, we turn our attention to the question of intelligence as a nonbinary process of knowledge production and consumption. We outline the debates around intelligence across psychology, philosophy, and computer science, among others, and offer Hayles's distinction between cognition and consciousness as a working vocabulary for thinking about the complex and evolving entanglement of human and technological modes of producing meaning.

Smart Spaces

The notion of smart in geography has been employed to describe a range of interventions and development strategies globally. As many have

discussed, the smart moniker has been used to describe both initiatives focused on adopting new digital technologies—embedding sensors in space and collecting and analyzing data in near real time (Kitchin 2014)—as well as those development strategies focused on attracting skilled workers in the "knowledge economy." In this sense, smart operates as a branding technique for emerging forms of digitalized public–private management—reflecting what Söderström, Paasche, and Klauser (2014) referred to as a form of "corporate storytelling" and Hollands (2008) called "urban labeling." Recognizing this, many scholars have ignored the term while focusing attention on the sociotechnical assemblages or political economic logics that it represents.

Geographers have explored how evolving digital systems—constituted by "coded objects, infrastructures, processes, and assemblages" (Kitchin and Dodge 2005, 170)—have come to play increasingly vital roles in the production of urban space and everyday life. Ash (2017) described the way "smart objects"—from light bulbs to watches and toothbrushes—generate "phases" or "space-times, around which human and non-human life is organized" (16). To describe the agentive capacities of these objects and assemblages, however, scholars have typically employed a constellation of terms interchangeably—including intelligence, sentience, and thinking—with very little operational discussion of their meaning (Beer 2007; Shepard 2011; Albino, Berardi, and Dangelico 2015; Picon 2015). For example, in writing about the rise of "sentient" spaces, Crang and Graham (2007) argued that the spatial diffusion of information processing means that "we not only think of cities but cities think of us" (789, italics added). Amin and Thrift (2016) similarly described urban sociotechnical systems as "thinking" entities. Both went on to offer detailed analyses of spatialized information processing, but their use of terms like *thinking* and *sentient* lacks specificity and deeper theorization. This ambiguous terminology blurs the distinctions between human and technical agency and modes of being.

When not clearly theorized, this ambiguous terminology reinforces problematic and hotly contested conceptions of intelligence. Perhaps worse, it reduces "thinking" to information processing. For geographers, this reduction poses a significant challenge to a robust theorization of intelligence and its application to emerging smart discourses and spaces. Put another way, this theory of intelligence—referred to as the "cognitivist paradigm" (Hayles 2017) or as the "executive theories" (Hardcastle 1995)—constitutes a direct relationship between information on the one hand and knowledge, intelligence, and consciousness on the other. In this model, the latter are theorized as the direct product of the former. The human brain is compared to a computer and thought is understood as the manipulation of abstract symbols inputted from an exterior world (Thompson 2007). Dupuy (2009) called this model the "mechanization of the mind," an approach that reproduces a mind–body dualism long deconstructed by feminist and queer theorists (Grosz 1994; Murray and Sullivan 2012).

This cognitivist approach to intelligence has been robustly critiqued by psychologists and cognitive biologists, among others (Hardcastle 1995). Even so, it continues to exert significant academic and applied influence, driving research programs in artificial intelligence (AI) and hopes or fears of a coming "technological singularity"—a moment of unprecedented technological growth marked by the rise of "thinking" machines operating well beyond human capacity—among some in the technologist community (Shanahan 2015). When scholars passively accept—or fully embrace—terms like smart, intelligent, or sentient and then focus on information processing systems, they help reinforce a problematic conception of intelligence at the heart of debates in computer science, psychology, and philosophy.

Similarly, Rose (2017) critiqued smart city research in geography for undertheorizing the differentiated forms of posthuman agency, as the agential capacities of technical objects become the primary object of analysis. Relying on Stiegler, Rose (2017) suggested that we must "theorize (digital) posthuman agency by thinking it as always already (digitally) sociotechnical" (789). Rose (2017) asserted that geographic scholarship must move away from the human as a "supplement" to digital life. Put another way, "geographers must … reconfigure their understanding of digitally mediated cities and acknowledge both the reinventiveness and the diversity of urban posthuman agency" (Rose 2017, 789). We cannot, as she intimated, fall into the trap of thinking that machine intelligence is the only intelligence of the future or that human intelligence will simply be augmented by new machinic intelligent systems. Our

spaces are overly complicated by the constitutive posthuman relations of sociotechnological life to assert such reductions.

Intelligence, Spatialized Cognition, and Consciousness

Given this lingering critique of intelligence in the conversation around smart spaces, we turn to debates surrounding the nature of human and nonhuman intelligence. In this, we employ Hayles's (2017) distinction between cognition and consciousness to build a vocabulary for thinking about the abilities of emerging spatialized information processing systems and their relationships to differentiated human forms of intelligence.

The Question of Intelligence

Within the fields of psychology and cognitive science, there is little agreement as to the nature of human intelligence. In relation to the debates over the cognitivist model discussed previously, psychologists have long questioned the role of consciousness, bodily processes, environment, and culture in relation to something called intelligence. Theories of general intelligence posit a correlation of relative individual strengths and weaknesses across an array of cognitive functions, with humans and other creatures falling somewhere along a continuum of intelligence. This theory holds that one's place on the continuum can be quantitatively measured through IQ tests and similar mechanisms.

Of course, such notions of intelligence have long been contested. Within psychology, Gardner (2011) rejected theories of general intelligence, arguing for a theory of multiple intelligences, distinguishing among eight to ten different intelligences, from logical–mathematical and verbal–linguistic to bodily–kinesthetic and visual–spatial intelligences. Others have critiqued theories of general intelligence for their racial biases and role in historical and contemporary projects of colonialism, eugenics, and various forms of scientific racism (Smedley and Smedley 2005). Recently, autism rights activists and psychologists have called for the recognition of "neurodiversity," deconstructing conceptions of "normal" neurological behavior and challenging the pathologization of autism, attention deficit hyperactivity disorder, dyslexia, and other common "mental disorders" (Armstrong 2015).

Such debates around intelligence become more complicated when claims of intelligence are extended to nonhumans. Hayles (2017), for example, discussed the debates around plant intelligence in which Brenner et al. (2006) drew homologies between animal neurology and plants' mechanisms for responding to environments. Although the topic is controversial among plant scientists and neurobiologists alike, Hayles (2017) pointed out that the controversy is more about the use of the word *intelligent* than it is about the science of plant behavior. Hayles (2017) saw in such work a "double intent to draw upon the cachet of 'intelligence' as an anthropocentric value while simultaneously revising the criteria for what constitutes intelligence" (19); in other words, what traits and attributes constitute intelligence and whether or not the term can and should be extended to nonhumans and to what extent. Brenner et al.'s (2006) critics draw a clear line between animal forms of intelligence—based on their neurological similarities to humans that see the brain as the site of intelligence—and responsive plant behavioral mechanisms not connected to a neurological structure.

Similar debates across computer science have found renewed attention in relation to AI. The field of AI is remarkably diverse and composed of multiple working theories as to the nature of intelligence and the route to engineering an AI. The ability to build an AI depends essentially on one's definition of intelligence. Many in the field have set "human-level intelligence" as a benchmark—essentially that an AI would be able to perform most or all human functions equal to or better than an "average" human. One of the primary approaches to human-level AI has been to engineer faster and more complex information processing systems. Recently, this approach has led to major advancements in machine learning. This approach to AI is most directly influenced by cognitivist notions of intelligence. A competing approach to AI has focused on brain emulation and neural networks. This approach locates intelligence in physical processes and structures in the brain—specifically in the electric signaling of neurons—and aims to emulate those processes in artificial systems. In all of these systems, computer scientists have begun to shy away from the term AI, preferring instead *intelligence*

augmented to describe the ways machine learning algorithms or even neural networks extend human intelligence but do not replace it. Although this reframing problematically reasserts the binary of human and nonhuman, it also suggests the possibility of a theorization based in coconstitution and not a binary logic.

Reflecting on these debates, several scholars have highlighted the possibility for multiple forms of intelligence, not directly comparable to a metric of "humanness" (Shanahan 2010, 2015). These scholars argue for an expanded notion of intelligence as multidimensional and diverse while decentering the human as the ultimate metric. As new technological apparatuses are created and used in new ways, we do not know what kinds of intelligences might emerge or are emerging. It is therefore important that we leave the question of intelligence open and shift attention to the diversity of interpretive and agential capabilities emergent within complex systems. This shift has clear ethical and analytical implications for how we understand the agential capacities of nonhumans, including technical apparatuses. To explore the diversity of intelligences across human and nonhuman actors, we reflect on the distinction that Hayles (2017), Shanahan (2010), and others have drawn between cognition and consciousness. This distinction allows us to think more specifically about the multiplicity of intelligences and their entanglements and interdependencies.

More-Than-Human Cognition

Drawing on sources across cognitive biology and philosophy, Hayles (2017) defined cognition as "a process that interprets information within contexts that connect it with meaning" (22). This means reframing cognition as a process rather than an attribute or ability residing in a particular individual or entity. Hayles (2017) further described this as "dynamic unfolding within an environment in which its activity makes a difference" (25). Cognitive roboticist Shanahan (2010) similarly argued that "[c]ognition has arisen because it beneficially modulates a creature's behaviour … it intervenes in the sensorimotor loop by means of which the creature interacts with its physical and social environments" (3). In both of these definitions, cognition is a process of interaction with an environment in which an entity processes sensory information and responds according to set goals—whether a biological drive for survival or a programmed objective.

Understood in this way, cognition "becomes a pervasive activity among humans, animals, and technical devices, with many different kinds of agents contributing to a rich ecology of collaborating, reinforcing, contesting, and conflicting interpretations" (Hayles 2017, 213). Understanding technical apparatuses of data input and algorithmic knowledge production as processes of cognition calls into question how they relate to other forms of spatialized cognition. Smart city data flows represent only one aspect of information processing in cities; they need to be understood in their larger entanglements. This requires an attentiveness to the various forms of information that might not be captured by technical devices and translated into digital data streams as well as the affective relationships among devices, bodies, and spaces that escape digital capture.

Yet, most geographies of information processing in smart spaces have shown little engagement with the questions of embodied sensory input and knowledge production long explored by feminist theorists (Davidson and Milligan 2004; Kwan 2007). How does, for example, the spatial knowledge produced through an Internet of Things (IoT) sensor network relate to the knowledge produced through embodied sensory experience? More important, how are these becoming entangled and enmeshed in new ways? To address these questions, Mitchell (2004) conceived of networked infrastructure and smart devices as extensions of the body, writing, "Not only are these networks essential to my physical survival, they also constitute and structure my channels of perception and agency—my means of knowing and acting upon the world. … They are as crucial to my cognition as my neurons" (61). Similarly, artistic and transfeminist political interventions highlight the way in which information and sensory data flow through entangled relationships among bodies, spaces, and technologies (Jones 2006; Egaña and Solà 2016). Such interventions call for an attentiveness to the differential experiences of embodiment in smart spaces, as well as to the role of bodies (of workers, residents, etc.) in the production, maintenance, and everyday functions of smart infrastructures.

Within geography, Amoore's work on algorithms—in collaboration with others—offers an excellent example of the relationship between technical and "human" forms of cognition, exploring how algorithms

reorient forms of perception—bringing to the fore previously imperceptible patterns (Amoore and Piotukh 2015; Amoore and Raley 2017; Amoore 2018). Reflecting on the entanglement of embodied action and the proliferation of algorithmic procedures, Amoore and Raley (2017) argued that "to draw attention to the embodied actions of algorithms, then, is precisely to reflect on how the already broad cognitive function of thought is distributed and extended through algorithms" (5). Amoore (2018) thus called to "extend attention beyond the data centre and into the spatialities of perception itself" (16)—or, as Hayles (2017) might argue, into the broader "cognitive ecology." By carefully tracing the operations and entanglements of complex infrastructures, data assemblages, and algorithms, this sort of scholarship decenters the human in cognition. Yet, by maintaining a focus on embodiment and forms of perception, the human is not simply dissolved into an amorphous milieu of agential entities but rather is understood as differentially coconstituted in complex spatial and temporal relations with myriad nonhuman others.

Hayles (2017) thus recognized a range of differentiated cognitive capacities across different entities and assemblages, writing: "On the technical side are speed, computational intensity, and rapid data processing; on the human side are emotion, an encompassing world horizon, and empathic abilities to understand other minds" (140). Hayles (2017) highlighted the entanglement of human and nonhuman cognition, and she also reinforced a narrow, normative conception of "human" cognition that calls for further critique. In exploring the distinct capacities of a diversity of cognitive agents, Hayles (2017) raised the question of consciousness and its role in human forms of cognition. We turn to this question of consciousness to build from but also push beyond Hayles's account to draw attention to the diversity of human modes of being and knowing.

Consciousness

Hayles did not dissolve human intelligence into an undifferentiated category of information processing. Instead, she highlighted the role of human consciousness, asserting the cognitive capacities of a range of actors. Hayles (2017) argued, for example, that "there is no technical agency without humans, who design and build the systems, supply them with power and maintain them, and dispose of them when they become obsolete" (32). Her argument builds on the work of Simondon (2017), who saw humans as the assemblers of technical systems, even as these systems operate according to distinct logics not directly controlled by a human inventor. Similarly, Stiegler (1998) highlighted the importance of the social and material processes of production and reinvention in which humans play a key role, simultaneously remaking their environments and themselves as subjects.

Hayles's (2017) notion of consciousness does not reinstate a sovereign human subject as the rational director of technical assemblages understood as tools of human will. Rather, she recognized the entanglement of consciousness in spatialized cognitive assemblages and reflected on the ability to interact with those assemblages in different ways. Significantly, this recognizes human consciousness as fundamentally embodied and embedded and thus in constant interaction with complex material realities—reflecting what Varela, Thompson, and Rosch (1991) termed an "enactive approach" to questions of mind and being. These cognitive assemblages—both technical information processing and embodied human forms of sensing—fundamentally shape modes of awareness. This is why Hayles distinguished between nonconscious cognition and consciousness, arguing that consciousness requires various forms of nonconscious cognition to process and filter sensory input. She wrote, "Mediating between material processes and modes of awareness, nonconscious cognition provides a crucial site where intra-actions connect sensory input from the internal and external environments ('events') with the emergence of the subject ('entities')" (Hayles 2017, 75). Whereas recent advances in neurobiology highlight the importance of nonconscious cognition in human life, Hayles (2017) reflected on the entanglement of "internal" nonconscious cognition with the rapidly expanding realm of "external" or "technical" nonconscious cognitions, recognizing that both shape modes of awareness.

Recognizing the multiple forms of agency exercised by nonhumans in the continual constitution of consciousness, this approach also stresses the role of conscious forms of human agency in shaping complex technological systems. As Hayles (2017) explained, "Effective modes of intervention seek for inflection points at which systemic dynamics can be decisively transformed to send the cognitive assemblage in a different direction" (203). Even as

cognitive assemblages shape consciousness, they are also the object of conscious intervention or, as Rose (2017) and Stiegler (1998) suggested, "reinvention." If we understand the cognitive functions being carried out by technical apparatuses and the ways in which they reorient human forms of perception and cognition within broader cognitive assemblages, then we open up space for conscious reflection on its workings and our differential relationships to it.

Reflecting on the ways in which posthuman forms of consciousness might be aware of and intervene in complex cognitive assemblages raises the questions of ethics and politics and opens up possibilities for imagining alternative technological futures. As Hayles (2017) asked, "How should we reimagine contemporary cognitive ecologies so that they become life-enhancing rather than aimed toward dysfunctionality and death for humans and nonhumans alike?" (141). Yet, reflecting on the possibilities for conscious forms of reinvention within such cognitive ecologies requires a move beyond Hayles to an attentiveness to how the human is differentiated in its complex entanglements with nonhumans—and the ways in which evolving cognitive assemblages rework those differentiations (Rose 2017). This includes a recognition of neurodiversity and the ways in which different human cognitions and consciousnesses might intra-act differently in complex social and material environments (Armstrong 2015). This also requires a critique of how access to "technological" knowledge and claims to "expertise" are policed and entangled in the reproduction of race, gender, class, age, and ability, as well as other markers of difference in the everyday power relations of militarized techno-capitalism. We might thus understand critical posthuman consciousnesses as ongoing embodied experiments with alternative modes of differential becoming with an array of cognitive agents.

A growing literature across digital geographies points toward expanded conceptions of intelligence, highlighting the entanglement of technological cognitions with differentiating forms of human experience, embodied sensing, and conscious reflection. Wilmott (2016), for instance, offered an ethnographic account of embodied experiences of spatial big data in Hong Kong and Sydney, highlighting how they challenge logics of calculability and profitability. Likewise, Pink and Fors (2017) argued that "while self-tracking technologies might appear on the surface to belong to a quantified world of measurement … they participate considerably in how people 'feel' or sense in their everyday environments" (376). Such accounts do not privilege one form of intelligence over another but rather point to the entanglement of multiple ways of knowing and producing meaning that operate according to a diversity of logics. Recognizing this diversity of possible relationships to technological systems presents opportunities for imaging and building alternative futures beyond the logics of the smart discourse. Lynch (2019), for example, examined how activists in a grassroots movement around "technological sovereignty" in Barcelona consciously reflect on the evolving role of digital technology in everyday life and experiment with alternative social practices for managing and negotiating these relationships.

Conclusion

This article expands and clarifies the vocabulary for geographers writing about the capacities and agencies of spatialized information processing systems. It offers a specificity to the kinds of operations performed by such systems and their relationships to a broader milieu, or what Hayles (2017) called a "cognitive ecology." Within this framing, we highlight the role of "posthuman agency" (Rose 2017) and the possibilities for forms of conscious posthuman reflection and interaction in complex technosocial entanglements. This is not meant to reinstate a rational human subject at the center; it is to recognize the complex interplay among a diversity of cognitive actors while also calling for critical reflection on the part of posthuman subjects.

Reframing debates over smart spaces to more actively engage theories of intelligence offers a number of benefits to geography. First, it connects smart city debates to discussions of the differentiated and entangled capabilities and agencies of a range of human and nonhuman actors. Second, it offers greater analytical specificity as to what kinds of processes are occurring and which actors are involved. Third, it helps geographers move beyond the smart city as a set of policy prescriptions and urban management techniques to think more about the spatialization of a range of technical devices and apparatuses. Fourth, it highlights the entanglements of emerging digital devices not only with differentiated human processes and forms of agency but also with biological forms of cognition. Finally, it raises the question of consciousness and the possibilities for collective

reflection and intervention into processes of technological development and implementation.

This intervention points toward several areas for future inquiry. On the one hand, there is a need for more empirical explorations of the multiplicity of social, political, and economic possibilities emergent in the evolving "cognitive ecology"—and of the work of social movements, hackers, cyberfeminists, and everyday individuals and collectives that creatively explore those possibilities. On the other hand, by resituating the smart city in relation to broader debates over intelligence and cognition, there emerges an opportunity for geographers to become more relevant to the discussions around AI, machine learning, and robotics. As these technologies develop and become increasingly ubiquitous, there is a need for a robust geographic theory that helps think about the spatial dimensions and entanglements of smart spaces and the various intelligences at play in everyday posthuman life.

ORCID

Casey R. Lynch ⓘ http://orcid.org/0000-0001-8839-876X

References

Aitken, S. C., and E. M. Bjorklund. 1988. Transactional and transformational theories in behavioral geography. *The Professional Geographer* 40 (1):54–64. doi: 10.1111/j.0033-0124.1988.00054.x.]

Albino, V., U. Berardi, and R. M. Dangelico. 2015. Smart cities: Definitions, dimensions, performance, and initiatives. *Journal of Urban Technology* 22 (1):3–21. doi: 10.1080/10630732.2014.942092.

Amin, A., and N. Thrift. 2016. *Seeing like a city*. Malden, MA: Wiley.

Amoore, L. 2018. Cloud geographies: Computing, data, sovereignty. *Progress in Human Geography* 42 (1):4–24. doi: 10.1177/0309132516662147.

Amoore, L., and V. Piotukh. 2015. Life beyond big data: Governing with little analytics. *Economy and Society* 44 (3):341–66. doi: 10.1080/03085147.2015.1043793.

Amoore, L., and R. Raley. 2017. Securing with algorithms: Knowledge, decision, sovereignty. *Security Dialog* 48 (1):3–10. doi: 10.1177/0967010616680753.

Armstrong, T. 2015. The myth of the normal brain: Embracing neurodiversity. *AMA Journal of Ethics* 17 (4):348–52. doi: 10.1001/journalofethics.2015.17.4.msoc1-1504.

Ash, J. 2017. *Phase media: Space, time and the politics of smart objects*. London: Bloomsbury.

Batty, M. 2013. Big data, smart cities and city planning. *Dialogues in Human Geography* 3 (3):274–79. doi: 10.1177/2043820613513390.

Beer, D. 2007. Thoughtful territories: Imagining the thinking power of things and spaces. *City* 11 (2):229–38. doi: 10.1080/13604810701395845.

Brenner, E. D., R. Stahlberg, S. Mancuso, J. Vivanco, F. Baluska, and E. Van Volkenburgh. 2006. Plant neurobiology: An integrated view of plant signaling. *Trends in Plant Science* 11 (8):413–19. doi: 10.1016/j.tplants.2006.06.009.

Crang, M., and S. Graham. 2007. Sentient cities: Ambient intelligence and the politics of urban space. *Information, Communication & Society* 6 (10):789–817. doi: 10.1080/13691180701750991.

Davidson, J., and C. Milligan. 2004. Embodying emotion sensing space: Introducing emotional geographies. *Social & Cultural Geography* 5 (4):523–32. doi: 10.1080/1464936042000317677.

Dupuy, J. P. 2009. *On the origins of cognitive science: The mechanization of the mind.* Cambridge, MA: The MIT Press.

Egaña, L., and M. Solà. 2016. Hacking the body: A transfeminist war machine. *TSQ: Transgender Studies Quarterly* 3 (1–2):76–82.

Gardner, H. 2011. *Frames of mind: The theory of multiple intelligences.* 3rd ed. New York: Basic Books.

Grosz, E. 1994. *Volatile bodies: Toward a corporeal feminism.* Bloomington: Indiana University Press.

Hardcastle, V. G. 1995. A critique of information processing theories of consciousness. *Minds and Machines* 5 (1):89–107. doi: 10.1007/BF00974191.

Hayles, N. K. 2017. *Unthought: The power of the cognitive nonconscious.* Chicago: University of Chicago Press.

Hollands, R. G. 2008. Will the real smart city please stand up? *City* 12 (3):303–20. doi: 10.1080/13604810802479126.

Kitchin, R. 2014. The real-time city? Big data and smart urbanism. *GeoJournal* 79 (1):1–14. doi: 10.1007/s10708-013-9516-8.

Kitchin, R., and M. Dodge. 2005. Code and the transduction of space. *Annals of the Association of American Geographers* 95 (1):162–80. doi: 10.1111/j.1467-8306.2005.00454.x.

Kwan, M.-P. 2007. Affecting geospatial technologies: Toward a feminist politics of emotion. *The Professional Geographer* 59 (1):22–34. doi: 10.1111/j.1467-9272.2007.00588.x.

Jones, C. A., ed. 2006. *Sensorium: Embodied experience, technology, and contemporary art.* Cambridge, MA: The MIT Press.

Leszczynski, A. 2016. Speculative futures: Cities, data, and governance beyond smart urbanism. *Environment and Planning A: Economy and Space* 48 (9):1691–1708. doi: 10.1177/0308518X16651445.

Lynch, C. 2019. Contesting digital futures: Urban politics, alternative economies, and the movement for technological sovereignty in Barcelona. *Antipode.* Advance online publication. doi: 10.1111/anti.12522.

Marvin, S., and A. Luque-Ayala. 2017. Urban operating systems: Diagramming the city. *International Journal of Urban and Regional Research* 41 (4):84–103. doi: 10.1111/1468-2427.12479.

Mitchell, W. 2004. *Me++: The cyborg self and the networked city.* Cambridge, MA: The MIT Press.

Murray, S., and N. Sullivan. 2012. *Somatechnics: Queering the technologisation of bodies.* Farnham, UK: Ashgate Publishing.

Picon, A. 2015. *Smart cities: A spatialized intelligence.* New York: Wiley.

Pink, S., and V. Fors. 2017. Being in a mediated world: Self-tracking and the mind–body–environment. *Cultural Geographies* 24 (3):375–88. doi: 10.1177/1474474016684127.

Pötzsch, H. 2015. The emergence of iBorder: Bordering bodies, networks, and machines. *Environment and Planning D: Society and Space* 33 (1):101–18. doi: 10.1068/d14050p.

Rose, G. 2017. Posthuman agency in the digitally mediated city. *Annals of the Association of American Geographers* 107 (4):779–93. doi: 10.1080/24694452.2016.1270195.

Shanahan, M. 2010. *Embodiment and the inner life: Cognition and consciousness in the space of possible minds.* Oxford, UK: Oxford University Press.

Shanahan, M. 2015. *The technological singularity.* Cambridge, MA: The MIT Press.

Shepard, M. 2011. *Sentient city: Ubiquitous computing, architecture, and the future of urban space.* Cambridge, MA: The MIT Press.

Simondon, G. 2017. *On the mode of existence of technical objects.* Minneapolis: University of Minnesota Press.

Smedley, A., and B. D. Smedley. 2005. Race as biology is fiction, racism as a social problem is real: Anthropological and historical perspectives on the social construction of race. *American Psychologist* 60 (1):16–26. doi: 10.1037/0003-066X.60.1.16.

Söderström, O., T. Paasche, and F. Klauser. 2014. Smart cities as corporate storytelling. *City* 18 (3):307–20. doi: 10.1080/13604813.2014.906716.

Stiegler, B. 1998. *Technics and time 1: The fault of Epimetheus.* Palo Alto, CA: Stanford University Press.

Thompson, E. 2007. *Mind in life: Biology, phenomenology, and the sciences of mind.* Cambridge, MA: Harvard University Press.

Varela, F. J., E. Thompson, and E. Rosch. 1991. *The embodied mind: Cognitive science and human experience.* Cambridge, MA: MIT Press.

Wilmott, C. 2016. Small moments in spatial big data: Calculability, authority and interoperability in everyday mobile mapping. *Big Data & Society* 3 (2):1–16. doi: 10.1177/2053951716661364.

Exploding the Phone Book: Spatial Data Arbitrage in the 1990s Internet Boom

Will B. Payne and David O'Sullivan

This article examines 1990s Internet firm Zip2 as an early case study in the political economy of location-based services (LBS) and "smart" cities in the United States. Tesla CEO Elon Musk's first startup, Zip2, combined digital maps built using public spatial data with computerized Yellow Pages listings to create city directory Web portals for daily newspapers. Examining the technical affordances, labor practices, and revenue models of the company's two key data providers, NavTech and American Business Information, reveals the conditions that enabled Zip2 to successfully employ data arbitrage as a corporate strategy. Despite its significant limitations as a consumer-facing technology, Zip2 was acquired by Compaq for $300 million in 1999. The company's founders leveraged government investment in digital mapping, fortuitous shifts in copyright law, and anxiety among newspaper publishers about missing the Internet boom into a business that prefigured many of the use cases and revenue models of contemporary LBS and Web maps. *Key Words: computer navigation, data arbitrage, location-based services, smart cities, web mapping.*

本文以上世纪 90 年代的互联网公司 Zip2 为例，对美国基于位置服务 (LBS) 和"智慧"城市的政治经济学进行了初步研究。Zip2是特斯拉 CEO 埃隆·马斯克创建的第一家初创公司，通过使用公共空间数据构建数字地图，结合电脑黄页中的信息，为日报创建城市目录门户网站。该公司有两个关键数据供应商 —— NavTech 和 American Business Information，通过研究这两家公司的技术可承受性、劳动实践和收入模型，揭示了 Zip2如何将数据套利变成一项成功的公司战略。尽管作为一家以消费者为主导的技术型企业，Zip2 还存在着很大的局限性，但还是在 1999 年被康柏以 3 亿美元的价格收购。面对政府对数字地图的投资、版权法变化的不确定性以及报纸出版商对错过互联网繁荣的焦虑，公司创始人极富前瞻性地成功抓住机遇，迅速发展成为一家具有现代 LBS 和网络地图企业，昭示了未来的更多应用领域和收入模式。*关键词: 计算机导航、数据套利、基于位置的服务、智慧城市、网络制图。*

Este artículo examina la firma Zip2 de la Internet de los años 1990 a manera de estudio de un caso temprano en la economía política de los servicios basados en localización (LBS) y de ciudades "inteligentes" en los Estados Unidos. La primera empresa emergente del CEO de Tesla Elon Musk, Zip2, combinó los mapas digitales elaborados usando datos espaciales públicos con los listados computarizados de las Páginas Amarillas, con el fin de crear para los periódicos diarios portales Web con el directorio de la ciudad. El examen de las posibilidades técnicas, las prácticas laborales y los modelos de ingresos de los dos proveedores claves de datos, NavTech y American Business Information, revela las condiciones que habilitaron a Zip2 para utilizar exitosamente el arbitraje de datos como estrategia corporativa. Pese a sus limitaciones significativas como tecnología que enfrenta al consumidor, Zip2 fue comprada por Compaq por 300 millones de dólares en 1999. Los fundadores de la compañía apalancaron la inversión gubernamental en mapeo digital, por cambios fortuitos en la ley de derechos de autor (copyright) y la ansiedad entre los publicistas de periódicos por perderse la bonanza de Internet en un negocio que anticipó muchos de los casos de uso y modelos de ingresos de los LBS contemporáneos y de los mapas de la Web. *Palabras clave: arbitraje de datos, ciudades inteligentes, mapeo de la Web, navegación en la computadora, servicios basados en localización.*

Making cities "smart" requires ongoing investment in communications infrastructure and data acquisition beyond the capabilities of even the largest technology companies. The current market for commercial location-based services (LBS) and digital maps, accessed across billions of Internet-connected devices, depends on access to the U.S. Air Force's Global Positioning System (GPS) infrastructure, at an annual cost of over $700 million (Rankin 2016). Even before the contemporary boom in companies commoditizing location data (Wilson 2012), there has been an abiding entanglement between state and corporate efforts to map, index, and profit from urban space. As Mattern (2017) noted, "Cities and media have historically served as one another's 'infrastructures'" (xxv). For-profit urban directories have helped to order U.S. cities since the nineteenth century, pioneering practices like house numbering adopted and extended by governments to make space legible for state power and capitalist speculation (Scott 1998; Rose-Redwood 2006). The commercial introduction of digital maps and online city directories from the early 1980s to the late 1990s is a key moment in the development of what former MIT Media Lab Director William Mitchell (1996) called the "city of bits." Excavating the political economy of location-based startups in the 1990s Internet boom illuminates the key role of government investment in spatial data production, fortuitous changes in copyright law, and competitive pressures in the publishing industry in laying the groundwork for contemporary smart cities and "sharing economy" ventures.

Online city directory publisher Zip2 set a template for later LBS and Web maps like Yelp, Foursquare, and Google Maps, despite significant technical and business constraints that prevented the company from building a lasting consumer brand. Zip2's sale to Compaq's AltaVista search engine division for $307 million in 1999, shortly after a failed merger with rival city directory Citysearch, enabled its young cofounder Elon Musk to invest in PayPal, where a new generation of LBS entrepreneurs like Yelp founder Jeremy Stoppelman acquired the money and connections to start their own ventures. Zip2's outsized value to investors was based on good timing and an "opportunistic hack": Musk's ability to connect spatial data collected by NavTech with off-the-shelf digitized business directories purchased from American Business Information (ABI; Chafkin 2007). Musk recognized the substantial use value of allowing Internet users to query this combined database to discover local businesses. More important, he exploited the additional discrepancies in exchange value afforded by the growing Internet bubble and newspaper publishers' fear of disintermediation shutting them out of online advertising.

In marketing Zip2 to legacy media companies, Musk and his partners performed a kind of data arbitrage common to contemporary technology companies investing in "smart cities" projects, from Google subsidiary Sidewalk Labs' attempts to instrument entire neighborhoods in New York City and Toronto for data collection to startups seeking to apply big data methods to real estate valuation (Mattern 2017; Shaw 2018). We define *data arbitrage* as a business strategy in which technologists connect data sets from distinct sources to realize an emergent value not reflected in the combined valuation of the underlying data. Where conventional arbitrageurs exploit "opportunities exposed when different evaluative devices yield discrepant pricings" across space and time, in data arbitrage the key linkage is between different contexts of data production and use (Beunza and Stark 2004, 374). Burt (1992) identified how successful entrepreneurs identify arbitrage opportunities and new business models by filling "structural holes" between economic actors in distinct networks. In data arbitrage, the synthesized data set itself bridges this discrepancy between prices, contexts, and industries.

The widespread practice of venerating (predominantly white and male) technology moguls like Musk as solely responsible for their success obscures the social factors that allowed them to accumulate billions of dollars. Musk was not an uncommonly talented programmer or executive; in creating Zip2, he combined the advantages of his class position, educational background, and gender presentation in seeking venture capital funding with the ability to secure inexpensive access to spatial routing data and electronic business listings. This access was only possible through a confluence of public investments in spatial data collection, fortuitous regulatory decisions, and emerging technical affordances. Examining the technical arrangements, labor practices, and revenue models of Zip2's two key data providers, NavTech and ABI, reveals the conditions that enabled the company to successfully deploy data arbitrage as a business strategy, despite its limitations as a consumer technology compared to later LBS and Web maps.

"What's Become of the Yellow Pages"

Popular accounts of Elon Musk's business success typically begin with X.com, the online bank he founded in 1999 that merged with Peter Thiel and Max Levchin's Confinity to form PayPal. The company raised over $61 million in a 2002 initial public offering (IPO), before being acquired by eBay the same year for $1.5 billion in one of the first major deals after the collapse of the dot-com bubble in 2000. A "PayPal Mafia" of former executives went on to start companies like LinkedIn, Palantir, Yelp, and YouTube. Musk's involvement in PayPal, however, and his later ventures like Tesla, SpaceX, and OpenAI, all stemmed from the successful sale of his first startup, Zip2 (Chafkin 2007).

In 1995, after interning in Silicon Valley during his final summer at the University of Pennsylvania's Wharton School, Musk was debating between starting a PhD in physics at Stanford and building an "online city guide that would help people navigate the geographic world via the virtual one" with his brother Kimbal (Vance 2015). Venture capitalists in Toronto had declined to fund the concept; Musk claims that one investor "[threw] a phone book at us and shout[ed], 'You think this is ever going to be replaced?'" (Anderson 1999, D9). Peter Nicholson, a banker whom Musk had worked for in Canada, advised him to pursue the idea in the San Francisco Bay Area, where "there was this crazy Internet thing going on, and … people will pay big money for damn near anything" (Vance 2015, 431). After watching Netscape go public, Musk dropped out during his first week at Stanford to work on the "Virtual City Navigator" concept with Kimbal and Greg Koury, a Toronto real estate developer and family friend. The brothers' South African mining engineer father Errol Musk contributed $28,000 to allow the brothers to rent a small office in Palo Alto, California, where they also lived for several months. The Musks' idea for an online city directory was not unique; a number of software engineers, urban planners, and academics were exploring the potential of re-creating and augmenting physical space in code, in addition to direct competitors like Citysearch, founded in Los Angeles in 1995 (Gelernter 1993; Mitchell 1996).

By March 1996, the group had hired a small team and amassed $3.6 million of funding, led by Mohr Davidow Ventures. The company changed its name from Global Link Information Network to Zip2 and hired professional managers to allay investors' concerns about Musk's mercurial leadership style. Zip2's first slogan, "What's Become of the Yellow Pages," answered an implied question about how business information would be accessed in the Internet age. The company's back-end geographic information system (GIS) enabled spatially restricted searches and driving directions, allowing users to "zip" from place to place. A magazine article before Zip2's launch makes this value proposition clear with its title— "Zip2 Links Yellow Pages to GIS"—and its conclusion that the company's digital map with turn-by-turn directions was its "key selling feature" for publishing partners and sponsors (Tribute et al. 1996). According to Kimbal Musk, securing the spatial data to power Zip2's system was straightforward. He and Elon called digital mapping company NavTech, "and they gave us the technology for free" (Vance 2015, 38). Elon then wrote some straightforward Java code linking these maps to the listing data for a prototype system to show funders and publishers (Tribute et al. 1996).

Unleashing the TIGER

Consumer digital navigation was barely a decade old when the Musks founded Zip2. Stanford engineering PhD Barry Karlin and University of California, Berkeley, engineering graduate Galen Collins founded Karlin & Collins (later Navtech, then Navteq) in 1984. The previous year, Atari founder Nolan Bushnell had launched Etak, storing road network data on cassette tapes and visualizing it on arcade-style vector graphics monitors affixed to the dashboard (Edwards 2017). In an enduring visual trope for digital navigation systems, Bushnell's team borrowed the notched triangle spaceship image from Atari's hit arcade game *Asteroids* to represent the driver's current location. Although Etak's technology was impressive, Karlin saw that the system's cost and the difficulty of interpreting a spartan, static map limited its target audience (Elmer-DeWitt 1987). In designing an alternative system, Karlin drew on Bell Labs research demonstrating that turn-by-turn directions were more effective than static maps (Streeter et al. 1985). NavTech's first product, DriverGuide, was an ATM-sized kiosk sold to car rental companies and hotels in the Bay Area for $12,000 each; drivers would select a starting point and destination, and the machine printed out a list of directions in under a minute (Freeman 1987). When Zip2 contacted NavTech to license their spatial data, the company was growing quickly, on its way to a 2004 IPO with a market capitalization of $3.5 billion. Since 2015, the company has been owned by German automotive companies BMW, Audi, and Daimler AG as HERE; it remains one of the largest mapping companies globally.

Like Uber and other contemporary technology giants, NavTech pursued a strategy of aggressive growth over steady profit, losing hundreds of millions of dollars in its quest to become the ubiquitous spatial data provider for navigation. The company's effort to map the Bay Area alone cost over $3 million, including hiring staff to drive the road network and describe everything they saw on Dictaphones (Pederson, Derdak, and Simley 2005). At the 1987 North American Cartographic Information Society (NACIS) meeting, Karlin (1987) presented a paper on "the race to build the definitive electronic street map," which he shrewdly believed would be "the basis for products and services whose range is now only beginning to be imagined" (15).

Instead of licensing existing commercial road maps, which NavTech's engineers found to not be accurate enough for turn-by-turn navigation, the company looked to more authoritative sources, including the digital Topologically Integrated Geographic Encoding & Referencing (TIGER/Line) files being prepared by the U.S. Census Bureau for the 1990 decennial census. In a 1987 job ad, the company called for candidates with cartography degrees and five years of digital mapping experience, at that point difficult to find outside government and academia (Karlin & Collins, Inc. 1987). Etak had already hired digital cartography expert Marvin White from the U.S. Census, where he had helped develop the Geographic Base File/Dual Independent Map Encoding (GBF/DIME) format for storing street address data (Chrisman 2006; Edwards 2017). NavTech also developed proprietary GIS software to correct for perspectival distortions in aerial photographs and combine them with TIGER data and city and county planning maps (Rosenberg 1997).

Sperling (1995) argued that "the TIGER database may be the most important 'data' file from the 1990 Census" (377), with wide-ranging commercial applications analogous to the civilian spin-offs of Cold War military contracting, not least the transition from the closed, government-owned ARPANET system to an open but increasingly privatized Internet (Abbate 2000; Light 2003). More than 2,000 Census Bureau employees and contractors prepared GBF/DIME files for the 1980 Census, at a cost of over $200 million, around 20 percent of the Bureau's budget; building out the TIGER database was similarly expensive (Marx 1986). When finished, the 1990 database cost $90,150, several orders of magnitude cheaper than building a national digital map from scratch (Carbaugh and Marx 1990). Crucially, the Bureau left the development of any support software

or data products to private companies like NavTech, which owned the copyright to any improved maps they built using public data.

As NavTech expanded, the company downplayed its reliance on Census data. An online FAQ in 2000 explained that NavTech was the only digital map "supplier that does not use the highly inaccurate TIGER file," instead relying on (also public) "USGS [U.S. Geological Survey] quad sheets (for positional accuracy) combined with aerial photographs and field collected and verified data" (NavTech 2000). Nevertheless, without massive public investment in spatial data, NavTech's ambition would have been unachievable. Leszczynski (2012) noted that "the 'roll-out' of market-based regimes of geographic information governance" in the United States has not signaled "the eclipsing of the state by the market" but instead been characterized by a "co-occurrence" of state and market-based geographic information governance (84). Far from being a slow-moving monolith "crowding out" innovation, what Mariana Mazzucato calls the "entrepreneurial state" sets the agenda and "crowds in" corporate investment, enabling audacious technical feats like GPS and the Internet that even patient investors could never justify (Mazzucato 2015; see also Abbate 2000; Light 2003; Rankin 2016). As Karlin suspected, a navigable digital street map of the United States would prove immensely valuable to companies like NavTech and Zip2, aided by a fortuitous Supreme Court ruling freeing telephone directory listings from copyright protection just as TIGER data were released to the private sector.

Exploding the Phone Book

To make the map data provided by NavTech useful to end users and advertisers, Musk purchased an off-the-shelf CD-ROM business directory for the Bay Area, published by Omaha-based ABI, for a few hundred dollars (Vance 2015). The commodification of local business data in electronic form was the product of technical, legal, and economic developments from the mid-1970s on, a confluence of events in which ABI (later Infogroup, valued at over $400 million by 2010) played a major role. After receiving an MBA from the University of Nebraska in 1971, Indian immigrant Vinod Gupta took a marketing job at mobile home manufacturer Commodore Corporation, where he was asked to put together a list of every mobile home dealer in the country. As a regulated monopoly, AT&T was required to publish directories of local subscribers free of charge, allowing Gupta to use Commodore's Wide Area Telephone Service (WATS) long-distance subscription to order all 4,800 Yellow Pages books published nationally and store them in his garage. Gupta then spent $10,000 hiring people to type these data into computers, soon leaving Commodore and recouping his expenses by marketing the list to other dealers (Graves 1995).

By 1986, ABI had digitized the entire set of Yellow Pages directories for the United States. The same year, regional Bell operating company NYNEX (formed in AT&T's legally mandated corporate breakup in 1984) released the first CD-ROM telephone directory for New York City, selling it to large companies and government agencies like the FBI and the IRS for $10,000 (Shapiro and Varian 1998). Telephone companies also pursued litigation against ABI and competitors like Dun & Bradstreet to maintain control over the copyright of their listings. In response, the emerging telemarketing industry, aided by lower long-distance rates and more granular demographic data on consumers (Goss 1995), pushed for cheaper access to listing data. When ABI unveiled its own listing CD-ROM in 1990, priced at $1495 for an annual lease, the company's ongoing revenue were at risk due to a series of pending court cases on telephone directories' copyright status (Hardin 1990; Miller 1990). The U.S. Supreme Court decided the most important of these cases, *Feist Publications, Inc., v. Rural Telephone Service Co.*, in 1991, with significant consequences for media and mapping businesses. Kansas-based Rural Telephone Service Co. refused to allow private directory company Feist Publications to republish Rural subscribers' information. In a move paralleling the fictitious "trap streets" cartographers have added to maps to assert copyright claims, Rural Telephone had inserted twenty-eight fake names into its phone book, four of which appeared in Feist's directory (M. W. Miller 1990). In a unanimous decision, the Court overturned the "sweat of the brow" doctrine, which had stipulated that effort in compilation alone merited copyright protection, to argue that facts like a subscriber's name, address, and telephone number are not protected unless a derivative work uses the same "selection and arrangement" as the original (Feist 1991).

Following the *Feist* decision, there was a rush to digitize and market the newly public data held in phone directories. In 1992, NYNEX executive James Bryant left to found his own company, Pro CD

(Shapiro and Varian 1998). Bryant tried licensing optical character recognition (OCR) software from a German firm, but the cheaper paper U.S. directory printers used ruled out this automated solution. Instead, he set up a bidding process among data entry firms, using the Nantucket phone book as a test case. Beijing University of Aeronautics and Astronautics won with a $2 million bid, compared to $40 million from a Denver-based company (Rosenberg 1995). The university's all-female team of student workers then typed up more than 70 million listings from every phone book published in the United States for $3.50 a day (Shapiro and Varian 1998). Pro CD's dispassionate cost-cutting strategy is stark evidence of the intersection between race, gender, and geography in the global technology market, with feminized and often outsourced labor in data processing and electronics manufacturing systematically undervalued compared to male-coded work in "higher order" programming and design (Hicks 2017; Elwood and Leszczynski 2018). With marginal production costs at under a dollar per CD-ROM, Bryant was able to drop prices drastically and still ensure a profit; competition from other providers sent prices falling further. Although ABI's data entry jobs were still based in Omaha, their practices for cutting labor costs prefigure other aspects of the contemporary "gig economy." The company granted agents flexible hours in return for accepting constant surveillance and wages conditional on meeting aggressive daily quotas, like completing seventy phone calls per hour to verify business details ("Business in the Telemarketing Capital" 1994).

Digital directory data had become a cheap commodity by the mid-1990s, but combined with NavTech's base map, it became a crucial selling point for Zip2's users and partners. In 1998, a *Washington Post* article compared Zip2 to Web map rivals Mapquest and Maps on Us, a Lucent Technologies product built on Etak data (Pegoraro 1998). Spun off from AT&T in 1996, Lucent included the Bell Labs division responsible for the wayfinding research that Barry Karlin commercialized a decade earlier. The *Post* analysis rated Mapquest as the best Web map overall but gave Zip2 credit for its vast ABI-powered business directory, beating the phone company at its own game.

"Groupware for the City"

Zip2 might have joined the ranks of failed dotcom startups in the 2000 bust if its executives had not found a group of deep-pocketed partners concerned about missing the Web revolution: daily newspaper publishers. The double-digit profit margins in print advertising at the time were clearly threatened by the rise of digital publishing. By 2000, the year that San Francisco–based online classifieds startup Craigslist expanded outside the Bay Area, newspaper classified ad spending hit its historic peak of $19.6 billion (Reinan 2014). Zip2's success in attracting publishing partners was enough to see it through several high-profile launches of newspaper-branded city portals and a failed merger with Citysearch before its fortuitous acquisition by Compaq's AltaVista division in 1999.

The company's flagship site New York Today, a collaboration with Zip2 investor *New York Times*, launched in June 1998. The site's general manager, Dan Donaghy, called it "groupware for New York City," overselling its limited interactivity but gesturing toward visions of LBS realized a decade later (Broersma 1998). The site's launch press release envisions an affluent (and presumably white) audience desirable to advertisers. Proposed use cases included searches for classic cars and new Calvin Klein linens to replace a set soiled by spilled cabernet ("Zip2 Announces Breakthrough" 1998). An invitation to find "quaint gourmet cafes you adore for leisurely reading and sipping espresso" is more characteristic of contemporary LBS, although the five-mile default search radius might indicate a more suburban user base than the marketing implied. Although it appears that this feature was never implemented, the release also claims that users would be able to read other patrons' recommendations, several years before LBS like Dodgeball, Foursquare, and Yelp would allow users to read and write reviews from mobile devices (Payne 2018).

New York Today's shortcomings as a consumer Web site and a media property are typical of the weaknesses of early city directories. The promised interactivity was far beyond the capabilities of most businesses, even in the heart of Silicon Valley or Manhattan. A Zip2 press release touts the ability for consumers to ask car dealerships about pricing using the company's "Internet-to-fax gateway," a necessary hack given that most businesses lacked Web sites or e-mail addresses at the time ("Zip2 Announces World-Class Technology Platform" 1996). Another factor was the tangle of criss-crossing deals between newspapers, phone companies, and tech startups. New York Today launched shortly after the failure of the New Century Network (NCN), a joint digital venture between nine newspaper companies, including *The New York Times* and Knight-Ridder (Peterson 1998). According to a *BusinessWeek* reporter at the time, the network had "a name most of its owners disliked" and "a mission nobody understood"; the network disbanded in March 1998 after spending over $27 million (Dugan 1998). NCN members had signed a dizzying array of side deals, with the Tribune Company and AOL collaborating on Digital City, the Times Mirror Company investing in Zip2 rival Citysearch, and even Knight-Ridder, one of Zip2's largest investors along with Japanese conglomerate SoftBank (Uber's biggest shareholder as of 2018), launching its own Real Cities Network.

In addition to confusing consumers and sponsors, the profusion of online city directories diluted the relatively small pool of online advertising dollars at the time, making it impossible to sustain the aggressive growth projections that investors demanded. There were some successes; early national advertising partners like Nissan paid for banner advertising and "enhanced listings" on the Zip2 Yellow Pages site ("Zip2 Announces World-Class Technology Platform" 1996), and Zip2's newspaper partners sold priority positioning in search results to local businesses (Tribute et al. 1996). According to an *AdWeek* analysis in 1998, though, businesses were increasingly interested in online marketing, but the pace of their spending "can't possibly keep up with the rate at which publishers are building city guides" (Warner 1998). In 1997, whereas local advertisers spent over $70 billion nationally, they only spent around $90 million online. City directory boosters would eventually run into the reality that neither the technology, the user base, nor the advertisers were sufficient to sustain the industry at the valuations their funders had anticipated.

Conclusion

Arbitrage is by nature a short-term strategy, reliant on the temporary discrepancy of values across contexts. Venture-backed startups similarly depend on the ability of founders to demonstrate future potential and quickly realize outsized returns in an IPO or sale. Although enriching founders like the Musk brothers, and gesturing toward technical innovations and business cases that later LBS companies like Yelp, Foursquare, and Google Maps would more fully exploit, the 1990s city directory boom failed to produce any lasting consumer brands to rival contemporaries like eBay and Amazon. In 1998,

Microsoft's Sidewalk.com and AOL Digital Cities both laid off editorial staff even while trying to expand to more markets (Warner 1998). Citysearch outlasted most of its competitors, but by 2010 it had fallen behind Yelp in monthly unique visitors, paralleling Mapquest's gradual eclipsing by Google Maps (C. C. Miller 2010).

The vision of cities made perfectly navigable and visible online, partial and speculative in Zip2's heyday, is becoming ever more real today, built on a foundation of quasi-public data sets, cheap and free labor, and surveillance-based advertising. In addition to their mobile-native technology for reading and writing information about urban space, challengers like Yelp and Foursquare were able to drastically reduce the editorial costs that earlier city directories had faced by relying on "free labor" from volunteer contributors to keep their listing data and reviews current (Terranova 2000; Payne 2018). Although contemporary LBS have a firmer advertising revenue base than Zip2, new waves of "smart cities" startups deploy similar data arbitrage strategies to attract financing and spur acquisition by tech giants like Google, Microsoft, and Amazon.

Acknowledgments

Thank you for the very helpful comments from the two anonymous reviewers and the "Smart Spaces and Places" special issue and "Methods, Models, and GIS" Section Editor Ling Bian. An earlier version of this work was presented at the 2018 Annual Meeting of the American Association of Geographers in the series titled "After the Smart City? The State of Critical Scholarship Ten Years On"; thank you to the session organizers, participants, and attendees for their role in clarifying and broadening the ideas in this article and to the Berkeley Center for New Media for contributing to travel costs for the conference. Paul Duguid at the University of California, Berkeley's School of Information also provided valuable feedback on the article.

Funding

The writing and revision of this article was partially supported by the Townsend-Global Urban Humanities Fellowship at the University of California, Berkeley.

References

Abbate, J. 2000. *Inventing the Internet.* Cambridge, MA: MIT Press.

Anderson, M. 1999. Dotcom 'concept stocks' of dubious value to Canada. *The Ottawa Citizen,* November 22:D9.

Beunza, D., and D. Stark. 2004. Tools of the trade: The socio-technology of arbitrage in a Wall Street trading room. *Industrial and Corporate Change* 13 (2):369–400. doi: 10.1093/icc/dth015.

Broersma, M. 1998. *New York Times* debuts enhanced city guide. Accessed November 28, 2018. https://www.zdnet.com/article/new-york-times-debuts-enhanced-city-guide/.

Burt, R. 1992. *Structural holes: The social structure of competition.* Cambridge, MA: Harvard University Press.

Business in the telemarketing capital of the world. 1994. *Telemarketing* 12 (7):42–43.

Carbaugh, L. W., and R. W. Marx. 1990. The TIGER system: A census bureau innovation serving data analysts. *Government Information Quarterly* 7 (3):285–306. doi: 10.1016/0740-624X(90)90026-K.

Chafkin, M. 2007. Entrepreneur of the year, 2007: Elon Musk. Accessed November 28, 2018. https://www.inc.com/magazine/20071201/entrepreneur-of-the-year-elon-musk.html.

Chrisman, N. R. 2006. *Charting the unknown: How computer mapping at Harvard became GIS.* Redlands, CA: ESRI Press.

Dugan, I. J. 1998. New media meltdown at new century. *BusinessWeek.* Accessed November 28, 2018. https://www.bloomberg.com/news/articles/1998-03-22/new-media-meltdown-at-new-century.

Edwards, B. 2017. The untold story of Atari founder Nolan Bushnell's visionary 1980s tech incubator. Accessed November 29, 2018. https://www.fastcompany.com/3068135/the-untold-story-of-atari-founder-nolan-bushnells-visionary-1980s-tech-incubator.

Elmer-DeWitt, P. 1987. Driving by the glow of a screen. *Time,* April 20:63.

Flwood, S., and A. Leszczynski. 2018. Feminist digital geographies. *Gender, Place & Culture* 25 (5):629–44. doi: 10.1080/0966369X.2018.1465396.

Feist. 1991. *Feist Publications, Inc. v. Rural Telephone Service Co.*

Freeman, M. 1987. *Electronics Times* discusses an alternative to maps for drivers. *Electronics Times,* July 16:14

Gelernter, D. 1993. *Mirror worlds: Or the day software puts the universe in a shoebox…How it will happen and what it will mean.* Oxford, UK: Oxford University Press.

Goss, J. 1995. "We know who you are and we know where you live": The instrumental rationality of geodemographic systems. *Economic Geography* 71 (2):171–98. doi: 10.2307/144357.

Graves, J. M. 1995. Building a fortune on free data. *Fortune,* February 6.

Hardin, S. 1990. American Business Disk. *CD-ROM World,* December 1:36.

Hicks, M. 2017. *Programmed inequality: How Britain discarded women technologists and lost its edge in computing.* Cambridge, MA: MIT Press.

Karlin, B. W. 1987. The North America street map: Developing a navigable database. In *NACIS VII program and abstracts*, ed. J. Sutherland, 15. Atlanta, GA: North American Cartographic Information Society.

Karlin & Collins, Inc. 1987. Advertisement for geographic information systems analyst. *Los Angeles Times*, July 19:I46.

Leszczynski, A. 2012. Situating the geoweb in political economy. *Progress in Human Geography* 36 (1):72–89. doi: 10.1177/0309132511411231.

Light, J. 2003. *From warfare to welfare: Defense intellectuals and urban problems in Cold War America*. Baltimore, MD: Johns Hopkins University Press.

Marx, R. W. 1986. The TIGER system: Automating the geographic structure of the United States census. *Government Publications Review* 13 (2):181–201. doi: 10.1016/0277-9390(86)90003-8.

Mattern, S. 2017. *Code and clay, data and dirt: Five thousand years of urban media*. Minneapolis: University of Minnesota Press.

Mazzucato, M. 2015. *The entrepreneurial state: Debunking public vs. private sector myths*. New York: Anthem.

Miller, C. C. 2010. Citysearch gives content to the competition. *The New York Times*, May 5:B3.

Miller, M. W. 1990. See "L" for lawyers: Phone firms, rivals scrap over who owns directory names. *The Wall Street Journal*, November 19:B1.

Mitchell, W. J. 1996. *City of bits: Space, place, and the Infobahn*. Cambridge, MA: MIT Press.

NavTech. 2000. Business applications: FAQ. Accessed November 28, 2018. https://web.archive.org/web/20001218033200/http://www.navtech.com:80/busapps/bus_faq.html.

Payne, W. B. 2018. Crawling the city. *Logic* Issue 4 (Spring):161–70.

Pederson, J. P., T. Derdak, and J. Simley, eds. 2005. *International directory of company histories*. Detroit, MI: St. James Press.

Pegoraro, R. 1998. Drive ways. *The Washington Post*, January 30:70.

Peterson, I. 1998. Newspapers end network for web sites. *The New York Times*, March 11:D6.

Rankin, W. 2016. *After the map: Cartography, navigation, and the transformation of territory in the twentieth century*. Chicago: University of Chicago Press.

Reinan, J. 2014. How Craigslist killed the newspapers' golden goose. Accessed November 28, 2018. https://www.minnpost.com/business/2014/02/how-craigslist-killed-newspapers-golden-goose/.

Rosenberg, R. 1995. Let your CD-ROM do the walking. *The Boston Globe*, February 1:42.

Rosenberg, R. 1997. Mapping out a new idea. *The Boston Globe*, February 17:39.

Rose-Redwood, R. S. 2006. Governmentality, geography, and the geo-coded world. *Progress in Human Geography* 30 (4):469–86. doi:10.1191/0309132506ph619oa

Scott, J. 1998. *Seeing like a state: How certain schemes to improve the human condition have failed*. New Haven, CT: Yale University Press.

Shapiro, C., and H. R. Varian. 1998. Versioning: The smart way to sell information. *Harvard Business Review* 107 (6):106–14.

Shaw, J. 2018. Platform real estate: Theory and practice of new urban real estate markets. *Urban Geography* 1–28. Advance online publication. doi: 10.1080/02723638.2018.1524653.

Sperling, J. 1995. Development and maintenance of the TIGER data base: Experiences in spatial data sharing at the U.S. Census Bureau. In *Sharing geographic information*, ed. H. Onsrud and G. Rushton, 377–96. Rutgers, NJ: Rutgers University Press Center for Urban Policy Research.

Streeter, L. A., D. Vitello, and S. A. Wonsiewicz. 1985. How to tell people where to go: Comparing navigational aids. *International Journal of Man-Machine Studies* 22 (5):549–62. doi: 10.1016/S0020-7373(85)80017-1.

Terranova, T. 2000. Free labor: Producing culture for the digital economy. *Social Text* 18 (2):33–58. doi: 10.1215/01642472-18-2_63-33.

Tribute, A., S. Edwards, R. Rossello, B. Drennan, and C. Fischer. 1996. Western Litho, AIII, Zip2, Pakon star at Nexpo. *Seybold Report on Publishing Systems* 25 (19):26.

Vance, A. 2015. *Elon Musk: Tesla, SpaceX, and the quest for a fantastic future*. New York: HarperCollins.

Warner, B. 1998. Online sprawl. *Adweek* 39 (21):10.

Wilson, M. W. 2012. Location-based services, conspicuous mobility, and the location-aware future. *Geoforum* 43 (6):1266–75. doi: 10.1016/j.geoforum.2012.03.014.

Zip2 announces debut of breakthrough city guide platform. 1998. BusinessWire, press release, June 16.

Zip2 announces world-class technology platform that enables media companies to dominate local on-line advertising. 1996. PR Newswire, press release, September 30.

PART II
Analytical Smartness

Rethinking Spatial Tessellation in an Era of the Smart City

Jin Xing, ⓘ Renee Sieber, and Stéphane Roche

Smart cities frequently rely on vast sensor networks, such as traffic cameras and ventilation controllers. This requires that we rethink methods of spatial tessellation. As tessellation is becoming more dynamic, we often combine multiple tessellation methods and switch tessellation shapes frequently for different data collection and analytics. In this article, we review how tessellation works with the object and field geographic spatial models. To achieve the "smartness" within cities, this article introduces the dynamic tessellation approach as the initial solution. *Key Words: big data, sensor network, smart city, spatial tessellation.*

智慧城市的运行通常都离不开一个庞大的传感器网络，比如交通摄像头和通风控制器等。 这就要求我们重新思考空间网格细分方法。随着这种网格细分的变化日益频繁，我们经常要把多种网格细分方法相结合，或者为收集和分析不同的数据，在不同网格细分形状中进行切换。在本文中，我们回顾了如何将网格细分方法与对象和场地理空间模型相结合，还阐述了如何将动态网格切分法作为一个初始解决方案，实现城市内核的"智能"。关键词: *大数据，传感器网络，智慧城市，空间网格切分。*

Con frecuencia, las ciudades inteligentes dependen de vastas redes de sensores, tales como las cámaras de tráfico y controladores de ventilación. Tal circunstancia exige que repensemos los métodos de teselado espacial. En la medida en que éste se hace más dinámico, a menudo combinamos múltiples métodos de teselado y cambiamos sus formas para diferentes colecciones y análisis de datos. En este artículo, revisamos el modo como el teselado trabaja con el objeto y con los modelos geográficos espaciales de campo. Para lograr la "inteligencia" dentro de las ciudades, este artículo presenta el enfoque del teselado dinámico como la solución inicial. *Palabras clave: big data, ciudad inteligente, red de sensores, teselado espacial.*

It is difficult to craft a single characterization of the smart city (Ching and Ferreira 2015), but it tends to be dominated by sensors, from Internet of Things (IoT)-connected lampposts to cell phones. Smart sensor networks rely on geolocations; for example, to interpolate heat islands from unevenly distributed temperature sensors. Sensors can produce geospatial big data sets (e.g., transportation, weather, noise level, air quality, and energy consumption), usually as point data. A case-by-case tracking of individual communication is onerous, with a huge number and high variety of sensors, so we often bound and aggregate the spatial extent of mobility. For example, mobile payments (i.e., smartphones and facial recognition) have been enabled via geofencing, so users can enjoy the convenience without worrying about the security of each transaction. If a transaction is detected at a location outside the jurisdictional bounding, then the payment is suspended and a fraud alert is issued. In the new field of platial science, point data from social media are often less analyzable and therefore less valuable until they are aggregated (Abdoullaev 2011). Because so many of the sensors that comprise smart cities are mobile, this compels us to consider more dynamic forms of analysis.

A single bounding box could tell us about the target object, but it also limits our understanding of its interconnection with objects outside the box. Limiting aggregation to a single bounding box could result in considerable information loss by ignoring complex interactions among sensors. Typically, spatial tessellation partitions geographic space into zones or tiles that are continuous and nonoverlapping across overall space; these tessellations allow us to subsequently define neighborhoods of geospatial data collection and analytics (Kuhn 2012). Tessellations need not be seamless across the spatial extent, but bounding boxes have been considered insufficient in smart city research (Calderoni, Maio, and Palmieri 2012). Another advantage of tessellation is its inherent topology (e.g., relations among zones), because a single bounding box cannot provide enough topological information.

For any given area, tessellations are usually static (Gold 2016a). Smart cities require dynamism in spatial tessellations of smart sensor networks that transform the interconnection among people, infrastructures, devices, organizations, and space (Bakıcı, Almirall, and Wareham 2013). Smart city sensors have become pervasive. Smart homes rely on continuous communication among various sensors to schedule heating and cleaning services; a smart vehicle obtains real-time information crowdsourced from mobile sensors (primarily other cars) to avoid potential collisions. Sensors, whether static (i.e., an IoT lamppost) or dynamic (i.e., a cell phone in a car), interact with geographic space at different resolutions and extents. Sensors in smart homes operate at the centimeter level across ranges of several meters, but smart electricity grids cover thousands of square kilometers. Complications induced by sensor dynamism (movement and scale) and diversity need to be addressed systematically before we can achieve the smartness in cities.

Over the years, numerous spatial tessellation methods have been explored. Lee, Li, and Li (2000) categorized spatial tessellation by mathematical and conceptual (real-world and database) abstractions of space. Miles (1972) also proposed a mathematical classification of tessellation methods, including rectangular tessellation, isotropic tessellation, anisotropic tessellation, and Voronoi diagram (VD; Aurenhammer 1991). In other reviews, Boots, Okabe, and Sugihara (1999) focused on various regular geometric shapes before concluding that the irregularly shaped VD was preferable. Gold (2016a) updated the review with respect to topology and data structures and followed this update with temporal and viewpoint changes in tessellations (Gold 2016b). There has been little research done on the specific case of smart cities. For instance, Calabrese, Ferrari, and Blondel (2015) examined base station planning for cell phone roaming. They considered the location of base stations but not the mobility of the cell phone. Khan, Ghamri-Doudane, and Botvich (2015) examined mobility tessellations of self-driving vehicles, considering the vehicle but not the varied and scalar interactions among tessellations.

The contribution of this article is twofold. First, we highlight how traditional spatial tessellation is challenged when we introduce sensor networks—the hardware of smart cities. Then we present our new framework to tessellate the smart city, which accommodates topology, temporality, and nested spaces. We conclude in the last section.

Methods to Tessellate the Smart City

Figure 1 shows that smart cities can be characterized as numerous interconnected sensor networks, which record geolocation-dependent information. Whether closed-circuit television (CCTV) cameras or unmanned aerial vehicle zones, sensors work in a spatial extent and many involve some kind of ubiquitous coverage. Hence, connected devices in the city can benefit from tessellation. The increasing dynamic nature implicated by these smart sensor networks requires tessellations varying by areal extent, resolution, and time. In smart transportation infrastructure, for instance, traffic sensors record the speed and location of individual vehicles at given locations (as depicted in Figure 1), whereas smart waste management and smart grids monitor the aggregated quality of service at neighborhood resolutions (Abdoullaev 2011). Tessellations in the smart city can provide (1) appropriate and varying spatial resolutions and extents for geolocation data collection and analytics; (2) aggregated patterns to offer more useful information than examining individual objects; and (3) appropriate accuracy tolerances due to the frequent movement of sensors (Hernández-Muñoz et al. 2011).

Many considerations in the city involve zones, which, although useful, can prove inadequate compared to tessellations. They are often connected to existing statistical boundaries according to a single feature (e.g., census tracts), which could be insufficiently flexible for positioning and tracking sensors. Zones might have blind spots; for example, a Global Positioning System (GPS) signal might not be available among skyscrapers. Smart city spatial tessellation should be able to handle multiple criteria, especially topology, to generate dynamic tessellations. We consider existing tessellations and their benefits, costs, and emerging challenges.

Field-Based Tessellation

Field-based tessellations discretize space through grids; this type of tessellation is often used with pixel

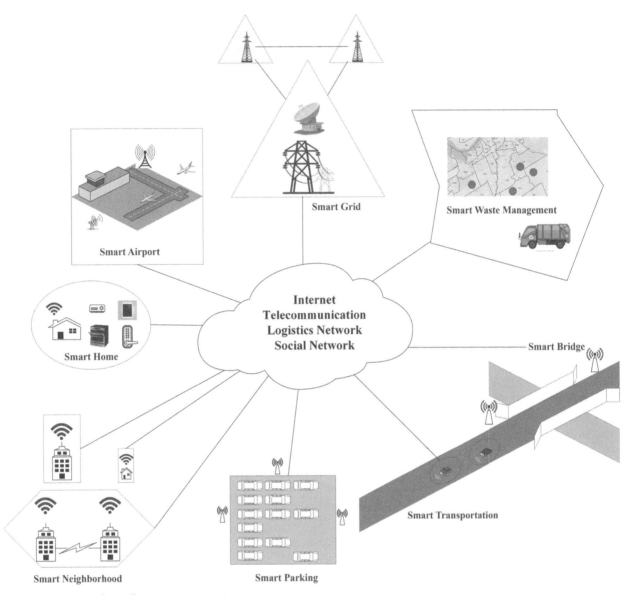

Figure 1. Various spatial tessellations in smart cities.

data, such as remotely sensed imagery. In smart cities, field-based tessellations employ regular or irregular geometric shapes to represent the spatial extent and neighborhood in which people, devices, institutions, infrastructures, and services function (Nam and Pardo 2011).

Field-based tessellation can be composed of regular or irregular shapes. The rectangular grid is the most frequently used approach (Kim et al. 2012). Buildings and city blocks are usually represented as rectangles on the map. Triangles (Daganzo and Knoop 2016) and hexagons (Feick and Robertson 2015) are less common but increasingly used, as with Loskutov, Zyrin, and Vukolov's (2016) smart power grid design or Uber's hexagonal hierarchical

spatial index (see https://eng.uber.com/h3/). Most urban features are irregular. As suggested in Figure 1, garbage collection is based on irregular boundaries of local communities. Administrative and political urban areas can be tessellated with irregularly shaped polygons (Laurini 2014). It is important to accommodate irregularly shaped territories because jurisdictions will impose their own political and legal regimes on smart sensor networks.

Field-based tessellation cannot tackle the increasing dynamics of smart sensor networks. The field model does not consider the variance in sensor coverage, which could result in the same tessellation for a light motion sensor (i.e., small extent) and a flooding sensor (i.e., large extent). Heterogeneity in spatiotemporal resolution represents another problem for field-based methods. A speed sensor needs to record individual vehicles at fine spatial resolution and short duration, whereas a larger spatiotemporal cell would be preferred for the structural condition of a bridge. Field tessellations are useful to account for the interaction among sensors because, broadly put, a field approach tessellates the background underneath which sensors function. Whether the background has been sensed properly (e.g., view angle, obstacles, distance, and privacy), however, cannot be answered using field-based tessellation alone. This is why researchers and practitioners are drawn to object-based tessellations.

Object-Based Tessellation

The object-based tessellation decomposes geographic space according to an attribute(s) of objects, especially their sensing location and extent. For example, the power consumption of buildings is aggregated as part of a smart electricity grid; sensors in airports not only track individual airplanes but also monitor their surrounds for risk management. WiFi coverage is routinely represented as circles or dodecahedrons, and hexagons are used to represent mesh networks of IoT devices (Centenaro et al. 2016). Here the objects determine the tessellation shapes and sizes.

To tessellate objects, we can aggregate and split. Object aggregation means grouping similar objects to achieve high intergroup similarity and reduce intragroup similarity (Logesh et al. 2018). Object aggregation can be accomplished via nearest neighbor (Yaqoob et al. 2017), optimization (X. Q. Li et al. 2007), density grouping (Vieira et al. 2010), agglomerative clustering (Rosenthal and Strange 2003), and fuzzy aggregation (Moreno et al. 2017) methods. An object-based splitting approach tends to decompose space using the dissimilarity among objects. Decomposition can be accomplished in various ways, such as K-means (Anderson 2009), graph-cut (Shi and Malik 2000), edge detection (Maini and Aggarwal 2009), and region dividing (Wu et al. 2008). Because sensors dictate the tessellation, object-based tessellation can accommodate heterogeneity (i.e., sensor capability, spatial resolution, and spatial extent) among numerous sensors. Object-based tessellations are seamless by

feature but not spatially (Calabrese, Ferrari, and Blondel 2015).

VD has attracted particular attention because it offers an efficient method to tessellate space by explicitly aggregating like objects and then connecting those aggregations (via seeds) to their surrounding space (i.e., field). The construction of seeds allows VD to merge thematic attributes (e.g., cars in a traffic jam) and the corresponding spatial neighborhood (e.g., extent of the traffic jam). The default VD employs Euclidean distances, but smart sensor networks could combine various metrics (e.g., area, shape, and boundary) in VD. To tackle heterogeneous sensor capabilities, convex or weighted VD has also been explored (Stergiopoulos and Tzes 2010). Xing and Sieber (2014) combined Euclidean distance and spectral distance with the sweepline algorithm to decompose big remote sensing images, which minimized the splitting of objects. A single VD cannot handle complex interactions among numerous sensors, as shown in Figure 1.

Graphical Model–Based Tessellation

Graphical models can represent complex dependence among objects as a probabilistic network built over multidimensional feature space. Geography has a long history using graphical models for network analysis. In road networks, the nodes and edges of graphical models are static in space; nodes are intersections and edges are line segments (roads). Sakakibara, Kajitani, and Okada (2004) extended this approach for potential disaster region detection via partitioning road networks into discrete sub graphs. Graphical models can be used for areas, which begin to resemble spatial tessellation (e.g., Zhu and Yu 2010). Krishnamurti and Roe (1979) first illustrated how to use graphical models to split geographic space into nested regular geometric shapes as a way to improve on architectural plans in three-dimensional visualization software. Some modern graphical tessellation algorithms include spanning tree (Clemens 2018), Markov random field (MRF; Liebig et al. 2014), and Bayesian network (Stassopoulou, Petrou, and Kittler 1998). Graphical tessellations differ from object tessellations in the former's emphasis on topological relations of nodes.

With smart sensor networks, sensors are represented as nodes and their spatial and topological relations as edges and edge weights. Notably, sensing zones vary in spatial extents, and there is no one node attribute or edge cost that will cover complex spatial interactions among sensors. Once determined, graphical models can represent the overlay and connectivity among sensors as the reliability, cost, throughput, or number of edges, which are fundamental in modeling smart sensor networks. This approach has not been explored for dynamic tessellation but shows promise for smart cities.

Dynamic Tessellation for Smart Cities

Smart cities depend on sensor networks to manage resource utilization, municipal services, and information exchange among various sensing platforms and surrounding areas (Ching and Ferreira 2015). Fields, objects, and graphs can address single static tessellations, but these methods alone are insufficient to address the intensive interaction and frequent movement of sensors. We propose a dynamic tessellation method to handle the increasing dynamics within sensor networks. This method is implemented as a nested graphical model, with each tile generated using object, field, or graphical model–based tessellation methods.

The dynamic tessellation has two components: (1) uneven and nested and (2) temporal tessellations. The first component generates tiles of various shapes and sizes to build a graph of graphs for tessellations. The temporal tessellation integrates spatio-temporal graphical models to tackle the mobility of sensors within sensor networks.

Uneven and Nested Graphical Tessellations

Uneven and nested graphical tessellations refer to tessellating given space into heterogeneous sizes and shapes of tiles with interconnections and overlaps among them. The sensing zone of the class of objects is encoded as a list of nodes with their spatial relationship represented as edges in the graph. IoT sensors coordinate with each other, such as the smart lamppost communicating with other lampposts, and smart parking might interact with lampposts (e.g., increase luminosity). These tessellations could afford overlay operations among the sensing zones to capture complex interactions (Fadel et al. 2015). Using nodes to represent tiles allows us to identify overlay areas (e.g., collision areas of smart vehicles), which

could provide advanced flexibility in smart sensor networks. According to Gold (2016a), graph edges offer an effective approach in characterizing the complex topological relationships and spatial distances among tile nodes. In addition, graphs of graphs can serve to "fill in the gaps" of noncontinuous sensing zone tiles. Consequently, seamless tessellation becomes less necessary and no longer constrains the graphical tessellation.

Definition 1. *Dynamic tessellation (DT) is a graph of graphs $<G, E>$, where G is a collection of graphs $<g_1, \ldots, g_i, \ldots, g_n>$ and E represents possible geographic information science operations (OP) among them. Each graph g_i is defined as $<Nodes, Edges, Tessellation\_Method>$. Nodes are a list of tiles, whereas Edges encode their topological and spatial relationships as arrays. Tessellation_Method is a constant in each graph, which could be either a field, an object, or a combination.*

$$DT = \{\langle Nodes_1, Edges_1, Tessellation_{method}1 \rangle, \ldots,$$

$$\langle Nodes_n, Edges_n, Tessellation_{method}n \rangle\},$$

$$\begin{bmatrix} N/A & \cdots & OP_{1,n} \\ \vdots & \ddots & \vdots \\ OP_{1,n} & \cdots & N/A \end{bmatrix}. \quad (1)$$

Uneven tessellations are discovered as the integration or overlay among two or more *DTs*. For example, if g_1 is a hexagon field tessellation and g_2 is an object-based vehicle radar sensing tessellation, then their overlay ($OP_{1,2} = Overlay$) highlights potential collision tiles. There exist numerous methods to compare graphs and graph of graphs (see Dale and Fortin 2010). To assess the similarity (distance) of two graphic tessellations, we employ various centrality distances (Pignolet et al. 2015).

Rule 1. *The distance between two DTs is given by differences of the centrality sum of each node graphs:*

$$D_f(DT_1, DT_2) = \sum_{i=1}^{m} f(g_i) - \sum_{j=1}^{n} f(g_j). \quad (2)$$

The function $f(g)$ stands for the calculation of graph centrality, which could be the sum of either degree centrality or betweenness centrality in each node graph. This rule not only defines the distance between DTs but also determines whether we can nest DTs: One graph becomes a node of another graph if their D_f is below a given distance threshold

(Lv et al. 2016). Where sensor tiles can be hierarchically ordered, tessellations can be represented as hierarchical graphs. The various sensors for the same object (e.g., sensors on a smart vehicle) can be encoded as a single node to investigate their communication with objects in other tiles. Smart sensor networks are then formulated as a routing problem across these graphs (Dragomir 2012). An example can be found at the optimization of multilevel power supply in the smart electricity grid (Liserre, Sauter, and Hung 2010). Multiple graphs can also be combined via graph embedding techniques (Xiong, Power, and Callan 2017). We also note that DT follows the concept of spatial graphing (Dale and Fortin 2010), because the location of nodes (tiles), the length (distance), and weights (topological information) of edges are very important in sensor networks.

Temporal Graphical Tessellations

Smart cities rely increasingly on real-time decision making. Autonomous vehicles must monitor other vehicles and people around them, identify potential overlaps, and take corresponding actions in a timely manner. Temporal graphical tessellation can be used to determine sensing zones in real time; Ibrahim et al. (2016) transformed scheduling into a routing problem across the graphs with real-time constraints. Katrakazas et al. (2015) introduced real-time scheduling algorithms into the spatial tessellation process to facilitate decision making in autonomous vehicles. We propose borrowing from movement theorists (Dodge 2016) who speak of path velocity, direction, acceleration, and sinuosity, which we can encode as edge weight methods in tessellation.

Traffic jams illustrate temporal dynamics as expressed via sensor networks: They can appear, expand, shrink, and vanish. Using Definition 1, the weights of edges could indicate the directional and speed changes of the traffic jam boundaries within each graphical tessellation (Sugiyama et al. 2008). To address the temporal dynamics of smart sensors, a key points–based tracking graphical model was investigated by Messing, Pal, and Kautz (2009). Temporal models have been further investigated alongside graphical tessellation, with increasing interest in semantic modeling of mobility tessellations (Diggle 2013).

We extend Definition 1 by designating the temporal change of tessellations as an array (see Dodge 2016) information within edges. Temporal information becomes a sequence of graphical tessellations, which means employing temporal interpolation to guarantee that $D_f(DT_{t1}, DT_{t2})$ falls within a predefined centrality distance with given periods (L. Li and Revesz 2004).

Rule 2. *Temporal interpolation of dynamic tessellation is afforded by a graph sequence of DTs based on (1), which satisfies* $\left\|\frac{D_f(DT_{t1},DT_{t2})}{T}\right\| < \varepsilon$, *and ε is the centrality distance threshold within given periods.*

$D_f(DT_1, DT_2)$ also can be applied to measure the magnitude of temporal changes among DTs.

Conclusion

Spatial tessellation has been given insufficient attention in smart cities. Current data collection and analytics largely depend on rectangular grids or individual object boundaries that fail to capture the dynamics of smart sensor networks. Because smart sensor networks are increasingly needed to ensure interoperability among sensors, we introduce dynamic spatial tessellation based on graphical models. We hope that this article will attract further research interest.

Funding

This work was funded by Mitacs, MITACS-IT09684.

ORCID

Jin Xing ⓘ http://orcid.org/0000-0001-5693-3414

References

Abdoullaev, A. 2011. Keynote: A smart world: A development model for intelligent cities. Paper presented at the 11th IEEE International Conference on Computer and Information Technology (CIT), Pafos, Cyprus, August 31.

Anderson, T. K. 2009. Kernel density estimation and K-means clustering to profile road accident hotspots. *Accident Analysis & Prevention* 41 (3):359–64. doi: 10.1016/j.aap.2008.12.014.

Aurenhammer, F. 1991. Voronoi diagrams—A survey of a fundamental geometric data structure. *ACM Computing Surveys* 23 (3):345–405. doi: 10.1145/116873.116880.

Bakıcı, T., E. Almirall, and J. Wareham. 2013. A smart city initiative: The case of Barcelona. *Journal of the Knowledge Economy* 4 (2):135–48. doi: 10.1007/s13132-012-0084-9.

Boots, B., A. Okabe, and K. Sugihara. 1999. Spatial tessellations. *Geographical Information Systems* 1:503–26. doi: 10.2307/2687299.

Calabrese, F., L. Ferrari, and V. D. Blondel. 2015. Urban sensing using mobile phone network data: A survey of research. *ACM Computing Surveys* 47 (2):25. doi: 10.1145/2655691.

Calderoni, L., D. Maio, and P. Palmieri. 2012. Location-aware mobile services for a smart city: Design, implementation and deployment. *Journal of Theoretical and Applied Electronic Commerce Research* 7 (3):15–87. doi: 10.4067/S0718-18762012000300008.

Centenaro, M., L. Vangelista, A. Zanella, and M. Zorzi. 2016. Long-range communications in unlicensed bands: The rising stars in the IoT and smart city scenarios. *IEEE Wireless Communications* 23 (5):60–67. doi: 10.1109/MWC.2016.7721743.

Ching, T. Y., and J. Ferreira. 2015. Smart cities: Concepts, perceptions and lessons for planners. In *Planning support systems and smart cities*, ed. S. Geertman, J. Ferreira, Jr., R. Goodspeed, and J. Stillwell, 145–68. Cham, Switzerland: Springer.

Clemens, J. 2018. Spanning tree modulus: Deflation and a hierarchical graph structure. PhD diss., Kansas State University.

Daganzo, C. F., and V. L. Knoop. 2016. Traffic flow on pedestrianized streets. *Transportation Research Part B: Methodological* 86:211–22. doi: 10.1016/j.trb.2015.12.017.

Dale, M. R. T., and M. J. Fortin. 2010. From graphs to spatial graphs. *Annual Review of Ecology, Evolution, and Systematics* 41:21–38. doi: 10.1146/annurev-ecolsys-102209-144718.

Diggle, P. J. 2013. *Statistical analysis of spatial and spatio-temporal point patterns*. Chapman and Hall/CRC.

Dodge, S. 2016. From observation to prediction: The trajectory of movement research in GIScience. In *Advancing geographic information science: The past and next twenty years*, ed. H. Onsrud and W. Kuhn, 123–36. Needham, MA: GSDI Association Press.

Dragomir, G. 2012. A spatial-temporal data model for choosing optimal multimodal routes in urban areas. In *2012 IEEE International Conference on Intelligent Computer Communication and Processing (ICCP)*, ed. J. Domingue, A. Galis, A. Gavras, T. Zahariadis, D. Lambert, F. Cleary, P. Daras, S. Krco, H. Müller, M.-S. Li, H. Schaffers, et al. 37–40. IEEE. doi: 10.1109/ICCP.2012.6356158.

Fadel, E., V. C. Gungor, L. Nassef, N. Akkari, M. A. Malik, S. Almasri, and I. F. Akyildiz. 2015. A survey on wireless sensor networks for smart grid. *Computer Communications* 71:22–33. doi: 10.1016/j.comcom.2015.09.006.

Feick, R., and C. Robertson. 2015. A multi-scale approach to exploring urban places in geotagged photographs. *Computers, Environment and Urban Systems* 53:96–109. doi: 10.1016/j.compenvurbsys.2013.11.006.

Gold, C. 2016a. Tessellations in GIS: Part I—Putting it all together. *Geo-Spatial Information Science* 19 (1):9–25. doi: 10.1080/10095020.2016.1146440.

Gold, C. 2016b. Tessellations in GIS: Part II—Making changes. *Geo-Spatial Information Science* 19 (2):157–67. doi: 10.1080/10095020.2016.1182807.

Hernández-Muñoz, J. M., J. B. Vercher, L. Muñoz, J. A. Galache, M. Presser, L. A. H. Gómez, and J. Pettersson. 2011. Smart cities at the forefront of the future internet. In *The future Internet assembly*, 447–62. Berlin: Springer. doi: 10.1007/978-3-642-20898-0_32.

Ibrahim, M. S., S. Muralidharan, Z. Deng, A. Vahdat, and G. Mori. 2016. A hierarchical deep temporal model for group activity recognition. In *Proceedings of the IEEE Conference on Computer Vision and Pattern Recognition*, ed. E. Mortensen and K. Saenko, 1971–80. Las Vegas: The Computer Vision Foundation.

Katrakazas, C., M. Quddus, W. H. Chen, and L. Deka. 2015. Real-time motion planning methods for autonomous on-road driving: State-of-the-art and future research directions. *Transportation Research Part C: Emerging Technologies* 60:416–42. doi: 10.1016/j.trc.2015.09.011.

Khan, J. A., Y. Ghamri-Doudane, and D. Botvich. 2015. Inforank: Information-centric autonomous identification of popular smart vehicles. In *2015 IEEE 82nd Vehicular Technology Conference (VTC2015-Fall)*, 1–6, Boston: IEEE.

Kim, S. A., D. Shin, Y. Choe, T. Seibert, and S. P. Walz. 2012. Integrated energy monitoring and visualization system for smart green city development: Designing a spatial information integrated energy monitoring model in the context of massive data management on a web based platform. *Automation in Construction* 22:51–59. doi: 10.1016/j.autcon.2011.07.004.

Krishnamurti, R., and P. O. N. Roe. 1979. On the generation and enumeration of tessellation designs. *Environment and Planning B: Planning and Design* 6 (2):191–260. doi: 10.1068/b060191.

Kuhn, W. 2012. Core concepts of spatial information for transdisciplinary research. *International Journal of Geographical Information Science* 26 (12):2267–76. doi: 10.1080/13658816.2012.722637.

Laurini, R. 2014. A conceptual framework for geographic knowledge engineering. *Journal of Visual Languages & Computing* 25 (1):2–19. doi: 10.1016/j.jvlc.2013.10.004.

Lee, Y. C., Z. L. Li, and Y. L. Li. 2000. Taxonomy of space tessellation. *ISPRS Journal of Photogrammetry and Remote Sensing* 55 (3):139–49. doi: 10.1016/S0924-2716(00)00015-0.

Li, L., and P. Revesz. 2004. Interpolation methods for spatio-temporal geographic data. *Computers, Environment and Urban Systems* 28 (3):201–27. doi: 10.1016/S0198-9715(03)00018-8.

Li, X. Q., S. M. Jiao, X. P. Zhang, S. Q. Zhu, and Z. F. Du. 2007. PSO spatial clustering with obstacles constraints. *Computer Engineering and Design* 24:32.

Liebig, T., N. Piatkowski, C. Bockermann, and K. Morik. 2014. Predictive trip planning-smart routing in smart cities. In *EDBT/ICDT Workshops*, ed. K. Selçuk Candan, S. Amer-Yahia, N. Schweikardt, V. Christophides, and V. Leroy, 331–38. Brussels: Schloss Dagstuhl–Leibniz-Zentrum fuer Informatik.

Liserre, M., T. Sauter, and J. Y. Hung. 2010. Future energy systems: Integrating renewable energy sources into the smart power grid through industrial electronics. *IEEE Industrial Electronics Magazine* 4 (1):18–37. doi: 10.1109/MIE.2010.935861.

Logesh, R., V. Subramaniyaswamy, V. Vijayakumar, X. Z. Gao, and V. Indragandhi. 2018. A hybrid quantum-induced swarm intelligence clustering for the urban trip recommendation in smart city. *Future Generation Computer Systems* 83:653–73. doi: 10.1016/j.future.2017.08.060.

Loskutov, A., D. Zyrin, and V. Vukolov. 2016. Analysis of different topological structures of electric power distribution for the cities and elaboration of method of reducing value of the short-circuit current based on algorithm of dividing network. In *2016 2nd International Conference on Industrial Engineering, Applications and Manufacturing (ICIEAM)*, ed. IEEE staff, 1–6. Chelyabinsk, Russia: IEEE.

Lv, T., H. Gao, X. Li, S. Yang, and L. Hanzo. 2016. Space-time hierarchical-graph based cooperative localization in wireless sensor networks. *IEEE Transactions on Signal Processing* 64 (2):322–34. doi: 10.1109/TSP.2015.2480038.

Mainetti, L., L. Patrono, and A. Vilei. 2011. Evolution of wireless sensor networks towards the internet of things: A survey. In *SoftCOM 2011, 19th International Conference on Software, Telecommunications and Computer Networks*, 1–6. Split, Croatia: IEEE.

Maini, R., and H. Aggarwal. 2009. Study and comparison of various image edge detection techniques. *International Journal of Image Processing* 3 (1):1–11.

Messing, R., C. Pal, and H. Kautz. 2009. Activity recognition using the velocity histories of tracked keypoints. In *2009 IEEE 12th International Conference on Computer Vision*, ed. M. Okutomi, S. Baker, T. Pajdla, and H. Zha, 104–11. Kyoto, Japan: IEEE.

Miles, R. E. 1972. The random division of space. *Advances in Applied Probability* 4:243–66. doi: 10.2307/1425985.

Moreno, M. V., F. Terroso-Sáenz, A. González-Vidal, M. Valdés-Vela, A. F. Skarmeta, M. A. Zamora, and V. Chang. 2017. Applicability of big data techniques to smart cities deployments. *IEEE Transactions on Industrial Informatics* 13 (2):800–809. doi: 10.1109/TII.2016.2605581.

Nam, T., and T. A. Pardo. 2011. Conceptualizing smart city with dimensions of technology, people, and institutions. In *Proceedings of the 12th Annual International Digital Government Research Conference: Digital Government Innovation in Challenging Times*, ed. J. Bertot, K. Nahon, S. Ae Chun, L. Luna-Reyes, and V. Atluri, 282–91. College Park, MD: ACM.

Pignolet, Y., M. Roy, S. Schmid, and G. Tredan. 2015. Exploring the graph of graphs: Network evolution and centrality distances. Preprint arXiv:1506.01565.

Rosenthal, S. S., and W. C. Strange. 2003. Geography, industrial organization, and agglomeration. *Review of Economics and Statistics* 85 (2):377–93. doi: 10.1162/003465303765299882.

Sakakibara, H., Y. Kajitani, and N. Okada. 2004. Road network robustness for avoiding functional isolation in disasters. *Journal of Transportation Engineering* 130 (5):560–67. doi: 10.1061/(ASCE)0733-947X(2004)130:5(560).

Shi, J., and J. Malik. 2000. Normalized cuts and image segmentation. *IEEE Transactions on Pattern Analysis and Machine Intelligence* 22 (8):888–905. doi: 10.1109/34.868688.

Stassopoulou, A., M. Petrou, and J. Kittler. 1998. Application of a Bayesian network in a GIS based decision making system. *International Journal of Geographical Information Science* 12 (1):23–46. doi: 10.1080/136588198241996.

Stergiopoulos, Y., and A. Tzes. 2010. Convex Voronoi-inspired space partitioning for heterogeneous networks: A coverage-oriented approach. *IET Control Theory & Applications* 4 (12):2802–12. doi: 10.1049/iet-cta.2009.0298.

Sugiyama, Y., M. Fukui, M. Kikuchi, K. Hasebe, A. Nakayama, K. Nishinari, S. Tadaki, and S. Yukawa. 2008. Traffic jams without bottlenecks—Experimental evidence for the physical mechanism of the formation of a jam. *New Journal of Physics* 10 (3):033001. doi: 10.1088/1367-2630/10/3/033001.

Vieira, M. R., V. Frias-Martinez, N. Oliver, and E. Frias-Martinez. 2010. Characterizing dense urban areas from mobile phone-call data: Discovery and social dynamics. In *2010 IEEE Second International Conference on Social Computing (SocialCom)*, ed. A. K. Elmagarmid and D. Agrawal, 241–48. Indianapolis, IN: IEEE.

Wu, Y. T., F. Y. Shih, J. Shi, and Y. T. Wu. 2008. A top-down region dividing approach for image segmentation. *Pattern Recognition* 41 (6):1948–60. doi: 10.1016/j.patcog.2007.11.020.

Xing, J., and R. Sieber. 2014. Sampling based image splitting in large scale distributed computing of earth observation data. In *2014 IEEE International Geoscience and Remote Sensing Symposium (IGARSS)*, ed. T. Lukowski, 1409–12. Quebec City, Canada: IEEE.

Xiong, C., R. Power, and J. Callan. 2017. Explicit semantic ranking for academic search via knowledge graph embedding. In *Proceedings of the 26th International Conference on World Wide Web*, ed. R. Barrett, R. Cummings, E. Agichtein, and E. Gabrilovich, 1271–9. Perth, Australia: International World Wide Web Conferences Steering Committee.

Yaqoob, I., I. A.T. Hashem, Y. Mehmood, A. Gani, S. Mokhtar, and S. Guizani. 2017. Enabling communication technologies for smart cities. *IEEE Communications Magazine* 55 (1):112–20. doi: 10.1109/MCOM.2017.1600232CM.

Zhu, W., and Q. Yu. 2010. Spatial chromatic tessellation: Conception, interpretation, and implication. *Annals of GIS* 16 (4):237–54. doi: 10.1080/19475683.2010.539983.

Understanding Place Characteristics in Geographic Contexts through Graph Convolutional Neural Networks

Di Zhu, ⓘ Fan Zhang, ⓘ Shengyin Wang, Yaoli Wang, Ximeng Cheng, ⓘ Zhou Huang, ⓘ and Yu Liu ⓘ

Inferring the unknown properties of a place relies on both its observed attributes and the characteristics of the places to which it is connected. Because place characteristics are unstructured and the metrics for place connections can be diverse, it is challenging to incorporate them in a spatial prediction task where the results could be affected by how the neighborhoods are delineated and where the true relevance among places is hard to identify. To bridge the gap, we introduce graph convolutional neural networks (GCNNs) to model places as a graph, where each place is formalized as a node, place characteristics are encoded as node features, and place connections are represented as the edges. GCNNs capture the knowledge of the relevant geographic context by optimizing the weights among graph neural network layers. A case study was designed in the Beijing metropolitan area to predict the unobserved place characteristics based on the observed properties and specific place connections. A series of comparative experiments was conducted to highlight the influence of different place connection measures on the prediction accuracy and to evaluate the predictability across different characteristic dimensions. This research enlightens the promising future of GCNNs in formalizing places for geographic knowledge representation and reasoning. *Key Words: big geodata, graph convolutional neural networks, place characteristic, place connection, spatial prediction.*

推断一个地理位置的未知属性既取决于观测的属性，也取决于其连接对象位置的特征。由于位置特征无具体结构且位置连接的指标是多元的，如果空间预测任务的结果可能受相邻区域绘图方式的影响，很难识别不同位置真正的相关性，也就很难将这些位置特征纳入其中。为了弥补这一缺陷，我们引入图卷积神经网络（GCNN）将位置建模为图谱，将每个位置确认为一个节点，将位置特征编码为节点特征，并将位置连接显示为连接节点的边。GCNN 通过优化图神经网络层之间的权重，来获取相关地理环境的知识。我们在北京城市区域设计了一项案例研究，根据观测到的属性和特定地理位置的连接，预测未观测到的位置特征。我们通过一系列比较实验，揭示了不同位置连接测量对预测准确性的影响，评估不同特征维度的可预测性。本研究表明，GCNN 在地理知识表述与推理的位置确认方面具有广阔的应用前景。 关键词: 大地理数据，图卷积神经网络，位置特征，位置连接，空间预测。

Inferir las propiedades desconocidas de un lugar se fundamenta tanto en los atributos observables como en las características de los lugares con los cuales aquel está conectado. Debido a que las características del lugar no están estructuradas y las métricas de las conexiones del lugar pueden ser diversas, es todo un reto incorporarlas en una tarea de predicción espacial, donde los resultados podrían afectarse por el modo como están delineados los vecindarios y donde la verdadera relevancia que hay entre los lugares es difícil de establecer. Para zanjar esta dificultad, introducimos el gráfico de las redes neurales convolucionales (GCNNs) para modelar los lugares como un gráfico, donde cada lugar es formalizado como un nódulo, las características del lugar son codificadas como rasgos nodales y las conexiones del lugar se representan como los bordes. Los GCNNs capturan el conocimiento del contexto geográfico relevante optimizando los pesos entre las capas del gráfico de redes neurales. Se diseñó un estudio de caso en el área metropolitana de Beijing para predecir las características no observadas del lugar con base en las propiedades observadas y las conexiones específicas del lugar. Se condujo una serie de experimentos comparativos para destacar la influencia de las medidas de diferentes conexiones del lugar sobre la exactitud de la predicción, y para evaluar la predictibilidad a través de diferentes dimensiones de las características. Esta investigación ilumina el prometedor futuro de los GCNNs para formalizar lugares para la representación y razonamiento del conocimiento geográfico. *Palabras clave: big geodata, características del lugar, conexión del lugar, gráfico de las redes neurales convolucionales, predicción espacial.*

This article has been republished with minor changes. These changes do not impact the academic content of the article.

The geographical concept of *place* is often used as "a portion of space" (Agnew and Duncan 2014) within which people carry out habitual aspects of their lives, such as recreation, work, and sleep (Goodchild 2011). The perception of a place is a comprehensive integration of location names, emotional feelings, and other properties (Shamai 1991; Adams and McKenzie 2013). The term *place characteristics* encompasses a broad range of properties used to depict a place that are important in describing the uniqueness of a specific environment. By place characteristics, we are referring not only to the features such as place names and types but also to the socioeconomic properties, human activities, and perceptions that can be measured, such as liveliness, greenness, or transport convenience (Rattenbury and Naaman 2009; Zhang et al. 2018).

Because places are not isolated but are connected to each other in many ways (Nystuen and Dacey 1961; Gould 1991; Noronha and Goodchild 1992), the contextual information for a place (i.e., its connection to other places) is crucial to understand its characteristics. These connections link a set of places to a network that indicates the predefined geographic contexts for the places (Kwan 2007). We use *place connections* to represent the measures between places, which could be both physical and social, such as distance, adjacency, and spatial interactions. Intuitively, the predictability of a place characteristic should be higher when choosing more appropriate connection measures. For example, the connection between places via taxi origin–destination flows can help identify the land-use characteristics of the places (X. Liu et al. 2016).

The prediction of a place's unknown characteristic relies on both the place's observed characteristics and the characteristics of the places to which it is connected. Despite various ways of measuring the connections, the "true causally relevant" contexts can be very complex (Golledge 2002; Kwan 2012). The prediction could be affected by how the geographic context is defined. A.-X. Zhu et al. (2018) suggested measuring the similarity of the locations' geographic configurations in the covariate space as the connections used for spatial prediction. In most cases, however, the researcher cannot be certain that the connections used in the study are appropriate. Further, as various place characteristics can now be sensed from multisource user-generated big geodata (Y. Liu et al. 2015; Jenkins et al. 2016; MacEachren 2017), it is even more difficult to incorporate both high-dimensional and unstructured places' characteristics and diverse place connections in a spatial prediction model. Attempts to simplify the knowledge of relevant contexts as predefined mathematical functions cannot adequately model its complex nature. Methods such as geographically weighted regression use kernel functions to consider connections but only focus on nearby observations in space (Fotheringham, Brunsdon, and Charlton 2003), which could be arbitrary when long-range relevance is nonnegligible and would require much effort to be applied to a non-Euclidean situation (Lu et al. 2014).

Recently, there has been a surge of interest in graph convolutional neural networks (GCNNs) for learning graph-structured data where the range of connection varies (Bruna et al. 2014). To effectively process the connection information, GCNNs generally follow an aggregation scheme where each node aggregates characteristics of its neighbors to learn a deep representation of the contextual information (Defferrard, Bresson, and Vandergheynst 2016). This powerful technique is able to capture both the long-range and short-range relationships through its neural network weights. Clearly, it is suitable for modeling a graph of connected places.

This research introduces the use of GCNNs to model connected places where each place is represented as a node, place characteristics are the node features to be computed, and place connections are represented as the graph edges. The graph convolution can effectively learn from the graph structures and node features to understand the place characteristics in a geographic context. The objective of this study was to investigate the feasibility of incorporating place connections to predict place characteristics. In a case study of the Beijing metropolitan area, we took advantage of GCNNs in formalizing, reasoning, and understanding places. Three scenarios were designed to consider different connection types. A series of comparative experiments revealed the influence of place connections on predicting place characteristics.

Methodology

Building the Place-Based Graph

Assuming a set of places P where each element p_i means a place, a place characteristic X can be represented as a feature vector $[x_1, x_2, ...]$ observed on P, where x_i denotes the values for the ith dimension of X. A set C includes various connection types between places, such as distance, topological adjacency, and spatial interactions. $c \in C$ refers to a certain connection metric. Then, a place-based graph $G = (V, E^{(c)})$ is constructed to connect places as a graph. Each place p_i is formalized as a node $v_i \in V$ in G, and the place characteristic X is encoded as the node feature $X_k \in X$ on every $v_k \in V$. The place connections in type c are represented as a set of edges $E^{(c)}$, where $e_{ij} = (v_i, v_j, a_{ij}) \in E^{(c)}$ is the edge between p_i and p_j and a_{ij} is the weight of the edge. Given c, $E^{(c)}$ represents the connection information in the place-based graph G.

The problem we addressed is illustrated in Figure 1. There are two kinds of place characteristics: X, which is easier to obtain (e.g., visual characteristic), and Y (e.g., functional characteristic), which is more difficult to obtain but is vital to residents' daily lives, activities, and perceptions. Often, we had the information of both visual and functional characteristics for certain sampled places but only the visual characteristics for unsampled places. Using the information collected, we were able to predict the functional characteristics for unsampled places based on their observed visual characteristics and the characteristics of their connected places given $E^{(c)}$. As illustrated in Figure 1, Y_1, Y_4, and Y_5 were the place characteristics to be predicted for the unsampled places p_1, p_4, and p_5, respectively, but we only had the ground truth Y_2 and Y_3 for the sampled places p_2 and p_3, respectively.

Predicting Place Characteristics Using GCNNs

The GCNN model was designed as a multilayer neural network structure to learn the node features and connection information of an input graph. By computing the differences between the model outputs and the expected output, a GCNN model iteratively updates its layer-wise neural network weights through a large amount of training and eventually approximates the predictability of place characteristics given certain place connections. The details of how to train a GCNN model can be found in the Appendix.

When using an m-layer GCNN model to predict the place characteristic Y of unsampled places (i.e., Y_1, Y_4, and Y_5 in Figure 1), the place characteristic X of both the sampled and unsampled places was used as the input node features, and then we selected a metric of place connections E to build the place-based graph. For each training iteration, the weights in the GCNN were optimized using a back-propagation method by computing the prediction loss between the outputs on p_2, p_3 and the observed Y_2, Y_3. A fully trained GCNN model would generate a final output of node features that are the most probable values for Y_1, Y_4, and Y_5 given the place connections.

Because the results could be affected by how the geographic contexts are defined, GCNNs can learn from both graph structures and node features to capture the "relevant context" underlying $E^{(c)}$ and to facilitate the prediction of unobserved Y. GCNNs fit a neural network model that considers place characteristics and connections among all of the places to make the most credible prediction. The advantage of using GCNNs is threefold. First, all of the places are connected as a graph, so the connections among unsampled places can be explicitly modeled. Second, GCNNs do not require a predefined kernel function (Fotheringham, Yang, and Kang 2017), so they could

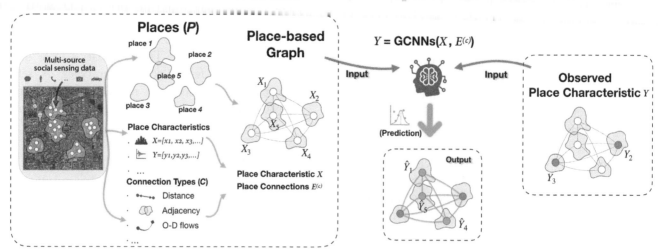

Figure 1. Building a place-based graph and using GCNNs to predict place characteristics. *Note:* O–D = origin–destination; GCNN = graph convolutional neural network..

learn both the long-range and short-range relationships in the place-based graph. Third, because a GCNN is designed as a nonparametric neural network, it does not assume any form of the spatial function to be fitted; for example, linear equations in most spatial regression models (Anselin 2010).

Case Study

The study area for this research was Beijing, the capital of China. The fast-developing economy has made this city a giant metropolitan area where the flexible behaviors of its residents have led to diverse functional characteristics common in urban places regarding dining, residences, transportation, business, and so on. Understanding the heterogeneous pattern of place characteristics and how it is formed in the complex geographic context is key to enhancing the quality of life for residents and building a smart city.

A case study was designed to leverage GCNNs in predicting the unobserved characteristics for places based on their observed characteristics and connection to other places. The boundaries of 203 places within the study area were identified as the study units. Two types of characteristics were collected for the places and represented as feature vectors. One was the visual features as the input characteristic X; the other was the functional features as the place characteristic Y to be predicted. Three scenarios of incorporating place connections—*self-only*, *adjacency*, and *spatial interaction*—were examined to build different place-based graphs. By randomly selecting some places as the sampled places and others as the unsampled places, experiments were conducted to evaluate (1) the influence of different place connection scenarios on the prediction accuracy and (2) how the predictability of Y varies across different dimensions.

Data Preparation

Delineating Place Boundaries. To predict the characteristics of these places, a data set with 243,065 points of interest (POIs) was collected in 2016 from Baidu, a major source of location data. Each POI was labeled with a place name that reflected residents' common perceptions of the location. To identify the boundaries of places, we adopted a kernel density estimation method to compute the POI densities for each place name in space (Wang, Liu, and Chen 2018) and to depict the intuitive boundary of the place name. For each place name, the corresponding POI kernel densities were normalized into [0, 1] as the membership functions indicating to what extent an area belongs to this place (Gao et al. 2017; Wu et al. 2019). Then, a place's boundary was delineated based on a membership threshold.

We adopted a threshold of 0.5 to delineate both the core and peripheral area of the place names (Wang, Liu, and Chen 2018). Figure 2 shows the 203 places extracted within the study area. These delineated polygons cover the urban areas where names are broadly known and used in the locals' daily lives. We used these places as the study units in the following experiments.

Quantifying Place Characteristics. Two kinds of place characteristics were collected. One is human perception of place locales (MacEachren 2017); that is, the visual characteristics. The other one is the land-use attributes of a place in terms of human activities (Lansley and Longley 2016); that is, the functional characteristics. These characteristics were treated as visual features and functional features in the GCNN model. Visual features and functional features characterized places from physical and social viewpoints, respectively. In this work, we used the visual features as the input to predict the functional features of places.

With regard to visual features, a total of 987,635 street view images taken in 2016 were collected from Tencent (Zhang et al. 2018). The images are photos describing the visual environment of the urban streets. We used a feature extractor (He et al. 2016) to derive place-based visual features from the images. The feature extractor has been proven efficient in scene perception modeling (Zhou et al. 2018). Each street view image was transformed into a 512-dimensional feature vector to represent human perceptions of a place. Visual features of a place were further obtained by taking the average of the vectors on all images within a place. Figure 3A shows that the visual features varied greatly for two places in the study area.

For functional features, we used a check-in data set collected from a social media platform named Sina Weibo. The data contained the annual number of check-ins for each POI. Within the study area, we obtained 4,099,016 check-in records for the 203

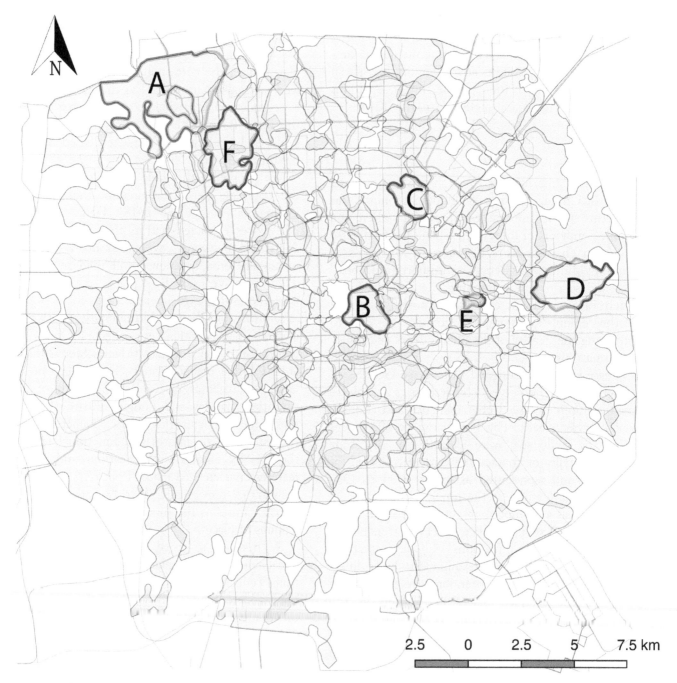

Figure 2. The 203 places in the Beijing metropolitan area. Some representative places are highlighted, including historic sites (A) the Summer Palace and (B) Tian'anmen; residential areas (C) Hepingli and (D) Shilipu; and commercial districts (E) Guomao and (F) Zhongguancun. For other maps in this article, we omit the north arrows and map scales for simplicity.

places. Check-in records with different activity labels were aggregated into seven functional characteristic types to represent seven different functional feature dimensions. Because the check-in intensities are heavy-tail distributed, we computed the logarithmic numbers of the check-in activities to the base 10 as the feature values to reduce the skewness of data. The seven functional feature dimensions included (along with their value range in parentheses) dining (0, 4.80], residence (0, 4.42], transport (0, 4.67], business (0, 4.72], recreation (0, 4.78], medical (0, 4.11], and outdoors (0, 4.24]. The spatial distribution of the functional features is plotted in Figure 3B.

Measuring Place Connections. To test the effect of place connections in predicting place characteristics, three scenarios of incorporating place

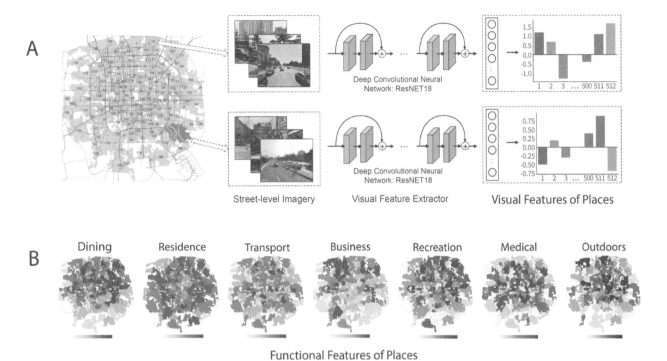

Figure 3. Place characteristics represented as feature values. (A) Visual features derived from street view images. (B) Seven functional feature dimensions sensed from social media check-in data; a darker hue denotes a higher value.

connections were examined to build different place-based graphs. The first scenario, self-only, considered no contextual information of a place, which means that we predicted the characteristics of the unsampled place without considering place connections. Self-only connection indicates that a place is connected only to itself. The second scenario, adjacency, incorporated connections in the prediction but was limited to only those connections between spatially adjacent places. The third scenario, spatial interaction, quantified both the short-range and long-range interaction volumes among places where the data were derived from a taxi origin–destination (O–D) data set (D. Zhu et al. 2017). Details about how these place connections were represented in matrix form can be found in the Appendix.

A GCNN Model to Predict Places' Functional Features

The GCNN model required some places to be the training places, and others were the test places. The training places were those sampled with both visual features X and functional features Y. The test places were places that have observed visual features X but do not have observations for functional features Y. The task of predicting Y for test places can be formulated as a semisupervised learning problem in the graph where we use the X of both training and test places but the Y of only training places to learn the place-based graph, as mentioned in the Methodology section.

Of the 203 places, we used 40 for training and 163 for testing. Figure 4A shows this initialization of the training and test places. A GCNN model was designed (Figure 4B) to predict the functional features for the test places. Seven submodels were developed, one for each of the seven functional feature dimensions. Each submodel was a two-layer GCNN that outputs the values for one functional dimension.

To report the prediction accuracy of a fully trained GCNN model, we used the lowest mean absolute percentage error (MAPE) on the test places. Let $N(T)$ be the total number of places in the test set T; the MAPE is calculated as $\frac{100\%}{N(T)} \sum_{i \in T} \frac{|\hat{Y}_i - Y_i|}{Y_i}$, where $\hat{Y}_i$ and Y_i are the predicted and ground truth feature value of place i, respectively.

Results and Discussion

The Influence of Place Connection Types

Three scenarios of place connections—self-only, adjacency, and spatial interaction—were used in our

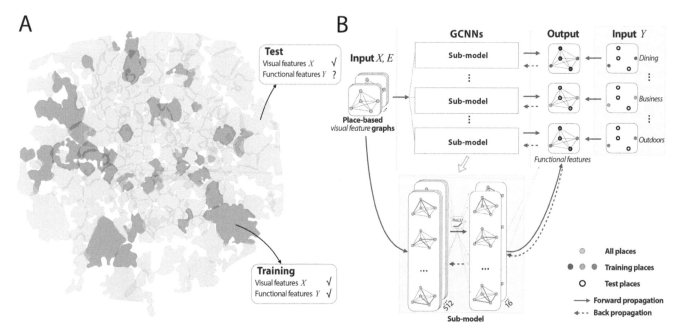

Figure 4. (A) An example of the training and test places initialization. (B) The GCNN model used to predict the seven functional feature dimensions for test places in a graph. *Note:* GCNN = graph convolutional neural network.

proposed GCNN model to predict the functional features of places. For each of the seven feature dimensions, place connections had a significant influence on the prediction accuracy. The GCNN model always achieved the best performance given the spatial interaction scenario. To highlight the influence of place connection metrics, we only report and discuss the results for the dining dimension in this section. Similar findings can be applied to the other six functional dimensions.

We simulated fifty different initializations of training and test places. The results of the fifty experiments on dining are plotted in Figure 5A. Self-only place connections show a median MAPE of 51.03 percent and a standard deviation of 0.0451; adjacency obtains the poorest median MAPE of 80.11 percent and a standard deviation of 0.1761; and the spatial interaction scenario displays the best overall accuracy with a median MAPE of only 16.24 percent and a standard deviation of 0.0148. Figure 5B shows the results for one of the fifty experiments. We found that neither self-only nor adjacency could achieve a predicted pattern similar to the real one, whereas for spatial interaction the GCNN achieved a predicted pattern strongly correlated with the real pattern.

The results of self-only indicated that using features alone cannot predict the functional features very well without considering the place connections.

Because we did not consider any contextual information in the GCNN model in the self-only scenario, the prediction worked very similarly to a multivariate nonspatial regression. The fact that adjacency had the worst result indicates that it is unwise to use Euclidean metrics such as adjacency to predict the functional features, because adjacency cannot accurately describe the meaningful connections between places. Moreover, as spatial interaction can better reflect human activities and the actual connections among places, it outperformed the other two scenarios in predicting places' functional features.

Variation across Functional Feature Dimensions

Because the performance of spatial interaction was better than the other two scenarios, the subsequent analysis focused on the spatial interaction scenario. When evaluating the prediction results of all seven functional feature dimensions, we conducted experiments based on multiple training ratios (from about 10 percent to 90 percent) of all of the places. Table 1 presents the results of the fifty parallel experiments. For each of the training ratios and feature dimensions, an average value of MAPE and a standard deviation of the fifty experiments were calculated for the test places to report the prediction accuracy and stability of the GCNN model, respectively.

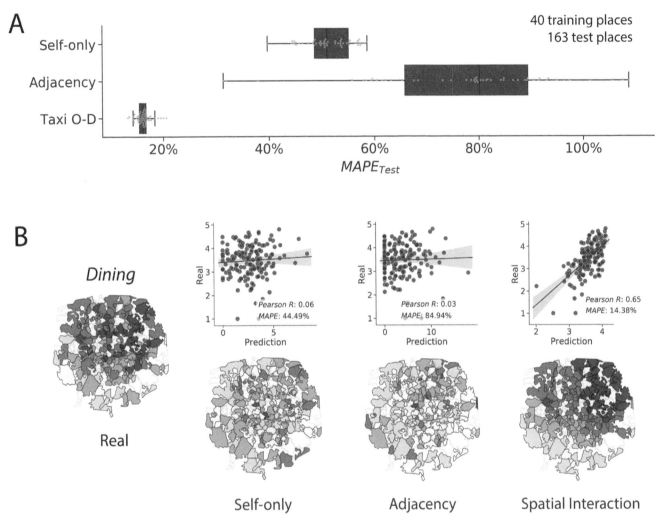

Figure 5. Different place connection scenarios resulted in different prediction accuracies. (A) The MAPE box plots of the fifty experiments on the dining dimension. (B) The predicted patterns on test places in one of the fifty experiments. Only the test places are colored to help interpret the model's performance. *Note:* O–D = origin–destination; MAPE = mean absolute percentage error.

According to the MAPEs in Table 1, dining and residence were better predicted than the other five dimensions, with average MAPEs lower than 20 percent. Medical was the most difficult dimension to predict and did not reach an average MAPE lower than 50 percent. We found that for a given training ratio, the accuracy remained stable, which means that the GCNN model can learn the knowledge for prediction even with only 10 percent observations. The standard deviation slightly rose when the training ratio was increased. This was because the test places of the fifty experiments were randomly selected but kept the same training ratio and the lesser test places led to a higher instability of the prediction. We checked the results of the best-fitted GCNN model under the 60 percent (120/203) training ratio for each functional feature dimension and visualized their

corresponding patterns on test places in Figure 6. First, dining and residence achieved quite good predicted patterns compared to the real patterns, indicating that taxi O–D flows can help in understanding these two functional characteristics. Second, business, recreation, and transport showed medium-level predictabilities, which implies that using taxi O–D flows as the place connections is still not satisfactory when trying to predict these three functional characteristics. Third, outdoors and medical had inferior predicted patterns, meaning that the predictability of these two characteristics is very low even when taxi O–D flows are used. Because the functional characteristics of places are closely related to residents' activity patterns, some activities, such as hanging around in a park or going to a local hospital, might have very little relationship with taxis (Gong et al. 2016).

Table 1. The predictions of different functional feature dimensions on the test places

# Train	# Test	Training ratio (%)	Average prediction accuracies of fifty parallel experiments in the spatial interaction scenario												
			Dining		Residence		Transport		Business		Recreation		Medical		Outdoors	
			MAPE (%)	SD	MAPE (%)	SD	MAPE (%)	SD	MAPE (%)	SD	MAPE (%)	SD	MAPE (%)	SD	MAPE (%)	SD
20	183	10	16.26	0.0157	1.94	0.0113	34.10	0.0214	33.32	0.0331	35.81	0.0270	51.24	0.0695	41.16	0.0430
40	163	20	16.31	0.0148	1.4	0.0146	34.72	0.0233	34.74	0.0417	36.56	0.0372	53.57	0.0636	42.18	0.0473
60	143	30	15.46	0.0171	1.42	0.0155	33.98	0.0252	34.42	0.0385	36.44	0.0320	52.32	0.0700	40.58	0.0305
80	123	40	15.32	0.0178	1.45	0.0209	33.72	0.0271	33.33	0.0288	35.72	0.0349	51.94	0.0664	41.56	0.0442
100	103	50	15.29	0.0225	1.93	0.0241	34.46	0.0312	33.97	0.0404	35.48	0.0351	51.88	0.0643	40.68	0.0406
120	83	60	15.08	0.0216	1.17	0.0311	34.76	0.0389	34.35	0.0395	35.13	0.0376	50.50	0.0677	40.73	0.0484
140	63	70	15.19	0.0273	1.99	0.0295	34.74	0.0474	33.62	0.0407	35.43	0.0511	53.58	0.0757	41.16	0.0586
160	43	80	15.57	0.0391	1.75	0.0446	33.76	0.0586	34.51	0.0587	36.79	0.0699	51.46	0.0850	41.41	0.0658
180	23	90	16.09	0.0590	1.62	0.0566	31.81	0.0858	33.61	0.0879	35.17	0.0962	50.84	0.1121	42.26	0.0942

Note: MAPE = mean absolute percentage error.

Additional data are needed to further investigate what kinds of spatial interactions are more appropriate in predicting these place characteristics.

As was implied in A.-X. Zhu et al. (2018) and Kwan (2012), the definition of explanatory covariates and geographic contexts could affect the findings on places. The predictability of a place's unknown characteristic could be affected by both the place's observed characteristics and the definition of its contextual places. Because our GCNN model was designed to capture the "relevant context" in its trainable neural network weights, the predictability of a task actually reflects to what extent a place is related to its connected places given a connection metric. The predictability could be higher when using suitable place connections and more informative explanatory characteristics, because the predictability is governed by the underlying relevance. For example, the results in Figure 5 reveal that the spatial interaction context is much more helpful in uncovering the relevance of places in terms of dining features than the adjacency context.

Conclusions

Predicting a place's characteristics is challenging, as the place is related to the characteristics of other places, given a measure of place connections. Researchers cannot be certain whether the connections adopted are appropriate to reflect the "relevant" geographic context of places. In addition, as the place characteristics sensed from big geodata could be unstructured and the metric of place connections could be diverse, it is difficult to incorporate both of them in a spatial model. This research introduces the idea of using GCNNs to bridge the gap by effectively learning from the place characteristics and place connections through its trainable neural network weights and helping explore the predictability of a place characteristic in a geographic context.

The place characteristics and place connections are integrated into a place-based graph. Then, a GCNN model is used to make a credible prediction for unobserved place characteristics in a graph of connected places. The advantage of GCNNs is threefold. First, all places are connected as a graph, so that the connections among unsampled places can be explicitly modeled. Second, GCNNs can learn both the short-range and long-range relationships in the place-based graph to capture the complex relevance in a

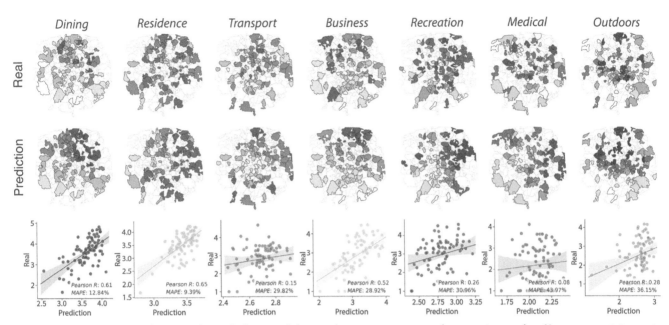

Figure 6. The best predicted pattern for each functional feature dimension using spatial interactions and a 60 percent training ratio. *Note:* MAPE = mean absolute percentage error.

geographic context. Third, the GCNN model is constructed as a nonparametric neural network and so does not assume any form of the spatial function to be fitted. An empirical study conducted in the Beijing metropolitan area showed that GCNNs are able to estimate the predictability of place characteristics in different geographic contexts.

Our contributions are as follows. First, we formalized places as a place-based graph to consider both place characteristics and place connections. Second, a semi-supervised GCNN model was proposed to predict a place's unobserved properties based on its observed properties and its contextual places. Third, the architecture of a GCNN model is flexible and powerful enough to encode both graph structure and node features, an ability that could be further exploited in future work. Moreover, with the recent advances in place-based knowledge graphs (Chen et al. 2018), introducing the knowledge of geographic contexts captured by black-box GCNN methods to such white-box methods could help uncover more geographic knowledge.

Acknowledgments

Our deep and sincere thanks go to Dr. Ling Bian, Dr. A-Xing Zhu, and the anonymous reviewers for their constructive comments, which greatly improved the content and clarity of this article. Author Yu Liu served as corresponding author for this article.

Funding

This research was supported by the National Key Research and Development Program of China (Grant 2017YFB0503602) and the National Natural Science Foundation of China (Grants 41625003, 41830645, 41771425, 41971331, and 41901321).

ORCID

Di Zhu ⓘ http://orcid.org/0000-0002-3237-6032
Fan Zhang ⓘ http://orcid.org/0000-0002-3643-018X
Ximeng Cheng ⓘ http://orcid.org/0000-0001-9923-7240
Zhou Huang ⓘ http://orcid.org/0000-0002-1255-1913
Yu Liu ⓘ http://orcid.org/0000-0002-0016-2902

References

Adams, B., and G. McKenzie. 2013. Inferring thematic places from spatially referenced natural language descriptions. In *Crowdsourcing geographic knowledge*, ed. D. Sui, S. Elwood, and M. Goodchild, 201–21. Springer.

Agnew, J. A., and J. S. Duncan. 2014. *The power of place: Bringing together geographical and sociological imaginations.* London and New York: Routledge.

Anselin, L. 2010. Thirty years of spatial econometrics. *Papers in Regional Science* 89 (1):3–25. doi: 10.1111/j.1435-5957.2010.00279.x.

Bruna, J., Z. Wojciech, S. Arthur, and L. Yann. 2014. Spectral networks and locally connected networks on graphs. In *International Conference on Learning Representations (ICLR2014)*, ed. Y. Bengio and Y. Lecun. Banff, AB, Canada: Conference Track Proceedings. Accessed December 10, 2019. http://arxiv.org/abs/1312.6203.

Chen, H., M. Vasardani, S. Winter, and M. Tomko. 2018. A graph database model for knowledge extracted from place descriptions. *ISPRS International Journal of Geo-Information* 7 (6):221. doi: 10.3390/ijgi7060221.

Defferrard, M., X. Bresson, and P. Vandergheynst. 2016. Convolutional neural networks on graphs with fast localized spectral filtering. In *Advances in neural information processing systems*, ed. D. D. Lee, M. Sugiyama, U. V. Luxburg, I. Guyon, and R. Garnett, 3844–52. Barcelona: Annual Conference on Neural Information Processing Systems. Accessed December 10, 2019. http://papers.nips.cc/paper/6081-convolutional-neural-networks-on-graphs-with-fast-localized-spectral-filtering

Fan, R. K. C. 1997. *Spectral graph theory*. Providence, RI: American Mathematical Society.

Fotheringham, A. S., C. Brunsdon, and M. Charlton. 2003. *Geographically weighted regression: The analysis of spatially varying relationships*. Hoboken, NJ: Wiley.

Fotheringham, A. S., W. Yang, and W. Kang. 2017. Multiscale geographically weighted regression (MGWR). *Annals of the American Association of Geographers* 107 (6):1247–65. doi: 10.1080/24694452.2017.1352480.

Gao, S., K. Janowicz, D. R. Montello, Y. Hu, J. Yang, G. McKenzie, Y. Ju, L. Gong, B. Adams, and B. Yan. 2017. A data-synthesis-driven method for detecting and extracting vague cognitive regions. *International Journal of Geographical Information Science* 31 (6):1245–71. doi: 10.1080/13658816.2016.1273357.

Golledge, R. G. 2002. The nature of geographic knowledge. *Annals of the Association of American Geographers* 92 (1):1–14. doi: 10.1111/1467-8306.00276.

Gong, L., X. Liu, L. Wu, and Y. Liu. 2016. Inferring trip purposes and uncovering travel patterns from taxi trajectory data. *Cartography and Geographic Information Science* 43 (2):103–14. doi: 10.1080/15230406.2015.1014424.

Goodchild, M. F. 2011. Formalizing place in geographic information systems. In *Communities, neighborhoods, and health*, 21–33. Berlin: Springer.

Gould, P. 1991. Dynamic structures of geographic space. In *Collapsing space and time: Geographic aspects of communication and information*, ed. S. D. Brunn and T. R. Leinbach, 3–30. London: HarperCollins.

He, K., X. Zhang, S. Ren, and J. Sun. 2016. Deep residual learning for image recognition. In *Proceedings of the IEEE Conference on Computer Vision and Pattern Recognition*, ed. Computer Vision Foundation, 770–78. Las Vegas: IEEE. Accessed December 10, 2019. https://www.cv-foundation.org/openaccess/content_cvpr_2016/papers/He_Deep_Residual_Learning_CVPR_2016_paper.pdf

Jenkins, A., A. Croitoru, A. T. Crooks, and A. Stefanidis. 2016. Crowdsourcing a collective sense of place. *PLoS One* 11 (4):e0152932. doi: 10.1371/journal.pone.0152932.

Kipf, T. N., and M. Welling. 2017. Semi-supervised classification with graph convolutional networks. *arXiv Preprint*. arXiv:1609.02907.

Kwan, M. P. 2007. Mobile communications, social networks, and urban travel: Hypertext as a new metaphor for conceptualizing spatial interaction. *The Professional Geographer* 59 (4):434–46. doi: 10.1111/j.1467-9272.2007.00633.x.

Kwan, M. P. 2012. The uncertain geographic context problem. *Annals of the Association of American Geographers* 102 (5):958–68. doi: 10.1080/00045608.2012.687349.

Lansley, G., and P. A. Longley. 2016. The geography of Twitter topics in London. *Computers, Environment and Urban Systems* 58:85–96. doi: 10.1016/j.compenvurbsys.2016.04.002.

LeCun, Y., Y. Bengio, and G. Hinton. 2015. Deep learning. *Nature* 521 (7553):436–44. doi: 10.1038/nature14539.

Liu, X., C. Kang, L. Gong, and Y. Liu. 2016. Incorporating spatial interaction patterns in classifying and understanding urban land use. *International Journal of Geographical Information Science* 30 (2):334–50. doi: 10.1080/13658816.2015.1086923.

Liu, Y., X. Liu, S. Gao, L. Gong, C. Kang, Y. Zhi, G. Chi, and L. Shi. 2015. Social sensing: A new approach to understanding our socioeconomic environments. *Annals of the Association of American Geographers* 105 (3):512–30. doi: 10.1080/00045608.2015.1018773.

Lu, B., M. Charlton, P. Harris, and A. S. Fotheringham. 2014. Geographically weighted regression with a non-Euclidean distance metric: A case study using hedonic house price data. *International Journal of Geographical Information Science* 28 (4):660–81. doi: 10.1080/13658816.2013.865739.

MacEachren, A. M. 2017. Leveraging big (geo) data with (geo) visual analytics: Place as the next frontier. In *Spatial data handling in big data era*, ed. C. Zhou, F. Su, F. Harvey and J. Xu, 139–55. Berlin: Springer.

Noronha, V. T., and M. F. Goodchild. 1992. Modeling interregional interaction: Implications for defining functional regions. *Annals of the Association of American Geographers* 82 (1):86–102. doi: 10.1111/j.1467-8306.1992.tb01899.x.

Nystuen, J. D., and M. F. Dacey. 1961. A graph theory interpretation of nodal regions. *Papers of the Regional Science Association* 7 (1):29–42. doi: 10.1007/BF01969070.

Rattenbury, T., and M. Naaman. 2009. Methods for extracting place semantics from Flickr tags. *ACM Transactions on the Web* 3 (1):1. doi: 10.1145/1462148.1462149.

Shamai, S. 1991. Sense of place: An empirical measurement. *Geoforum* 22 (3):347–58. doi: 10.1016/0016-7185(91)90017-K.

Wang, S., Y. Liu, and Z. Chen. 2018. Representing multiple urban places' footprints from dianping.com data. *Acta Geodaetica et Cartographica Sinica* 47 (8):1105–13.

Wu, X., J. Wang, L. Shi, Y. Gao, and Y. Liu. 2019. A fuzzy formal concept analysis-based approach to uncovering spatial hierarchies among vague places extracted from user-generated data. *International Journal of Geographical Information Science* 33 (5):991–1016. doi: 10.1080/13658816.2019.1566550.

Zhang, F., B. Zhou, L. Liu, Y. Liu, H. H. Fung, H. Lin, and C. Ratti. 2018. Measuring human perceptions of a large-scale urban region using machine learning. *Landscape and Urban Planning* 180:148–60. doi: 10.1016/j.landurbplan.2018.08.020.

Zhou, B., A. Lapedriza, A. Khosla, A. Oliva, and A. Torralba. 2018. Places: A 10 million image database for scene recognition. *IEEE Transactions on Pattern Analysis and Machine Intelligence* 40 (6):1452–64. doi: 10.1109/TPAMI.2017.2723009.

Zhu, A.-X., G. Lu, J. Liu, C.-Z. Qin, and C. Zhou. 2018. Spatial prediction based on Third Law of Geography. *Annals of GIS* 24 (4):225–40. doi: 10.1080/19475683.2018.1534890.

Zhu, D., and Y. Liu. 2018a. Modelling irregular spatial patterns using graph convolutional neural networks. *arXiv Preprint*. arXiv:1808.09802.

Zhu, D., and Y. Liu. 2018b. Modelling spatial patterns using graph convolutional networks. In *10th International Conference on Geographic Information Science (GIScience2018)*. Germany: Schloss Dagstuhl-Leibniz-Zentrum fuer Informatik. doi: 10.4230/LIPIcs.GISCIENCE.2018.73

Zhu, D., N. Wang, L. Wu, and Y. Liu. 2017. Street as a big geo-data assembly and analysis unit in urban studies: A case study using Beijing taxi data. *Applied Geography* 86:152–64. doi: 10.1016/j.apgeog.2017.07.001.

Appendix

The use of traditional convolutional neural networks (CNNs; LeCun, Bengio, and Hinton 2015) could be problematic when the data are not structured in the regular spatial domain (Defferrard, Bresson, and Vandergheynst 2016). GCNNs are facilitated by the demand of generalizing well-established CNNs to the irregular spatial domain. Graph Fourier transform is needed to transform a place-based graph into the spectral domain (Fan 1997), such that a graph convolutional filter can be defined to learn the connection information among places and support the prediction of place characteristics.

Given a place-based graph $G = (V, E)$ with n places, let $\tilde{X}$ be the node feature matrix of place characteristic X and $\tilde{E} \in \mathbb{R}^{n \times n}$ be the matrix form of place connections' intensities, a general form of forward propagation $h(\cdot)$ between the lth and $(l + 1)$ th hidden layer in a GCNN is

$$\tilde{X}^{l+1} = h(\tilde{X}^l, \tilde{E}) = \sigma\left(\tilde{D}^{-\frac{1}{2}}\tilde{E}\tilde{D}^{-\frac{1}{2}}\tilde{X}^l W^l\right), \qquad (1)$$

where $\tilde{D}$ is the diagonal degree matrix with $\tilde{D}_{ii} = \sum_j \tilde{E}_{ij}$, W^l is the layer-wise weights in neural networks, and $\sigma(\cdot)$ is a certain kind of activation function, for which we choose ReLU activation in this work to facilitate the nonlinear function approximation.

The normalized Laplacian matrix $\tilde{D}^{-\frac{1}{2}}\tilde{E}\tilde{D}^{-\frac{1}{2}}$ contains the preset connection information among places, and the trainable weights W^l enable GCNNs to approximate the predictability of characteristics in the geographic context defined by $\tilde{E}$. Further explanations of Equation 1 can be found in Defferrard, Bresson, and Vandergheynst (2016), Kipf and Welling (2017), and D. Zhu and Liu (2018), where a Chebyshev polynomial was suggested to simplify and compute the weights in GCNNs. We do not include it here, as it is outside the scope of this article, but we do apply similar methods and design a trainable GCNN model in the study.

In the case study, three scenarios of place connections, self-only, adjacency, and spatial interactions, were examined as the $\tilde{E}$. In the self-only scenario, connections were represented as a diagonal matrix, with all diagonal elements equal to one. In the adjacency scenario, connections were represented as a binary matrix that contained the basic adjacency information of places; that is, adjacent places were connected and the others were not. In the spatial interactions scenario, the connections were represented as a dense symmetric matrix with each element indicating the number of taxi O–D flows for a pair of places.

Spatial Learning in Smart Applications: Enhancing Spatial Awareness through Visualized Off-Screen Landmarks on Mobile Devices

Rui Li

Smartphones have become a significant platform in everyone's daily lives. For example, maps and map-based services on smartphones bring great convenience for wayfinding. They affect users' spatial awareness, however, due to their small sizes. That impacted spatial awareness can lead to degraded spatial knowledge and disorientation. This study intends to address these issues associated with spatial learning on smartphones by adapting cartographic and cognitive theories and investigating a new design for presenting spatial information on smartphones that can support users' awareness of space. The design uses the distinctive identities of spatial locations beyond the mapped screen as landmarks and visualizes the identities and distances of landmarks in distance through visual variables. Following previous pilot studies, this study evaluates the effectiveness of using such a design on aspects related to spatial awareness. Results provide additional details on the advantage of using specific visual variables to enhance the acquisition of spatial knowledge and spatial orientation. Although smart devices are ubiquitous in everyone's lives, it is still important to address the cognitive issues between those devices and their users. This study provides evidence that design can further contribute to the improvement of map-based applications on smartphones, which provides convenience and enhances users' spatial learning of new places. *Key Words: off-screen landmarks, orientation, spatial awareness, visual variable, visualization.*

智能手机已经成为人们日常生活中的一个重要工具，比如手机上的地图和基于地图的各种服务为人们寻路导航带来了极大的便利。但它也存在一个弊端：尺寸较小会影响用户的空间感知。人的空间意识受到影响后，就会引发空间知识退化和定向障碍问题。本研究希望通过改变地图和认知理论，研究一种在智能手机上呈现空间信息、可以支持用户空间感知的全新设计，从而解决与智能手机空间学习相关的问题。该设计利用地图屏幕以外的空间位置的独特标识作为地标，通过视觉变量，显示远处地标的标识和距离。本研究在之前曾进行了几次试点研究，随后评估了使用这种设计在空间感知方面的效果。研究结果以更多相信的信息，说明了利用具体视觉变量强化空间知识获取和空间定位所具有的优势。虽然如今智能设备在我们的生活中无处不在，但解决设备与用户之间的认知问题，仍是一个重要环节。根据本研究获得的证据可以说明，可使用该设计进一步改进基于地图的智能手机应用程序，为用户提供便利，强化用户对新地点的空间学习。关键词：屏幕外地标、方位、空间感知、视觉变量、可视化。

Los teléfonos inteligentes se han convertido en una plataforma significativa en la vida diaria de toda la gente. Por ejemplo, los mapas y los servicios basados en mapas de los teléfonos inteligentes son altamente convenientes para trazar planes de viaje y saber cómo y a dónde ir. Sin embargo, estos aparatos afectan la conciencia espacial de los usuarios, debido a su pequeñez. Esa conciencia espacial impactada puede conducir a un conocimiento espacial degradado y a la desorientación. El presente estudio intenta abocar estas cuestiones asociadas con aprendizaje espacial, por medio de los teléfonos inteligentes, adaptando teorías cartográficas y cognoscitivas, e investigando un nuevo diseño para presentar información espacial en el teléfono celular, que puede soportar la conciencia de espacio del usuario. El diseño usa como hitos las distintas identidades de ubicaciones espaciales más allá de la pantalla cartografiada y visualiza las identidades y distancias de los hitos en distancia a través de variables visuales. Siguiendo estudios pilotos anteriores, este estudio evalúa la efectividad de usar tal diseño en aspectos relacionados con la conciencia espacial. Los resultados proveen detalles adicionales sobre la ventaja de usar variables espaciales específicas para dar robustez a la adquisición de conocimiento espacial y orientación espacial. Aunque los aparatos inteligentes son ubicuos en la vida cotidiana de todo el mundo, todavía es importante abocar los asuntos cognitivos entre tales aparatos y sus usuarios. Este estudio suministra evidencia de que el diseño puede contribuir en mayor grado a la mejora de las aplicaciones de los teléfonos inteligentes basadas en mapas, lo cual genera conveniencia y fortalece el aprendizaje espacial de nuevos lugares por los usuarios. *Palabras clave: conciencia espacial, orientación, referentes locacionales, variable visual, visualización.*

Smartphones are becoming an inevitable part of everyone's daily lives. Map-based applications on mobile devices are early examples of smart applications. Without referring to road maps to learn about a place, drivers or pedestrians can use those applications to navigate. Researchers have pointed out the shortcomings associated with using such applications. For example, Speake and Axon (2012) found that the tendency of users to rely on those applications leads to very weak cartographic literacy and spatial awareness. This issue is not a simple topic in one field but an interdisciplinary issue involving cartography, geographic information, spatial cognition, and computer science. This study intends to address the shortcoming associated with using map-based applications by suggesting a design for visualizing distant locations, which can potentially enhance users' spatial learning and their awareness of space. This study implements this design on a smartphone as a prototype and evaluates its effect on aspects related to spatial awareness in a real environment.

When maps are on the small screens of smartphones, they can only provide a very limited portion of the environment at a time. Researchers have pointed out that the limit of the small screen on mobile devices is associated with learning of a new environment and efficiency of wayfinding (see Dillemuth 2005; Ishikawa and Kiyomoto 2008; Willis et al. 2009). In particular, Dillemuth (2009) suggested that using maps on small screens in navigation tasks results in degraded spatial knowledge and accuracy. The represented small portions of the environment do not seamlessly integrate into a full mental representation of the larger environment. Other researchers have also suggested that the maps on small screens lead to poor attention and disorientation in an environment, because learned spatial knowledge is limited (Ishikawa and Kiyomoto 2008; Gardony et al. 2013). The very limited spatial knowledge and disorientation consequently result in poor spatial awareness.

Improving the outcome of spatial learning on smartphones is a continuing research challenge. The initial strategy is to overcome the limit of the small screen size for representing spatial information. It is important to note that smart technologies such as existing map-based applications are effective in some aspects. It is not the purpose of this study to suggest otherwise. Instead, this study

focuses on the degraded spatial awareness associated with using such applications, which can be a common issue between users and smart applications in the future, and then sheds light on potential design improvements that can achieve both effectiveness and users' spatial awareness. Klippel, Hirtle, and Davies (2010) suggested that positive spatial awareness is a result of learning structuring features in the environment and then organizing spatial knowledge using those learned structuring features. Therefore, representing information that facilitates the learning of structuring features in an environment by using landmarks is a likely solution. The goal of this study is to help a user using a smartphone become aware of a space that is much larger than the mapped area. To do so, this study first reviews methods addressing similar issues and then introduces a design to visualize locations beyond the mapped area. A behavioral experiment carried out in a real environment provides evidence to evaluate the effectiveness of designs using visual variables on spatial awareness. This study then presents the results and discusses the potential and considerations for future work.

Related Work

Before computers were widely accessible in the late twentieth century, paper maps were the main source for learning a new environment. Researchers investigated spatial learning using maps, the major source of acquiring spatial knowledge about an area, and suggested that maps support quick and accurate learning of spatial knowledge at the configurational level (Thorndyke and Hayes-Roth 1982; Richardson, Montello, and Hegarty 1999). Because the size of some paper maps limits the representation of spatial information, cartographers use the edge of maps to infer the names of nearby towns outside the mapped area in the format of strip maps. As a special type of route map, strip maps also show mostly routes and their connections. Their distinctive difference is that strip maps label names of nearby locations, indicating where the routes at the map edge lead. MacEachren (1986) reviewed strip maps and suggested that they contribute to the convenience and spatial learning of users when navigating a new environment. Figure 1 shows a portion of a strip map produced around 1921 by the Automobile Club of Southern California to imply nearby cities along

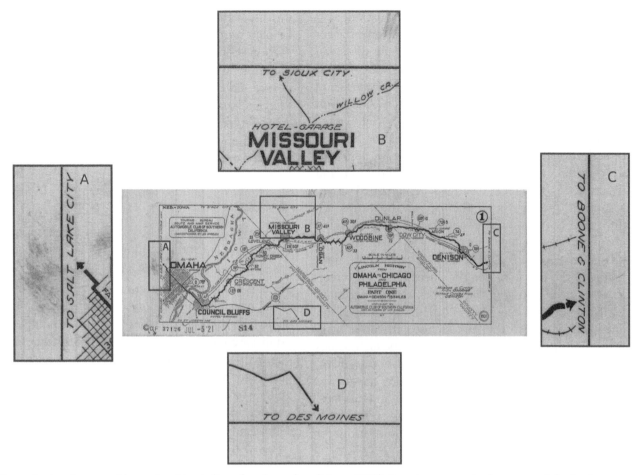

Figure 1. A strip map with nearby cities indicated at four edges (Moore 2018). In strip maps, only the names of destinations at each edge of the map are marked based on the connection of road networks. The zoomed views of the four edges show the directions to (A) Salt Lake City, (B) Sioux City, (C) Boone and Clinton, and (D) Des Moines.

Lincoln Highway. Examples of strip maps are also common in maps published by the American Automobile Association. This simple strip form contributes to easier acquisition of spatial knowledge. In addition, the implied locations at edges of the map help drivers to further establish their awareness of a new environment (MacEachren and Johnson 1987). As the representation of digital maps became mobile in the early twenty-first century, devices such as in-vehicle navigation systems and personal smartphones provide immediate access to digital maps. Because of the small screen size of those mobile devices, the degradation of spatial learning among users has become a topic that researchers started to investigate.

Adapting the design applied in strip maps on smartphones is a possible way to facilitate spatial learning, because those locations can serve as anchors in spatial knowledge to structure the learned

environment. These landmarks can also play an important role in spatial orientation and wayfinding (Raubal and Winter 2002; Winter et al 2000). Because locations of those landmarks might not be in the mapped area on a mobile device, these landmarks are termed "off-screen landmarks" in this study. Many approaches introduced in the field of computer science, however, focus on the locations off-screen but overlook the distinctive identities of those locations, which can serve as landmarks to facilitate spatial learning. For example, the halo approach (Baudisch and Rosenholtz 2003) uses partial circles of varying sizes at the edge of a mobile screen to imply different distances to those off-screen locations. If a partial circle is relatively flat, it indicates that the distance to the off-screen location is further than those partial circles that curve more. In addition to the issue that all off-screen locations are visualized using generic geometry in varying sizes while neglecting the identity of any location as a landmark, the geometry can also become an issue for users to infer distance when the shapes are too small or overlapped at the edge. Similar methods such as the scaled arrow approach (Burigat, Chittaro, and Gabrielli 2006) and the wedge approach (Gustafson et al. 2008) use different geometries for the same purpose. The scaled arrow approach uses the size and direction of the arrow, whereas the wedge approach uses partial triangles to indicate the direction and distance to off-screen locations. The scaled arrow approach directly shows the tip of the arrow for direction but the wedge approach needs users to complete a triangle mentally to estimate an off-screen location. Like the halo approach, both methods do not consider the identity of off-screen locations as potential landmarks. More recently, Gollenstede and Weisensee (2014) introduced an approach similar to the strategy used in earlier strip maps. This approach visualizes a network such as bus lines from a user's location to the edges of a screen. Similar to the previous methods using geometric shapes, this method uses generic lines and neglects the potential of using the identity of off-screen locations as landmarks. Figure 2 shows illustrations of these approaches.

In short, the approaches just discussed consider the visualization of distance and direction to an off-screen location. They do not, however, use the identification of a location (e.g., museum) that can serve as a cue for users to structure learned spatial knowledge. In terms of the important roles of landmarks, an earlier study (Li et al. 2014) used cartographic icons embedding the identities of off-screen locations

to indicate their directions on the edges of the mobile screen. This approach prevents user disorientation compared to a controlled version without icons of off-screen landmarks. This approach, however, did not visualize the distance to those off-screen locations. Building on the positive results of this earlier test, this study extends this approach and adds the visualization of distance in the design of those cartographic icons.

Design

Visual variables introduced in cartography (MacEachren 1995; Robinson et al. 1995) that can encode this quantitative information are considered in this study. Effective visual variables for presenting this type of information include size, color value, color saturation, fuzziness (blur), resolution, and transparency. Considering the similar representation between color value and color saturation as well as between resolution and fuzziness in designed icons, pilot studies to test visual variables used size, color value, and fuzziness (Li 2016). Color value did not show an advantage in users' perceptions of distance. Therefore, this study uses three visual variables: size, fuzziness, and transparency in design. Size represents the variation in area of an icon. The larger an icon is, the closer distance to a location it implies. Fuzziness represents the vagueness of boundary in a map symbol. The fuzzier (more blurred) an icon is, the further distance to a location it implies. Transparency is the third visual variable used in this study. Transparency represents the opaqueness of a

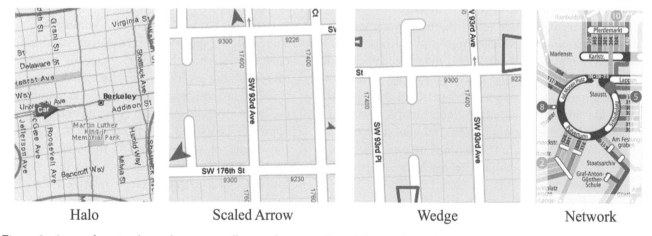

| Halo | Scaled Arrow | Wedge | Network |

Figure 2. Approaches visualizing distance to off-screen locations. From left to right: Halo approach, scaled arrow approach (Burigat, Chittaro, and Gabrielli 2006), wedge approach (Gustafson et al. 2008), and network approach (Gollenstede and Weisensee 2014).

symbol if features underneath the symbol are visible. The more transparent an icon is, the further distance to a location it implies.

Level of Measurement

After choosing the visual variables, the next consideration is the level of measurement for representing distance, either ratio or ordinal. Using size as an example, icons at the ratio level change in size proportionally based on an established ratio. Icons at the ordinal level only change based on defined thresholds of distances representing the concept of being close, intermediate, and far. Many previous approaches use different sizes of geometric shapes at the ratio level to infer distance. By introducing the design of icons using size at the ordinal level, the results could provide direct comparison with previous studies. The evaluation addresses the effect of level of measurement on the perception of distance (Li and Zhao 2017). Results suggest that those icons at the ordinal level contribute to better perception of distance than icons at the ratio level. This study uses all three visual variables at the ordinal level.

Visualizing Off-Screen Landmarks

The visual variables in a design icon vary in terms of the threshold of distance to a user's location. For distance up to 1 km, the distance to off-screen landmarks is in the threshold of being close. For distances between 1 km and 3 km, the distance to off-screen landmarks is considered intermediate. For distances further than 3 km, the distance to off-screen landmarks is in the threshold of being far. The decision on the threshold of distance is based on a pilot study for this this purpose (Zhao and Li 2016). The pilot study surveyed local residents to provide the most useful landmarks within the author's city and suggested the ten most frequently mentioned distant landmarks for the design of off-screen landmarks. Considering the distance from each off-screen landmark to the center of campus, the derived thresholds of distance resulted in three close off-screen landmarks, four intermediate off-screen landmarks, and three far off-screen landmarks. Furthermore, considering the screen size of a 5.2-in. screen with a resolution of $1{,}080 \times 1{,}920$ pixels, 90×90 pixels, 60×60 pixels, and 30×30 pixels

were assigned to close, intermediate, and far icons using size as the visual variable, which would not occupy a large portion of the screen and were legible to users. For the fuzziness visual variable, open-source software Inkscape version 0.92 (2017) was used to create blurred icons. In particular, icons that represent close off-screen locations had no blur. Icons representing intermediate distances were 3 percent blurred both horizontally and vertically. Icons representing far off-screen locations were 6 percent blurred in both directions. Chosen degree of fuzziness was examined for legibility and distinctiveness before testing. These icons were tested through an online platform for the effectiveness of perceiving the embedded information (Li 2016). In the design of using transparency, 100 percent, 75 percent, and 50 percent opacity were applied to icons indicating close, intermediate, and far off-screen locations. The effectiveness of these different levels of transparency was tested through an online platform (Li 2017). In addition to the applied visual variables, each icon for off-screen landmarks had a bounding box to indicate being off-screen, which on-screen landmarks did not have. If an off-screen icon is close to an on-screen landmark, the bold bounding box is the feature that distinguishes these two different types. When a user pans to an area where an off-screen landmark is, it is no longer off-screen. The icon of this landmark would appear at its actual location without the bounding box or varied visual variable to indicate an on-screen location. Likewise, when the location of an on-screen landmark is no longer on-screen because of a user zooming in or panning, the icon for this location would appear with a bounding box and corresponding visual variable representing its threshold of distance. When multiple icons overlap, only the closest icon will be visible. In addition, each edge of the screen would show a maximum of two icons of off-screen landmarks. Figure 3 shows the variation in icons using specific visual variables on a smartphone. Previous pilot studies have only tested these visual variables through an online platform. The lack of interaction through online testing did not provide a comprehensive understanding of roles of these visual variables in acquiring spatial knowledge in a real environment. It is crucial to carry out the evaluation in a real environment, which can provide further evidence for improving and introducing this design for smart applications.

Size   Fuzziness   Transparency

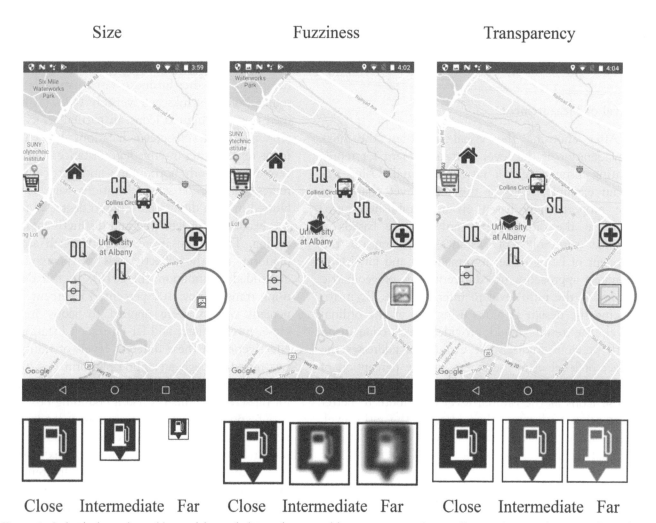

Close   Intermediate   Far   Close   Intermediate   Far   Close   Intermediate   Far

Figure 3. Individual visual variables used for symbolizing identity and location in icons for an off-screen location (gas station) are shown below each scenario, in which highlighted icons in circles showed icons using individual variables to imply the furthest off-screen distance in a developed prototype.

Methods

The experiment took place in the underground tunnel system on the author's campus. One can easily get lost in the tunnel system due to its lack of visual access to the outside and its extremely symmetric structure. In addition to these characteristics, this tunnel environment has very limited signage to provide wayfinding assistance, which is another reason that users of the tunnel can easily get lost. Although this type of environment might not be commonly seen in other places, it serves as an ideal testing environment for evaluating the effectiveness of the design. First, it is controllable to exclude influence from features in the environment. When features in an environment are distinctive, participants can easily use those features, such as colored doors or signs, as cues in wayfinding (Lynch 1960).

Because the color and appearance in this tunnel environment is generic, influence from environmental features is minimal, allowing the evaluation of the design based on the visual variables and the participant's spatial ability. Second, the symmetric structure of this tunnel reduced the likelihood of using the asymmetric environment structure as a cue to recognize and orient in this environment (Lynch 1960), which further minimizes the influence of the environment itself to the evaluation of the design prototype. If the effectiveness of the design prototype is noticeable in this evaluation, it could provide valuable support for users who do not use or notice environmental characteristics or provide additional support for spatial learning in other environments.

This study recruited students on campus who were unfamiliar with the tunnel system through on-campus flyers and listservs. In total, twenty-six students (thirteen male and thirteen female) signed up. Their ages ranged from eighteen to thirty (M = 22.08, SD = 3.02). Each participant received $10 at the end of the experiment as compensation. The main task consisted of the use of a prototype on a 5.2-in. Android smartphone showing the central part of the campus with icons of off-screen locations as well as on-screen landmarks in the mapped area (Figure 3). Each visual variable forms a specific scenario on this smartphone, so the experimenter can switch among scenarios during the experiment for each participant in a random order.

Procedure

Once a participant gave consent to the experimenter in the laboratory for this study, he or she was taken to the tunnel and then one of three selected locations to start the experiment. Each participant was taken to a location in a random order. The three selected locations were at different areas in the tunnel. There were no visible signs, maps, or access to outside at each location. Figure 4 shows the three selected locations. At each location, a participant was asked to estimate the direction and distance to an outside location before using the prototype. Each participant then used the prototype on the smartphone to learn about the surrounding and mapped locations. Participants could turn around and browse the immediate surroundings while using the prototype for functions such as zoom, pan, or click. When a participant was ready, he or she then estimated direction and distance to the same location at the same standing point and facing the same direction. The prototype recorded the user's interactions with it, including panning on the screen, zooming in or out of the map, and clicking on the visualized icons. After this estimation task, a questionnaire that consisted of ten questions asked participants for their judgment of distance among those off-screen landmarks. Once this task was complete at one location, the experimenter took the participant to the next location, repeating the same tasks using a different visual variable on the smartphone. Once a participant completed the tasks at the third location, he or she returned to the lab to complete spatial ability tests.

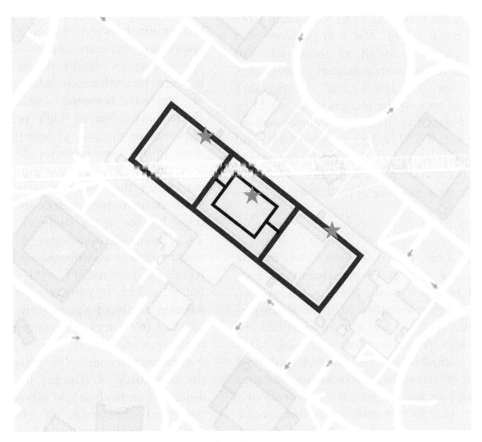

Figure 4. Three selected locations (stars) in the campus tunnel for this experiment.

The spatial ability tests used three psychometric tests to measure a participant's spatial ability, which is another factor that contributes to one's performance in space (Liben, Myers, and Kastens 2008). Each test addressed one specific aspect. The Paper Folding Test was used for testing visuospatial memory (Linn and Petersen 1985). The Mental Rotation Test (Vandenberg and Kuse 1978) was used for testing mental rotation ability. The Water Level Test was used for testing spatial perception (Liben and Golbeck 1980). A split using median score separated participants into two groups based on their spatial ability on these tests. If the average score of all three tests was above the split score, corresponding participants were in the higher spatial ability group. Likewise, participants with scores below the split score were in the lower spatial ability group. The purpose of splitting was to convert the continuous variables of spatial ability into categorical variables for the analysis of variance (ANOVA), because an individual participant's combined scores of spatial ability from three tests vary from each other. Splitting based on the median value can be arbitrary, but the split can be meaningful when a noticeable gap exists between the two groups and both median and mean score are very close (5.87 for median score and 5.94 for mean score). The average score of a participant's spatial ability in all three tests can range from zero to ten, with a median value of five. On an actual test, it is rare to find a participant's spatial ability at zero, so the median value would be slightly higher than five. The split groups based on the median value of 5.87 had an average of 4.65 and 7.65 in the lower spatial ability and higher spatial ability group, respectively. Using the mean split resulted in the same grouping of participants.

Limitations

There are some potential limitations in the design. First, the number of participants in each group is still relatively small. For this reason, this study did not employ a control group in which participants use the designed app without visualized off-screen landmarks. An earlier study (Li et al. 2014) supports this consideration, because results showed that visualized off-screen landmarks without distance information support spatial awareness in a real environment. The focus in this study is whether specific visual variables would lead to different effects. Increasing the number of participants can further

provide clarity for distinguishing the role of their spatial ability. Second, the base map used in the prototype displays map labels at various scales. Therefore, depending on the scale that a participant zooms to, displayed or disappeared map labels could potentially change the level of details on the screen. These changing labels could potentially influence the acquisition of spatial knowledge.

Results

Based on the design of the experiment, a repeated-measures ANOVA used each measure as a dependent variable and the visual variable as the within-subject variable. Spatial ability was the between-subject variable. Results represent three major categories of performance: the distance judgment, the estimations of direction and distance, and the user interactions.

Distance Judgment

In the tasks of distance judgment, participants' performance reflected their perceptions of distance to the closest landmark, the furthest landmark, the comparison between two on-screen landmarks, the comparison between one on-screen landmark and one off-screen landmark, and the comparison between two off-screen landmarks. The percentage of correct responses indicates their score. Participants achieved high percentages of correct responses in the task of selecting the closest landmark in all three scenarios: 88 percent in the size scenario, 85 percent in the fuzziness scenario, and 92 percent in the transparency scenario. Participants, however, performed differently in judging the furthest distance to the landmark among three scenarios. The repeated-measures ANOVA showed that visual variables contributed significantly to the correctness of response, $F(2, 48) = 3.69$, $p < 0.05$, partial $\eta^2 = 0.13$. In particular, participants in the size scenario exhibited significantly lower correctness of response (69 percent) in comparison to those in the fuzziness scenario (92 percent) and those in the transparency scenario (92 percent). Table 1 shows the correctness of response in judging the furthest distance to a landmark in all scenarios of this prototype and the result of the repeated-measures ANOVA. Spatial ability did not contribute significantly to the differences.

Table 1. Percentage of correct responses of participants selecting the furthest landmark in each scenario and the corresponding result of repeated-measures analysis of variance

Source	M	SD	N	df	F	p	Partial η^2
Between subjects				(error df = 24)			
Spatial ability				1	1.88	0.18	0.07
Lower SA	2.64	0.47	13				
Higher SA	2.80	0.41	13				
Within subjects				(error df = 48)			
Visual variable				2	3.69	<0.05	0.13
SA * V				2	1.61	0.20	0.06
Size	0.69	0.52	26				
Fuzziness	0.92	0.27	26				
Transparency	0.92	0.27	26				

Notes: SA = spatial ability; V = visual variable.

Table 2. Percentage of correct responses of participants comparing distance between one on-screen landmark and one off-screen landmark and the corresponding result of repeated-measures analysis of variance

Source	M	SD	N	df	F	p	Partial η^2
Between subjects				(error df = 24)			
Spatial ability				1	10.26	<0.01	0.30
Lower SA	2.54	0.49	13				
Higher SA	2.97	0.49	13				
Within subjects				(error df = 48)			
Visual variable				2	2.24	0.12	0.09
SA * V				2	1.70	0.19	0.07
Size	2.69	0.55	26				
Fuzziness	2.88	0.33	26				
Transparency	2.69	0.62	26				

Notes: SA = spatial ability; V = visual variable.

In the tasks of comparing relative distances between two visualized icons, spatial ability had a significant effect on the score of comparing the distance between one on-screen landmark and one off-screen landmark, $F(1, 24) = 10.26$, $p < 0.01$, partial $\eta^2 = 0.30$. As shown in Table 2, participants with lower spatial ability had lower scores compared to those with higher spatial ability regardless of visual variable. No main effect of spatial ability or visual variable was significant in the score for comparing the distance between two on-screen landmarks and the score for comparing the distance between two off-screen landmarks.

Estimations of Direction and Distance

At each testing location in the tunnel, participants estimated the direction and distance to an unseen outdoor location. The results showed that prior to using the prototype, participants' estimations did not differ between spatial ability. In the estimating tasks after using the prototype, however, spatial ability significantly contributed to the errors of estimated direction, $F(1, 24) = 10.24$, $p < 0.005$, partial $\eta^2 = 0.30$. Participants with higher spatial ability achieved fewer errors (°) in the estimation of direction than those with lower spatial ability. Table 3 shows the estimation errors before and after using the prototype and the result of the repeated-measures ANOVA. Visual variable did not contribute to the differences in estimations after using the prototype. It appeared that all visual variables contribute to the estimation of direction, because the accuracy of estimation of direction increased in all scenarios after using the prototype.

User Interactions

Recorded user interactions include clicking, panning, and zooming during the experiment. Results

Table 3. Estimation errors (°) between spatial ability groups before and after using the prototype in each scenario and the corresponding result of repeated-measures analysis of variance

Source	M	SD	N	df	F	p	Partial η^2
Between subjects				(error df = 24)			
Spatial ability				1	10.26	<0.01	0.30
Lower SA	47.64	1.60	13				
Higher SA	21.49	1.60	13				
Within subjects				(error df = 48)			
Visual variable				2	0.02	0.98	0.001
SA * V				2	0.52	0.60	0.02
Size	34.31	41.39	26				
Fuzziness	35.65	44.00	26				
Transparency	33.73	31.63	26				

Notes: SA = spatial ability; V = visual variable.

Table 4. Mean frequency of panning based on visual variable and spatial ability and corresponding result of repeated-measures analysis of variance

Source	M	SD	N	df	F	p	Partial η^2
Between subjects				(error df = 24)			
Spatial ability				1	3.79	0.06	0.14
Lower SA	22.00	21.47	13				
Higher SA	10.64	9.72	13				
Within subjects				(error df = 29.27)			
Visual variable				1.22	6.62	<0.05	0.22
SA * V				1.22	0.59	0.48	0.02
Size	22.42	23.29	26				
Fuzziness	14.23	15.02	26				
Transparency	12.31	13.87	26				

Notes: SA = spatial ability; V = visual variable.

showed that the visual variable significantly influenced the panning frequency, $F(1.22, 29.27) = 6.65$, $p < 0.05$, partial $\eta^2 = 0.22$, using Greenhouse–Geisser correction. Pairwise comparison showed that participants using the size scenario panned on average 22.42 times, compared to 14.23 times in the fuzziness scenario and 12.31 times in the transparency scenario. The spatial ability was only marginal in its effect on the panning frequency, $F(1, 24) = 3.79$, $p = 0.06$, partial $\eta^2 = 0.14$. Table 4 reports these results. Neither visual variable nor spatial ability had a significant effect on clicking or zooming.

Discussion

Acquisition of Spatial Knowledge

All three visual variables used in this study appear effective for representing off-screen landmarks. Some visual variables, however, seem more effective. In the selection of icon representing the closest on-screen location, participants all had relatively high accuracy across all scenarios. In the task of selecting the icon representing the furthest off-screen distance, the visual variables of fuzziness and transparency seemed more effective than size, even though the accuracy in size scenario is still relatively high (70 percent) compared to an earlier study on an online platform (Li 2016). In this earlier study, the visual variables size and fuzziness were compared with the visual variable color value. Participants could only see screenshots of the three scenarios and performed the same tasks. The visual variable color value led to the lowest accuracy of comparing distance involving off-screen locations (10 percent) and size led to slightly higher accuracy of 20 percent, which was much lower than the score in the real environment. In addition to the visual variable, spatial ability seems to differentiate participants' perceptions of distance. As reflected in the task of comparing the distance between a pair of icons, a participant's higher spatial ability further contributes to higher accuracy in comparing distance between one on-screen icon and one off-screen landmark (Table 2). In these tasks of comparing pairs of icons, the visual variable does not contribute differently to participants' performance, indicating that all three visual variables can clearly represent relatively close, intermediate, and far distances. The results are slightly different from those in the earlier study on an online platform. This is likely due to the fact that the interaction between users and the prototype benefits users' accuracy. A separate later section specifically addresses user interaction. In summary, for representing distance to off-screen locations, size, fuzziness, and transparency all have positive effects on perception of encoded distance. Fuzziness and transparency in particular lead to higher scores in comparison of the distance between off-screen location and the user's location.

Spatial Awareness

In addition to the perception of distance, another important goal of this study is to address the effect of using icons for off-screen locations on the enhancement of spatial awareness. The design of off-screen icons seems to have a positive effect on the enhancement of users' spatial awareness. At each task, participants need to align their mental representations with the actual environment to estimate the direction and distance to an unseen location outside the tunnel, which is not included in the design. This is reflected in the improvement of accuracy in the estimation of direction after using the prototype, which can directly contribute to one's orientation in space. Although the difference among visual variables is not significant, spatial ability is an important factor affecting the improvement of spatial awareness. In other words, participants with higher spatial ability show greater improvement in the estimated direction after using the prototype. Future study is necessary to clarify the specific contribution of spatial ability and visual variables to the improvement of spatial awareness.

Regarding the significant role of spatial ability in the performance related to their awareness of the space, participants could benefit from mental rotation and spatial perception abilities. In this study, participants were not familiar with the tunnel environment. After spending at least a year on campus, participants became aware of some known locations on campus, one of which (i.e., Science Library) was used in the estimation task. As introduced earlier, a participant's spatial ability in this study consists of the ability for mental rotation, visuospatial memory, and spatial perception. When a participant with better spatial ability can efficiently align the newly learned tunnel environment with his or her existing mental representations of the campus, the estimation of direction can be more accurate than for those who need more time to align the new environment with their mental representation. The results show the contribution of a participant's spatial ability to the improvement of his or her awareness of the environment but do not provide sufficient information to distinguish the influence of visual variables on the enhancement of spatial awareness. A follow-up study involving a comparison between the prototype and a controlled design is necessary to assess the contribution of visual variables and spatial ability.

User Interaction

Regarding user interaction, this study clearly shows that size seems to be a weaker visual variable, because it requires more effort of the user to learn the environment through a smartphone. Participants in the size scenario had to pan twice as much as participants in the fuzziness or transparency scenarios. In addition to the visual variables, participants' spatial abilities seem to have an influential role as well. As shown in the frequency of panning on the prototype (Table 4), participants with better spatial ability panned fewer times than those with weaker spatial ability. Visual variables do not seem to affect interactions including zooming and clicking.

Considering participants' performance in other categories, fuzziness and transparency seem more effective than size. For example, the performance in selecting the furthest icon is noticeably different between the size scenario and the other two scenarios. Without user interaction on an online platform, the accuracy of selecting the furthest location is very poor (Li 2016). The enabled interaction between a user and the prototype contributes to higher scores in acquiring distance knowledge. Although the score in the size scenario is still higher than that obtained from the online platform, the size leads to lower scores compared to fuzziness and transparency. Overall, it seems that the efficiency of size to visualize distance is not as good as that of fuzziness or transparency.

Conclusion

Overcoming the limit of small screens on smart devices to facilitate spatial learning is a continuing challenge. Its outcome can contribute to addressing the cognitive and behavioral concerns of users with their growing adaption of smart applications. This study introduces the design of icons that visualize the direction and distance of landmarks at off-screen locations on smartphones. Adapting the earlier finding of level of measurement for symbolization, this study introduces the design and its evaluation using three visual variables: size, fuzziness, and transparency. This study examines the effectiveness of these visual variables for symbolizing distance information, supporting spatial learning, and user interactions. This study further introduces the factor of spatial ability and evaluates its influence on the performance and interactions of users. The results in this study imply that all visual variables can contribute to spatial learning. Although the performance of participants does not show significant differences among visual variables, it is necessary to investigate participants' preferences for those visual variables in future studies. This could provide additional insight for suggesting visualization strategies on smart devices. The implementation of this design addresses issues such as overlapped and cluttered icons. It needs to consider issues such as possible conflict between icons and elements on the map layer. The base map in the prototype shows map labels. In future studies, it is necessary to introduce a controlled version of the prototype (e.g., maps with only on-screen landmarks or maps without labels) to further help clarify the contribution of each individual visual variable to performance. Furthermore, future studies should consider additional factors such as familiarity and users' spatial strategies to shed light on the effect of off-screen landmarks on smartphones in the wider scope of applications.

Acknowledgments

The author sincerely appreciates the editor and all anonymous reviewers for their comments that improved this article. The author thanks Andrew Kim and Jiayan Zhao for their assistance with this project.

Funding

The author thanks the Faculty Research Award Program (FRAP) of the University at Albany for providing support for this project.

References

Baudisch, P., and R. Rosenholtz. 2003. Halo: A technique for visualizing off-screen objects. Paper presented at CHI 2003, April 5–10, Ft. Lauderdale, FL.

Burigat, S., L. Chittaro, and S. Gabrielli. 2006. Visualizing locations of off-screen objects on mobile devices: A comparative evaluation of three approaches. Paper presented at the 8th Conference on Human–Computer Interaction with Mobile Devices and Services, Helsinki, Finland.

Dillemuth, J. A. 2005. Map design evaluation for mobile display. *Cartography and Geographic Information Science* 32 (4):285–301. doi: 10.1559/152304005775194773.

Dillemuth, J. A. 2009. Navigation tasks with small-display maps: The sum of the parts does not equal the whole. *Cartographica: The International Journal for Geographic Information and Geovisualization* 44 (3):187–200. doi: 10.3138/carto.44.3.187.

Gardony, A. L., H. A. Taylor, and T. T. Brunyé. 2016. Gardony map drawing analyzer: Software for quantitative analysis of sketch maps. *Behavior Research Methods* 48 (1):51–177. doi: 10.3758/s13428-014-0556-x.

Gollenstede, A., and M. Weisensee. 2014. Animated cartographic visualisation of networks on mobile devices. Paper presented at the 11th International Symposium on Location-Based Services, November 26–28, Vienna.

Gustafson, S., P. Baudisch, C. Gutwin, and P. Irani. 2008. Wedge: Clutter-free visualization of off-screen locations. Paper presented at CHI 2008, April 5–10, Florence, Italy.

Inkscape (version 0.92) 2017. http://www.inkscape.org.

Ishikawa, T., and M. Kiyomoto. 2008. Turn to the left or to the west: Verbal navigational directions in relative and absolute frames of reference. In *GIScience 2008: Lecture notes in computer science*, ed. T. J. Cova, H. J. Miller, K. Beard, A. U. Frank, and M. F. Goodchild, vol. 5266, 119–32. Berlin: Springer.

Klippel, A., S. Hirtle, and C. Davies. 2010. You-are-here maps: Creating spatial awareness through map-like representations. *Spatial Cognition & Computation* 10 (2–3):83–93. doi: 10.1080/13875861003770625.

Li, R. 2016. Effects of visual variables on the perception of distance in off-screen landmarks: Size, color value, and crispness. In *Progress in location-based services*

2016, ed. G. Gartner and H. Huang, 89–103. Cham: Springer.

Li, R. 2017. Off-screen landmarks on mobile devices: Interaction and perception of distance. Paper presented at the annual meeting of the American Association of Geographers, April 5–9, Boston.

Li, R., A. Korda, M. Radtke, and A. Schwering. 2014. Visualising distant off-screen landmarks on mobile devices to support spatial orientation. *Journal of Location Based Services* 8:166–78. doi: 10.1080/17489725.2014.978825.

Li, R., and J. Zhao. 2017. Off-screen landmarks on mobile devices: Levels of measurement and the perception of distance on resized icons. *KI-Künstliche Intelligenz* 31 (2):141–49. doi: 10.1007/s13218-016-0471-7.

Liben, L. S., and S. Golbeck. 1980. Sex differences in performance on Piagetian spatial tasks: Differences in competence or performance? *Child Development* 51: 594–97. doi: 10.2307/1129301.

Liben L. S., L. J. Myers, and K. A. Kastens. 2008. Locating oneself on a map in relation to person qualities and map characteristics. In *Spatial cognition VI. Learning, reasoning, and talking about space. Spatial cognition 2008. Lecture notes in computer science*, ed. C. Freksa, N. S. Newcombe, P. Gärdenfors, S. Wölfl, vol. 5248. Berlin, Heidelberg: Springer.

Linn, M. C., and A. C. Petersen. 1985. Emergence and characterization of sex differences in spatial ability: A meta-analysis. *Child Development* 56 (6):1479–98. doi: 10.1111/j.1467-8624.1985.tb00213.x.

Lynch, K. 1960. *The image of the city*. Cambridge, MA: MIT Press

MacEachren, A. M. 1986. A linear view of the world: Strip maps as a unique form of cartographic representation. *The American Cartographer* 13 (1):7–26. doi: 10.1559/152304086783900185.

MacEachren, A. M. 1995. *How maps work: Representation, visualization, and design*. New York: Guilford.

MacEachren, A. M., and G. B. Johnson. 1987. The evolution, application and implications of strip format travel maps. *The Cartographic Journal* 24 (2):147–58. doi: 10.1179/caj.1987.24.2.147.

Moore, R. 2018. 1920s road trip: The Lincoln Highway in strip maps. February 21. Washington, DC: Library of Congress Geography and Map Division. Accessed November 19, 2018. https://blogs.loc.gov/maps/2018/02/1920s-road-trip-the-lincoln-highway-in-strip-maps/.

Raubal, M., and S. Winter. 2002. Enriching wayfinding instructions with local landmarks. In *GIScience 2002: Lecture notes in computer science*, ed. M. J. Egenhofer and D. M. Mark, vol. 2478, 243–59. Berlin: Springer.

Richardson, A. E., D. R. Montello, and M. Hegarty. 1999. Spatial knowledge acquisition from maps and from navigation in real and virtual environments. *Memory & Cognition* 27 (4):741–50. doi: 10.3758/BF03211566.

Robinson, A. H., J. L. Morrison, P. C. Muehrcke, A. J. Kimerling, and S. C. Guptill. 1995. *Elements of cartography*. 6th ed. New York: Wiley.

Speake, J., and S. Axon. 2012. "I never use 'maps' anymore": Engaging with sat nav technologies and the implications for cartographic literacy and spatial awareness. *The Cartographic Journal* 49 (4):326–36. doi: 10.1179/1743277412Y.0000000021.

Thorndyke, P. W., and B. Hayes-Roth. 1982. Differences in spatial knowledge acquired from maps and navigation. *Cognitive Psychology* 14 (4):560–89. doi: 10.1016/0010-0285(82)90019-6.

Vandenberg, S. G., and A. R. Kuse. 1978. Mental rotations, a group test of three-dimensional spatial visualization. *Perceptual and Motor Skills* 47 (2):599–604. doi: 10.2466/pms.1978.47.2.599.

Willis, K. S., C. Hölscher, G. Wilbertz, and C. Li. 2009. A comparison of spatial knowledge acquisition with maps and mobile maps. *Computers, Environment and Urban Systems* 33 (2):100–110. doi: 10.1016/j.compenvurbsys.2009.01.004.

Winter, S., M. Tomko, B. Elias, and M. Sester. 2008. Landmark hierarchies in context. *Environment and Planning B: Planning and Design* 35 (3):381–98. doi: 10.1068/b33106.

Zhao, J., and R. Li. 2016. Visualizing distance objects on mobile phones: Choice of resizable icons. Paper presented at the AutoCarto 2016, September 14–16, Albuquerque, NM.

Assessing Mobility-Based Real-Time Air Pollution Exposure in Space and Time Using Smart Sensors and GPS Trajectories in Beijing

Jing Ma, Yinhua Tao, Mei-Po Kwan, ⓘ and Yanwei Chai

Using real-time data from portable air pollutant sensors and smartphone Global Positioning System trajectories collected in Beijing, China, this study demonstrates how smart technologies and individual activity-travel microenvironments affect the assessment of individual-level pollution exposure in space and time at a very fine resolution. It compares three different types of individual-level exposure estimates generated by using residence-based monitoring station assessment, mobility-based monitoring station assessment, and mobility-based real-time assessment. Further, it examines the differences in personal exposure to $PM_{2.5}$ associated with different activity places and travel modes across various environmental conditions. The results show that the exposure estimates generated by monitoring station assessment and real-time sensing assessment vary substantially across different activity locations and travel modes. Individual-level daily exposure for residents living in the same community also varies significantly, and there are substantial differences in exposure levels using different approaches. These results indicate that residence- or mobility-based monitoring station assessments, which cannot account for the differences in air pollutant exposures between outdoor and indoor environments and between different travel-related microenvironments, could generate considerably biased estimates of personal pollution exposure. *Key Words: indoor environment, real-time exposure to air pollution, smart technologies, travel modes, the uncertain geographic context problem.*

本研究运用可携式空气污染感应器与手机全球定位系统轨迹在中国北京所搜集的实时数据，在极细微的分辨路上展现智慧科技与个人的活动—旅行之微观环境，如何影响个人层级在时空中的污染暴露量。本研究比较通过运用以居住为基础的监测站评估、根据移动的监测站评估，以及根据移动的实时评估所产生的污染暴露估计。此外，本研究检视与横跨各种环境条件的不同活动场所和旅行方式有关的$PM_{2.5}$个人暴露量。研究结果显示，由监测站评估和实时感应评估所生产的暴露估计量，在不同的活动地点和旅行模式之间有着显著的差异。居住在相同社区中的个人层级每日暴露量差异显著，且运用不同的方法亦会得到显著的暴露层级差异。这些研究结果意味着，以居住或移动为基础的观测站评估，无法考量室内与室外环境、以及与旅行相关的微观环境之间的空气污染暴露的差异，因而可能产生个人污染暴露量的大幅偏误。关键词：*室内环境，实时空气污染暴露，智慧科技，旅行方式，不确定的地理脉络问题。*

Usando datos de tiempo real de sensores portátiles de contaminantes aéreos y de trayectorias de Posicionamiento Global de teléfonos inteligentes, recogidos en Beijing, China, este estudio demuestra cómo las tecnologías inteligentes y los microambientes de actividad viajera individual inciden sobre la evaluación de exposición a la contaminación a nivel de individuo en el espacio y en el tiempo, a resolución muy fina. El estudio compara tres tipos diferentes de estimativos de exposición a nivel individual, generados mediante el uso de evaluación de la residencia a partir de una estación de monitoreo, evaluación de estación de monitoreo basada en movilidad, y evaluación en tiempo real basada en movilidad. Además, se examinan las diferencias de exposición personal al $PM_{2.5}$ asociada con diferentes lugares de actividad y modos de viaje a través de varias condiciones ambientales. Los resultados muestran que los cálculos de exposición generados por evaluación con estaciones de monitoreo y la evaluación por sensación en tiempo real varían sustancialmente a través de diferentes localizaciones de actividad y modalidades de viaje. La exposición diaria a nivel de individuo para residentes que habitan en la misma comunidad también varía significativamente y

Color versions of one or more figures in the article can be found online at www.tandfonline.com/raag.

hay diferencias sustanciales en los niveles de exposición cuando se usan diferentes enfoques. Estos resultados indican que la residencia o las evaluaciones con estaciones de monitoreo basadas en movilidad, que no puede responder por las diferencias en exposición a contaminantes aéreos entre ambientes interno y externos y entre diferentes microambientes relacionados con viajes, podrían generar cálculos considerablemente sesgados de exposición personal a la contaminación. *Palabras clave: ambiente interior, exposición a la contaminación aérea en tiempo real, modos de viaje, problema de contexto geográfico incierto, tecnologías inteligentes.*

Air pollution and its adverse health effects are a major health concern worldwide. Health geographers and public health researchers have conducted pollution exposure assessment for decades. Many epidemiological studies, for instance, have examined people's exposure to air pollution and its health effects using air quality data obtained from stationary monitoring stations and a static residence-based approach. In these studies, people who live in the same area (e.g., in the same residential neighborhood) are assigned the same value of air pollution exposure derived from stationary and sparse monitoring stations (Jerrett et al. 2005; Hoek et al. 2013; Kumar et al. 2015). Using such geographic areas as contextual units, residence-based exposure assessments are suitable for large population-based studies and long-term monitoring of outdoor air pollution (Monn 2001; Steinle, Reis, and Sabel 2013).

Studies based on this static residence-based approach have several limitations, however, that might undermine the reliability of their findings. For instance, they ignore human mobility and thus do not take into account people's exposures to environmental contexts besides their residential neighborhood and variations in personal exposure for people living in the same residential neighborhood. They use sparse stationary monitoring station data that do not accurately capture the space–time dynamics of air pollution. The uneven geographic distribution of stationary and sparse monitoring stations might introduce considerable uncertainty during the exposure modeling process (Avery et al. 2010; Gray, Edwards, and Miranda 2013). Because individuals are mobile in space and time and often undertake activities outside of their residential neighborhood, personal air pollution exposure is influenced by the complex interactions between the spatiotemporal dynamics of air pollution and human mobility. The static residence-based approach ignores individuals' space–time behaviors and, as a result, its use could lead to biases in exposure assessments and health effect estimates (Lu and Fang 2015; Chen, Song, and Jiang 2018).

As some recent studies indicate, past research that ignores human mobility and uses sparse stationary monitoring station data could face two major methodological issues that are especially relevant when assessing individual exposure to air pollution: the uncertain geographic context problem (UGCoP) and the neighborhood effect averaging problem (NEAP). The UGCoP is the problem that the use of different delineations of spatiotemporal contexts could lead to different research findings about the health effects of environmental influences on individuals (Kwan 2012; J. Wang and Kwan 2018; Zhao, Kwan, and Zhou 2018; Kwan et al. 2019), and the NEAP is the problem that ignoring people's daily mobility and exposures to nonresidential contexts could lead to biased estimations of personal exposure and the neighborhood effect (Kwan 2018a, 2018b; Kim and Kwan 2019).

There are growing concerns with addressing these methodological issues through reducing potential biases and increasing the accuracy of pollution exposure assessment in recent years. This has contributed to the development of mobility-based dynamic exposure assessment that simultaneously considers the spatiotemporal variability of air pollutant concentrations and human mobility (Yoo et al. 2015; Park and Kwan 2017). Increasing the accuracy of pollution exposure assessment is important because it can help improve our understanding of the impacts of exposure to air pollutants (e.g., fine particulates) on human health and well-being, especially for vulnerable or disadvantaged groups such as young children, the elderly, pregnant women, and people with respiratory or cardiovascular diseases (Yoo et al. 2015).

The rapid development and widespread use of smart technologies—such as wearable location-aware devices like Global Positioning System (GPS) devices, smartphones, and air quality sensors—in recent years have greatly facilitated the development of mobility-based dynamic assessment of personal exposures to air pollution. These technologies have considerably advanced the acquisition of accurate data on human space–time behaviors at fine spatio-temporal resolutions (J. Wang, Kwan, and Chai 2018). They could provide detailed space–time information essential for personal exposure assessment and have been increasingly used in geographic and epidemiological research (Castell et al. 2017; Jerrett et al. 2017; Rai et al. 2017; Caryl et al. 2019; Kwan et al. 2019).

Using GPS devices or smartphones, the potential biases and inaccuracy in assessments of personal exposure to air pollution can be mitigated. First, human mobility can be taken into account in exposure assessments. For instance, individual-level information on travel modes and activity type, location, time, and duration can be spatiotemporally linked to pollution concentrations derived from stationary monitoring stations (Nazelle et al. 2013). Some research indicated that individual-level exposure estimates could differ substantially for residents living in the same neighborhood, mainly due to their different movement trajectories (Dons et al. 2011; Yoo et al. 2015; Park and Kwan 2017). Compared to the conventional static residence-based approach, such mobility-based monitoring station assessment provides improved personal exposure estimates by considering individuals' daily mobility (Kwan 2013).

This mobility-based monitoring station approach still has two limitations, however—that is, the dependence on the air quality data from a limited number of monitoring stations and inability to account for the differences in air pollutant concentrations between outdoor and indoor environments—that can also be mitigated by deploying smart technologies (Tang and Lu 2012). Because people generally spend most of their time in indoor environments, using only their outdoor exposures could lead to erroneous personal exposure assessments (Hoek et al. 2013). People's indoor pollution exposures might be independent of or quite different from the outdoor environments because it depends on specific indoor factors like ventilation, air conditioning, and the concentrations of harmful substances from sources like tobacco smoke, cooking, and heating with solid fuels or natural gas (Steinle, Reis, and Sabel 2013; Q. Wang et al. 2019). Indoor pollution exposure should thus be an important element in personal exposure assessments, and it has received increasing attention in research and policymaking in recent years (Zou et al. 2009; Steinle, Reis, and Sabel 2013). Moreover, there is great variability in air pollutant concentrations within

different transportation modes or travel microenvironments (e.g., traveling in a bus, subway, or a private car), which have been found to play an important role in personal exposure assessment (e.g., Adams et al. 2001; Kaur, Nieuwenhuijsen, and Colvile 2007; Huang et al. 2012; Cepeda et al. 2017).

Thus, to also take into account individuals' exposures to air pollution in specific indoor and travel microenvironments, smart sensing technologies (e.g., air pollutant sensors) can be used to directly measure personal real-time pollution exposure (Mead et al. 2013; Piedrahita et al. 2014). Portable air pollution sensors integrated with GPS can simultaneously monitor individuals' geographic locations and measure real-time pollutant concentrations in individuals' immediate surroundings at very fine spatiotemporal resolutions (Panis 2010; Lu and Fang 2015). Compared with residence-based or mobility-based monitoring station assessments, mobility-based real-time assessments (RTAs) can take into account the differences in personal exposures between indoor and outdoor environments and between different transportation modes or travel microenvironments, as well as the differences due to the use of individual protective measures (e.g., using an air purifier at home, reducing travel, or changing travel modes; Steinle, Reis, and Sabel 2013; Jiao, Xu, and Liu 2018). Note that some studies have observed considerable differences between personal real-time exposures derived from portable sensors and modeled exposures derived from stationary monitoring station data, thus indicating the need for the use of smart sensors in personal exposure assessment (Nieuwenhuijsen et al. 2015). Moreover, because smart sensors can provide real-time information on pollutant concentrations at very high spatiotemporal resolutions (ranging from one-second intervals to several minutes), the information they provide allows people, especially those already at risk, to make informed decisions to effectively avoid high pollution exposure (Kumar et al. 2015). Possibly due to the high cost and additional efforts involved in research using wearable tracking and sensing devices (Gerharz, Antonio, and Klemm 2009), however, studies that analyzed and compared personal mobility-based real-time air pollution exposure at very fine spatiotemporal resolutions in various indoor and outdoor environments have been limited to date, particularly in developing countries like China.

In this research, we use smart technologies and advanced geospatial methods to examine how individuals' space–time behaviors and the spatiotemporal dynamics of air pollutant concentrations in different microenvironments influence individual-level real-time exposure in space and time. Using data collected with GPS devices, activity-travel diaries, and wearable air pollutant sensors in Beijing from December 2017 to February 2018, we compared three different types of exposure estimates at both the activity and travel episode level and the individual level generated by residence-based monitoring station assessment, mobility-based monitoring station assessment, and mobility-based RTA. Further, we examined the differences in personal exposure to fine particulates between different activity places (e.g., homes, workplaces, shops, and outdoor locations) and different travel modes (e.g., walking, cycling, public transport, and private car) across various environmental conditions (e.g., on lightly or heavily polluted days) in Beijing, China. The results indicate that the exposure estimates obtained by assessments based on monitoring station data and real-time data vary substantially between different activity locations and travel modes. There are also substantial differences in personal exposure levels between estimates obtained from different approaches, suggesting that the residence- or mobility-based monitoring station assessments could generate considerably biased estimates of personal pollution exposure. This study illustrates how smart technologies can help us analyze and visualize personal pollution exposure in space and time at very high resolutions, which in turn helps address both the UGCoP and the NEAP.

Method

Study Area and Data

Air pollution is a major environmental problem and public health concern in Beijing. With an annual average PM$_{2.5}$ concentration of above $80 \, \mu g/m^3$ in 2015, the air quality in Beijing was among the worst in the world (Ma et al. 2017). Air pollutant concentrations vary spatially and temporally in Beijing, with frequent high pollution and persistent smog episodes in winter. This study focuses on the Meiheyuan community, which is located in the inner suburban area of northern Beijing, adjacent to the railway, highway,

the Fifth Ring Road, and some other pollution sources (Figure 1). Housing types in this community are diverse, including price-controlled commercial housing and low-rent housing for low- or medium-income residents and *danwei* housing for employees allocated by their work units at discounted prices (Y. P. Wang and Murie 2011). Using the Meiheyuan community as a case study, we aim to highlight the differences in individual-level pollution exposure obtained from different exposure assessments for residents living in the same residential neighborhood.

The survey was conducted in Meiheyuan from December 2017 to February 2018. Using a stratified sampling approach, we recruited a total of 117 residents aged eighteen to sixty years old to participate in the survey via six waves. Smart air pollutant sensors, GPS tracking devices, and activity-travel diaries were used together to collect data on individuals' space–time behaviors and real-time exposures to fine particulates (PM$_{2.5}$). Specifically, each participant was asked to carry a GPS-equipped smartphone and a portable air pollutant sensor all the time over a continuous forty-eight-hour period that covers a workday and a weekend day (e.g., Friday and Saturday or Sunday and Monday). The GPS trajectories and real-time PM$_{2.5}$ concentrations in each participant's immediate surroundings were recorded simultaneously at one-second intervals. Moreover, each respondent was asked to complete a questionnaire and a two-day activity-travel diary that collected information about their sociodemographic attributes, health status, activities, and travels (e.g., activity types, places visited, and travel modes used). Finally, the survey obtained valid data from 112 participants. Table 1 presents the sociodemographic characteristics of the participants. In addition, hourly data on PM$_{2.5}$ concentrations from the thirty-five stationary monitoring stations in Beijing were also collected for the corresponding survey days and used in this research.

Procedure

In this study, we aim to compare three different types of exposure estimates generated by using residence-based monitoring station assessment, mobility-based monitoring station assessment, and mobility-based RTA. For the residence-based approach, we assumed that participants are nonmobile and used their residential location as their single spatial location throughout the day (Figure 2A). We first used

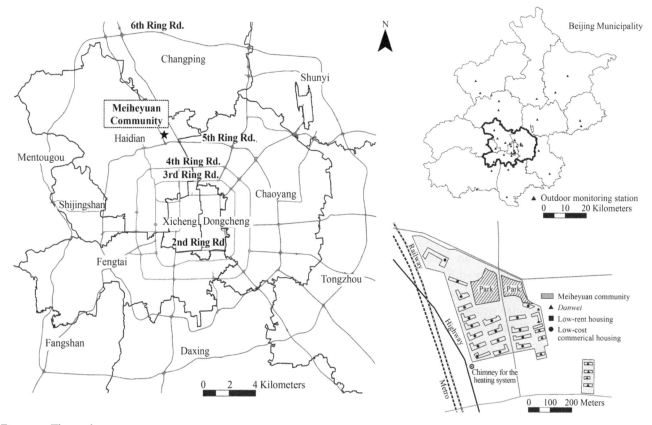

Figure 1. The study area.

Table 1. Key sociodemographic attributes of the survey participants

Variable	Description	N	Proportion (%)
Gender	Male	54	48.2
	Female	58	51.8
Age	18–30	20	17.9
	30–49	65	58.0
	50–60	27	24.1
Monthly income (RMB)	<3,000	13	11.6
	3,000–6,000	22	19.6
	6,000–10,000	22	19.6
	10,000–15,000	21	18.8
	15,000+	34	30.4
Housing type	*Danwei* housing	32	28.6
	Low-rent housing	29	25.9
	Price-restricted commodity housing	51	45.5

Notes: $N = 112$. RMB = renminbi, official Chinese currency.

the hourly air quality data from the thirty-five stationary monitoring stations in Beijing and kriging interpolation to create twenty-four hourly $PM_{2.5}$ concentration surfaces for each survey day. We then extracted the $PM_{2.5}$ concentration values at each participant's residential location for each hour and used their standardized sum as the person's static exposure estimates. In contrast, the mobility-based monitoring station assessment simultaneously takes into account each participant's activity-travel patterns in space and time, such as activity place, duration, and sequences, as well as the spatiotemporal dynamics of air pollution (Figure 2A). Using the hourly $PM_{2.5}$ concentration surfaces derived from the monitoring station data and GPS-integrated activity diary data, we extracted each participant's exposure to $PM_{2.5}$ by identifying the intersection between his or her daily movement trajectories and the twenty-four hourly layers of $PM_{2.5}$ concentrations. Individual exposure to air pollution is again obtained as the standardized sum of these twenty-four hourly exposures, which capture the variations in a person's pollution exposure due to changes in his or her location and in pollution concentrations in the environment. For these two assessments that rely on the data collected at the thirty-five stationary monitoring stations, we have conducted cross-validation to evaluate the kriging results and find

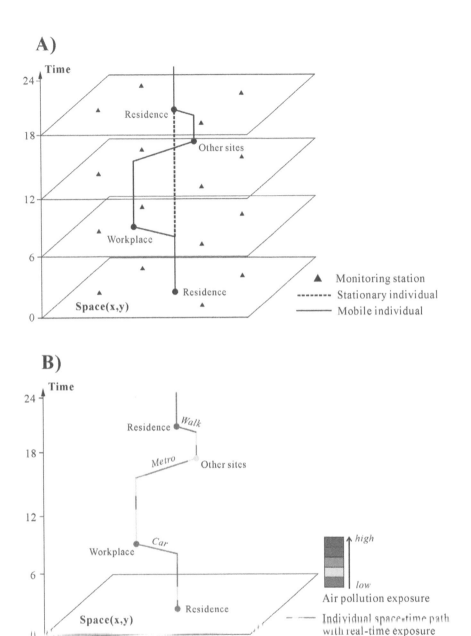

Figure 2. Personal exposure assessments: (A) residence-based or mobility-based monitoring station assessment and (B) mobility-based real-time assessment.

that this interpolation technique shows good performance ($R^2 = 0.94$).

To generate the mobility-based RTA, we extracted $PM_{2.5}$ concentrations from the air pollutant sensors (AirBeam from HabitatMap) and GPS trajectories from participants' smartphones, which were both logged at one-second intervals. This approach incorporates the real-time spatiotemporal interactions between air pollution and individual movement and the differences between indoor and outdoor exposures as well as between different travel modes or travel microenvironments (Figure 2B). Because low-cost portable

sensors might have lower specificity or sensitivity when compared to expensive monitoring station sensors (De Vito et al. 2009; Lewis and Edwards 2016), we conducted field testing and calibration to evaluate the performance of the portable sensors under different urban conditions using the colocation calibration technique, which was implemented as follows. First, the AirBeam sensors were placed near a stationary monitoring station, which was used to provide the reference data. Data on $PM_{2.5}$ concentrations were then collected from both the portable sensors and the stationary monitoring station at the same time over two

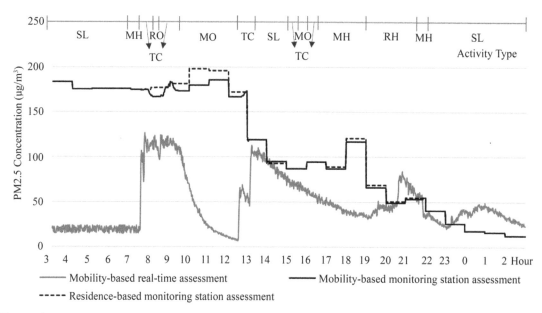

Figure 3. Temporal variations in exposure estimates at different activity locations on a weekend day, 14 January 2018. *Note:* SL = sleeping; MH = maintenance activity at home; RO = recreation activity outside home; MO = maintenance activity outside home; TC = travel by car; RH = recreational activity at home.

days with different environmental conditions. Using regression modeling, a strong correlation was found between the PM$_{2.5}$ concentrations obtained by the AirBeam sensors and monitoring station measurement ($R^2 = 0.86$). The portable sensors showed good performance during the field test, and there were no missing data during the colocation calibration period. Moreover, to compare each participant's daily average exposure to PM$_{2.5}$ between different assessments, we calculated the sum of the twenty-four hourly exposure estimates derived from the monitoring station assessments and the sum of the per second exposure estimates from the mobility-based RTAs. These different types of exposure sums were then standardized by the number of the respective measurement periods on a day.

Comparative Assessments of Personal Exposure in Space and Time

Comparing the Temporal Variations of Different Individual Exposure Assessments

Using the air pollution data and GPS trajectories, we first used one representative respondent to illustrate and compare the temporal variations in individual exposure estimates during each activity and travel generated by the three types of exposure

assessments described earlier. As shown in Figure 3, there was little variation between the static residence-based and mobility-based monitoring station estimates (MSEs) throughout the day, because the respondent spent much time at home on a weekend day when ambient air pollution was relatively high. In contrast, the variation between the mobility-based RTA and MSE was substantial, because the RTA accounted for the differences in exposure between indoor and outdoor environments and individual protective measures (e.g., using an air purifier at home). For instance, when sleeping and performing maintenance activities (e.g., eating meals) at home before 0 a.m., the respondent turned on an air purifier to avoid high exposure to pollution at home and, as a result, the real-time exposure was much lower than the monitoring station estimates. When traveling and performing recreational activities outside the home between 8 a.m. and 10 a.m., however, real-time exposure to pollution increased significantly but still was less than the MSE, mainly due to the differences between indoor and outdoor environments. In contrast, due to pollution from smoking and heating at home in the evening, the respondent's real-time exposure was higher than the MSE between 9 p.m. and 10 p.m. These results indicate that widely used monitoring station assessments tend to overestimate individual exposure levels when ambient air pollution is relatively high and, more

Table 2. Exposure estimates at different activity locations using different types of assessments

			Low daily concentration		Middle daily concentration		High daily concentration	
			M	SD	M	SD	M	SD
Indoor locations	Home	MSE	10.58	7.12	35.98	18.02	78.02	10.50
		RTA	45.33	50.24	47.64	38.55	65.91	15.11
	Workplace	MSE	9.41	2.26	23.77	11.78	79.06	8.28
		RTA	26.78	32.48	32.23	28.19	53.40	8.33
	Shops	MSE	9.36	1.17	25.40	10.95	96.60	9.91
		RTA	17.83	14.41	23.60	12.50	72.19	10.76
	Other	MSE	9.68	4.44	27.91	14.39	104.03	8.59
		RTA	31.26	31.74	44.68	47.06	71.56	10.10
Outdoor locations		MSE	9.14	1.72	27.83	9.65	83.65	10.65
		RTA	25.45	44.31	35.47	32.66	104.87	7.87

Notes: MSE = mobility-based monitoring station estimation; RTA = real-time assessment. Low daily concentration is daily mean $PM_{2.5}$ concentration $\leq 25 \mu g/m^3$; middle daily concentration is $25 \mu g/m^3 <$ daily mean $PM_{2.5}$ concentration $\leq 75 \mu g/m^3$; high daily concentration is $PM_{2.5}$ concentration $> 75 \mu g/m^3$.

important, individual activity-travel contexts or microenvironments have a significant impact on personal exposure to air pollutants ($PM_{2.5}$) in space and time.

Comparing Different Types of Exposure Estimates across Activity Places

Because air pollution concentrations vary between indoor and outdoor environments, we then compared the exposure estimates generated by mobility-based MSEs and RTAs at different activity locations. Further, to compare the variation across various environmental conditions, we divided the survey days into three categories based on the daily average $PM_{2.5}$ concentrations released by the Chinese Ministry of Environmental Protection. As shown in Table 2, there were obvious variations in $PM_{2.5}$ concentrations among different activity places, and the exposure estimates generated by MSE and RTA varied substantially across different activity locations. When the ambient air quality was good (i.e., with daily mean concentration of $PM_{2.5}$ less than $25 \mu g/m^3$), estimates from RTA were higher than MSE across different activity locations, particularly for residential locations, possibly due to the fact that RTA takes into account various sources of particulate matter pollution such as cooking, smoking, and traffic-related emissions in various indoor and outdoor environments. In contrast, on days with high levels of air pollution (i.e., with a daily mean $PM_{2.5}$ concentration higher than $75 \mu g/m^3$), estimates from RTA were lower than MSE in various indoor

locations but higher in outdoor environments. These results suggest that monitoring station assessments tend to underestimate personal exposure levels in indoor environments, particularly at home, when the ambient air quality is good and overestimate individual exposure in indoor environments when ambient air pollution levels are relatively high.

Comparing Different Types of Exposure Estimates by Travel Modes

Because travel modes might have a significant effect on personal exposure to fine particulates in different traffic microenvironments (Huang et al. 2012; Cepeda et al. 2017), we also compared different types of exposure estimates with respect to different travel modes across various environmental conditions. As shown in Table 3, variations in average $PM_{2.5}$ concentrations experienced from different travel modes were evident, and there were substantial differences in exposure estimates obtained from mobility-based MSEs and RTAs. Specifically, when ambient air quality was good, real-time exposures were higher than the estimates from MSE for various travel modes, such as walking, cycling, and public transport, whereas on days with serious air pollution, real-time exposures to $PM_{2.5}$ were much lower than those estimated by MSE. This suggests that pollution concentrations estimated with data from stationary monitoring stations, which fail to account for the differences in personal exposure in different travel microenvironments, are not appropriate surrogates for individual real-time $PM_{2.5}$ exposures on days

Table 3. Exposure estimates by different travel modes using different types of assessments

		Low daily concentration		Middle daily concentration		High daily concentration	
		M	SD	M	SD	M	SD
Car	MSE	9.05	2.28	30.04	14.45	114.34	44.88
	RTA	13.88	17.88	26.86	33.82	63.78	35.37
	Daily frequency	0.80	1.53	0.67	1.25	0.69	1.20
Bicycle/motorcycle	MSE	9.35	2.35	31.66	12.03	96.36	28.62
	RTA	16.85	14.01	36.19	29.63	63.94	37.72
	Daily frequency	1.15	1.97	0.76	1.87	0.59	1.05
Bus/subway	MSE	9.32	1.97	30.93	16.99	104.21	32.82
	RTA	28.05	42.87	30.04	19.52	64.04	28.06
	Daily frequency	0.24	0.79	0.65	1.04	0.52	1.00
Walk	MSE	8.88	1.33	29.50	15.47	101.33	35.40
	RTA	20.17	29.23	36.22	28.48	74.84	33.73
	Daily frequency	1.87	2.40	1.46	1.90	1.37	1.88
All	Daily frequency	4.05	2.59	3.55	2.22	3.17	2.18

Notes: MSE = mobility-based monitoring station estimation; RTA = real-time assessment. Low daily concentration is daily mean $PM_{2.5}$ concentration $\leq 25\,\mu g/m^3$; middle daily concentration is $25\,\mu g/m^3 <$ daily mean $PM_{2.5}$ concentration $\leq 75\,\mu g/m^3$; high daily concentration is $PM_{2.5}$ concentration $> 75\,\mu g/m^3$. Daily frequency refers to the average number of trips per day made by various transportation modes for the survey respondents.

with low or high air pollution. When ambient air pollution was moderate (i.e., with a daily mean concentration of $PM_{2.5}$ ranging from $25\,\mu g/m^3$ to $75\,\mu g/m^3$), however, there was little variation between exposure estimates obtained from RTA and MSE. Moreover, as shown in Table 3, with increasing air pollution, participants tend to decrease their daily trip frequencies significantly, especially for walking and cycling trips, and change their travel modes to motorized vehicles to avoid high exposure to air pollutants in heavily polluted environments.

Comparing Different Assessments of Individual-Level Exposure

Given the variations in activity and travel episode-level $PM_{2.5}$ concentrations in space and time, we further compared the differences in individual-level daily exposure estimates obtained from the three approaches: static residence-based assessment, mobility-based MSEs, and mobility-based RTAs. Figure 4 shows the three exposure estimates for each survey respondent on a workday and a weekend day based on the survey days. It indicates that the individual daily average exposure estimates obtained by assessments based on monitoring station data and real-time data vary substantially. By taking into account the interaction of air pollution and human mobility, the mobility-based MSE (white bars) showed some variations in individual daily exposure levels for

residents living in the same residential neighborhood. Because ambient air quality was relatively good with little spatiotemporal variation on some survey days or the respondents spent much time at home when ambient air pollution was relatively high, however, there were small differences between the residence-based assessment (dark lines) and mobility-based MSE (white bars). These results tend to support the argument that mobility-based stationary monitoring assessment might be similar to static residence-based estimates if air pollution concentrations have little spatiotemporal variation or when individuals have a low level of mobility (e.g., Yoo et al. 2015).

In contrast, when taking into account the differences in air pollutant concentrations between indoor and outdoor environments or in various microenvironments, there were substantial differences in individual daily exposure to $PM_{2.5}$ between MSE (white bars) and RTA (gray bars). For instance, on a workday (e.g., 25 December 2017) when ambient air quality was relatively good, the real-time $PM_{2.5}$ exposures for some respondents were much higher than the exposure levels obtained from MSE (Figure 4A). There were greater variations between individual-level real-time exposure estimates on the surveyed workdays from 15 December 2017 to 26 January 2018, ranging from $3.12\,\mu g/m^3$ to $144.04\,\mu g/m^3$, whereas mobility-based MSEs ranged from $7.64\,\mu g/m^3$ to $59.61\,\mu g/m^3$. In contrast, on a weekend day (e.g., 14 January 2018)

A)

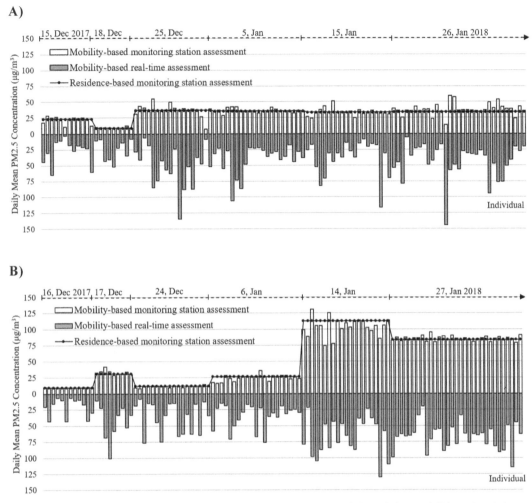

B)

Figure 4. Comparing different assessments of individual-level exposure estimates on (A) workdays and (B) weekend days.

when ambient air pollution was relatively high, individual daily exposures to $PM_{2.5}$ obtained from RTA were much lower than those obtained from MSE, but there were greater variations between individual real-time exposure estimates (Figure 4B). This is mainly because the RTAs take into account the differences in exposure levels between indoor and outdoor environments, the various sources of particulate matter pollution like smoking, cooking, and traffic-related emissions in individual-specific spatiotemporal microenvironments, as well as individual protective measures in a heavily polluted environment such as using an air purifier at home, reducing travel, or changing travel modes from nonmotorized modes to motorized vehicles to avoid high exposure to air pollution.

To better illustrate the spatiotemporal dynamics of air pollution and human mobility, the respondents' space–time trajectories were geovisualized in

3D and color-coded based on the real-time $PM_{2.5}$ exposures obtained from the air pollutant sensors. Further, we incorporated the interpolated hourly air pollution layers derived from data provided by the limited number of monitoring stations in the 3D geovisualizations for a typical workday and a weekend day. As shown in Figure 5, participants living in the same neighborhood but with different activity-travel patterns experienced different levels of $PM_{2.5}$ concentrations, and the interpolated hourly air pollution surfaces cannot accurately represent individual-level real-time exposures in space and time at a fine resolution. For instance, on the workday (15 January 2018) between 9 a.m. and 12 p.m., some respondents were exposed to high levels of pollution when traveling on the main roads, possibly due to traffic congestion and traffic-related emissions (Figure 5A). In the evening after 9 p.m. when ambient air quality deteriorated, the real-time $PM_{2.5}$

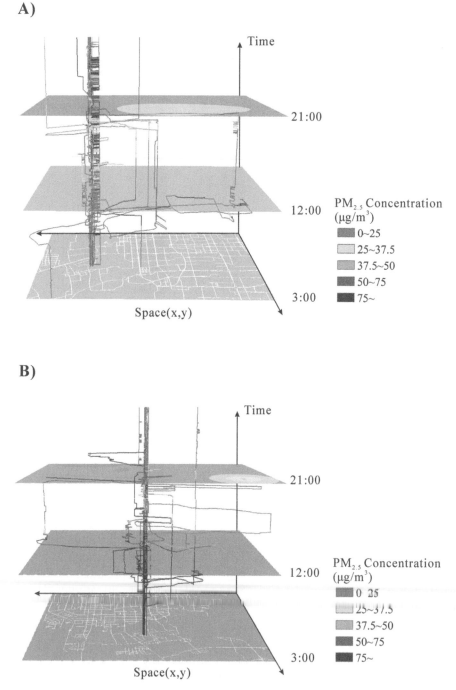

Figure 5. 3D geovisualization of respondents' space–time trajectories with real-time PM$_{2.5}$ concentrations from portable air pollutant sensors on (A) a workday (15 January 2018) and (B) a weekend day (14 January 2018). The interpolated hourly air pollution layers were derived from the static and sparse monitoring station data using the kriging interpolation method.

exposures of some respondents were very low (below 25 μg/m^3), possibly because they used an air purifier at home, whereas the real-time exposure estimates of some other participants were relatively high. The great variations in individual-level real-time exposures in space and time as well as the differences between real-time estimates and hourly monitoring station prediction surfaces were also evident on a typical weekend day (Figure 5B). Such differences should be taken into account in personal exposure assessments, particularly when estimating short-term exposure, which can directly cause acute symptoms (e.g., acute asthma), in epidemiological research.

Conclusion

Smart technologies such as smartphones and portable sensors could collect high-resolution geospatial data on individual space–time behaviors and real-time air pollution exposures. They offer great potential for reducing the errors in exposure measurement and increasing the accuracy of personal exposure estimates that are essential for epidemiological and geographic studies (Han and Naeher 2006; Jerrett et al. 2017). These smart technologies can play an important role in real-world environmental monitoring and facilitating intelligent individual decision making. They can help residents obtain valuable information and increase their real-time awareness of the surrounding environments, as well as make more "informed, responsible choices" to optimize their daily behavior, avoid exposure to high levels of pollution, and improve their quality of life (Gabrys 2014). Smart technologies facilitate and enable the development of smart cities, in which digital technologies are used to enhance people's life quality and well-being, reduce the cost of data collection and resource consumption, and engage more actively and effectively with urban residents (Hashem et al. 2016; Nieuwenhuijsen 2016).

With recent advancements in smart sensing technologies, our focus on exposure assessment has turned from people's residential locations to their daily movements, from space to space–time, and from static monitoring to real-time sensing (Kwan 2012, 2013). Using real-time data from smartphones and air pollutant sensors collected in Beijing, this study contributes to the environmental health and geographic literature by demonstrating how smart technologies and attention to individual activity-travel microenvironments enable us to analyze individual-level exposure in space and time at a very fine resolution. We compared three different types of exposure estimates generated by residence-based monitoring station assessment, mobility-based monitoring station assessment, and mobility-based real-time sensing assessment. Further, we examined the differences in personal exposures to fine particulates associated with different activity places (e.g., homes, workplaces, shops, and outdoor locations) and

different travel modes (e.g., walking, cycling, public transport, and private cars) across various environmental conditions in a realistic setting. The results showed that there were considerable variations in individual exposure to $PM_{2.5}$ in different activity places, and the exposure estimates generated by monitoring station assessment and real-time sensing assessment varied substantially between different activity locations. Because the monitoring station assessment is unable to take into account some important indoor and outdoor pollution sources, it could generate substantially biased estimates by under- or overestimating individual exposure at indoor locations, particularly at home and especially when ambient air quality is very good or very poor. Moreover, people's travel modes had a significant effect on individual exposure to air pollution, and the averaged $PM_{2.5}$ concentrations experienced when using different travel modes generated by different approaches varied substantially across various environmental conditions.

This study also found that individual-level daily average $PM_{2.5}$ exposures for residents living in the same community varied significantly, and there were substantial differences in exposure estimates obtained from different assessment methods. Real-time sensing technologies could greatly improve accuracy in the estimation of individual exposure to air pollution at very fine spatiotemporal resolutions, and they are superior to the static residence-based or mobility-based monitoring station assessments. This is mainly because smart sensing technologies take into account the differences in individual exposure between indoor and outdoor environments, various pollution sources (e.g., smoking, cooking, and traffic-related emissions) in person-specific spatiotemporal microenvironments, as well as the protective measures that people adopted in heavily polluted environments (e.g., using an air purifier at home, reducing travel, or changing travel modes). Using smart technologies, this study helps to improve our understanding of individual exposure to air pollution in space and time at very fine resolutions. It also shows how the UGCoP and the NEAP could be mitigated through deploying smart technologies to obtain more accurate estimates of individual exposures to environmental influences. Despite encouraging improvements in low-cost sensor technologies, however, more efforts are needed to conduct testing and calibration to evaluate sensor performance, improve the consistency and durability of sensing elements, and develop better devices to improve data accuracy and sensitivity (Kumar et al. 2015). As suggested by Lewis and Edwards (2016, 31), "Well-designed sensor experiments, that acknowledge the limitations of the technologies as well as the strengths, have the potential to simultaneously advance basic science, monitor air pollution—and bring the public along."

Acknowledgments

The authors are very grateful for the comments of the reviewers and the editor, which have helped improve the article considerably.

Funding

This work was supported by the National Natural Science Foundation of China (Grant Nos. 41529101, 41601148, and 41571144).

ORCID

Mei-Po Kwan http://orcid.org/0000-0001-8602-9258

References

Adams, H., M. Nieuwenhuijsen, R. Colvile, M. McMullen, and P. Khandelwal. 2001. Fine particle (PM$_{2.5}$) personal exposure levels in transport microenvironments, London, UK. *Science of the Total Environment* 279 (1–3):29–44. doi: 10.1016/S0048-9697(01)00723-9.

Avery, C. L., K. T. Mills, W. Ronald, K. A. Mograw, P Charles, R. L. Smith., et al. 2010. Estimating error in using residential outdoor PM$_{2.5}$ concentrations as proxies for personal exposures: A review. *Epidemiology* 118 (5):215–23. doi: 10.1097/EDE.0b013e3181cb41f7.

Caryl, F., N. K. Shortt, J. Pearce, G. Reid, and R. Mitchell. 2019. Socioeconomic inequalities in children's exposure to tobacco retailing based on individual-level GPS data in Scotland. *Tobacco Control.* Advance online publication. http://dx.doi.org/10.1136/tobaccocontrol-2018-054891

Castell, N., F. R. Dauge, P. Schneider, M. Vogt, U. Lerner, B. Fishbain, D. Broday, and A. Bartonova. 2017. Can commercial low-cost sensor platforms contribute to air quality monitoring and exposure estimates? *Environment International* 99:293–302. doi: 10.1016/j.envint.2016.12.007.

Cepeda, M., J. Schoufour, R. Freak-Poli, C. M. Koolhaas, K. Dhana, W. M. Bramer, and O. H. Franco. 2017. Levels of ambient air pollution according to mode of transport: A systematic review. *The Lancet Public Health* 2 (1):e23–e34. doi: 10.1016/S2468-2667(16)30021-4.

Chen, B., Y. Song, and T. Jiang. 2018. Real-time estimation of population exposure to PM$_{2.5}$ using mobile- and station-based big data. *International Journal of Environmental Research and Public Health* 15:573. doi: 10.3390/ijerph15040573.

De Vito, S., M. Piga, L. Martinotto, and G. Di Francia. 2009. CO, NO$_2$ and NO$_x$ urban pollution monitoring with on-field calibrated electronic nose by automatic Bayesian regularization. *Sensors and Actuators B: Chemical* 143 (1):182–91. doi: 10.1016/j.snb.2009.08.041.

Dons, E., L. I. Panis, M. Van Poppel, J. Theunis, H. Willems, R. Torfs, and G. Wets. 2011. Impact of time-activity patterns on personal exposure to black carbon. *Atmospheric Environment* 45 (21):3594–3602. doi: 10.1016/j.atmosenv.2011.03.064.

Fang, T. B., and Y. Lu. 2012. Personal real-time air pollution exposure assessment methods promoted by information technological advances. *Annals of GIS* 18 (4):279–88. doi: 10.1080/19475683.2012.727866.

Gabrys, J. 2014. Programming environments: Environmentality and citizen sensing in the smart city. *Environment and Planning D: Society and Space* 32 (1):30–48. doi: 10.1068/d16812.

Gerharz, L. E., K. Antonio, and O. Klemm. 2009. Applying indoor and outdoor modeling techniques to estimate individual exposure to PM$_{2.5}$ from personal GPS profiles and diaries: A pilot study. *Science of the Total Environment* 407 (18):5184–93. doi: 10.1016/j.scitotenv.2009.06.006.

Gray, S. C., S. E. Edwards, and M. L. Miranda. 2013. Race, socioeconomic status, and air pollution exposure in North Carolina. *Environmental Research* 126 (4):152. doi: 10.1016/j.envres.2013.06.005.

Han, X., and L. Naeher. 2006. A review of traffic-related air pollution exposure assessment studies in the developing world. *Environment International* 32 (1):106–20. doi: 10.1016/j.envint.2005.05.020.

Hashem, I., A. Targio, V. Chang, N. B. Anuar, K. Adewole, I. Yaqoob, A. Gani, E. Ahmed, and H. Chiroma. 2016. The role of big data in smart city. *International Journal of Information Management* 36 (5):748–58. doi: 10.1016/j.ijinfomgt.2016.05.002.

Hoek, G., R. M. Krishnan, R. Beelen, A. Peters, B. Ostro, B. Brunekreef, and J. D. Kaufman. 2013. Long-term air pollution exposure and cardio-respiratory mortality: A review. *Environmental Health* 12 (1):43. doi: 10.1186/1476-069X-12-43.

Huang, J., F. Deng, S. Wu, and X. Guo. 2012. Comparisons of personal exposure to PM$_{2.5}$ and CO by different commuting modes in Beijing, China. *Science of the Total Environment* 425:52–59. doi: 10.1016/j.scitotenv.2012.03.007.

Jerrett, M., A. Arain, P. Kanaroglou, B. Beckerman, D. Potoglou, T. Sahsuvaroglu, J. Morrison, and C. Giovis. 2005. A review and evaluation of intraurban air pollution exposure models. *Journal of Exposure Science & Environmental Epidemiology* 15 (2):185–204. doi: 10.1038/sj.jea.7500388.

Jerrett, M., D. Donaire-Gonzalez, O. Popoola, R. Jones, R. C. Cohen, E. Almanza, A. de Nazelle, et al. 2017. Validating novel air pollution sensors to improve exposure estimates for epidemiological analyses and citizen science. *Environmental Research* 158:286–94. doi: 10.1016/j.envres.2017.04.023.

Jiao, K., M. Xu, and M. Liu. 2018. Health status and air pollution related socioeconomic concerns in urban China. *International Journal for Equity in Health* 17 (1):18. doi: 10.1186/s12939-018-0719-y.

Kaur, S., M. Nieuwenhuijsen, and R. Colvile. 2007. Fine particulate matter and carbon monoxide exposure concentrations in urban street transport microenvironments. *Atmospheric Environment* 41 (23):4781–4810. doi: 10.1016/j.atmosenv.2007.02.002.

Kim, J., and M.-P. Kwan. 2019. Beyond commuting: Ignoring individuals' activity-travel patterns may lead to inaccurate assessments of their exposure to traffic congestion. *International Journal of Environmental Research and Public Health* 16 (1):89. doi: 10.3390/ijerph16010089.

Kumar, P., L. Morawska, C. Martani, G. Biskos, M. Neophytou, S. Di Sabatino, M. Bell, L. Norford, and R. Britter. 2015. The rise of low-cost sensing for managing air pollution in cities. *Environment International* 75:199–205. doi: 10.1016/j.envint.2014.11.019.

Kwan, M.-P. 2012. The uncertain geographic context problem. *Annals of the Association of American Geographers* 102 (5):958–68. doi: 10.1080/00045608.2012.687349.

Kwan, M.-P. 2013. Beyond space (as we knew it): Toward temporally integrated geographies of segregation, health, and accessibility. *Annals of the Association of American Geographers* 103 (5):1078–86. doi: 10.1080/00045608.2013.792177.

Kwan, M.-P. 2018a. The limits of the neighborhood effect: Contextual uncertainties in geographic, environmental health, and social science research. *Annals of the American Association of Geographers* 108 (6):1482–90. doi: 10.1080/24694452.2018.1453777.

Kwan, M.-P. 2018b. The neighborhood effect averaging problem (NEAP): An elusive confounder of the neighborhood effect. *International Journal of Environmental Research and Public Health* 15 (9):1841. doi: 10.3390/ijerph15091841.

Kwan, M.-P., J. Wang, M. Tyburski, D. H. Epstein, W. J. Kowalczyk, and K. L. Preston. 2019. Uncertainties in the geographic context of health behaviors: A study of substance users' exposure to psychosocial stress using GPS data. *International Journal of Geographical Information Science* 33 (6):1176–95. doi: 10.1080/13658816.2018.1503276.

Lewis, A., and P. Edwards. 2016. Validate personal air-pollution sensors. *Nature* 535 (7610):29–31. doi: 10.1038/535029a.

Lu, Y., and T. B. Fang. 2015. Examining personal air pollution exposure, intake, and health danger zone using time geography and 3D geovisualization. *ISPRS International Journal of Geo-Information* 4 (1):32–46. doi: 10.3390/ijgi4010032.

Ma, J., G. Mitchell, G. Dong, and W. Zhang. 2017. Inequality in Beijing: A spatial multilevel analysis of

perceived environmental hazard and self-rated health. *Annals of the American Association of Geographers* 107 (1):109–29. doi: 10.1080/24694452.2016.1224636.

Mead, M. I., O. A. M. Popoola, G. B. Stewart, P. Landshoff, M. Calleja, M. Hayes, J. J. Baldovi, et al. 2013. The use of electrochemical sensors for monitoring urban air quality in low-cost, high-density networks. *Atmospheric Environment* 70:186–203. doi: 10.1016/j.atmosenv.2012.11.060.

Monn, C. 2001. Exposure assessment of air pollutants: A review on spatial heterogeneity and indoor/outdoor/personal exposure to suspended particulate matter, nitrogen dioxide and ozone. *Atmospheric Environment* 35 (1):1–32. doi: 10.1016/S1352-2310(00)00330-7.

Nazelle, A., E. Seto, D. Donaire-Gonzalez, M. Mendez, J. Matamala, et al. 2013. Improving estimates of air pollution exposure through ubiquitous sensing technologies. *Environmental Pollution* 176:92–99. doi: 10.1016/j.envpol.2012.12.032.

Nieuwenhuijsen, M. 2016. Urban and transport planning, environmental exposures and health—New concepts, methods and tools to improve health in cities. *Environmental Health* 15 (Suppl. 1):38. doi: 10.1186/s12940-016-0108-1.

Nieuwenhuijsen, M., D. Donaire-Gonzalez, I. Rivas, M. Castro, M. Cirach., et al. 2015. Variability in and agreement between modeled and personal continuously measured black carbon levels using novel smartphone and sensor technologies. *Environmental Science & Technology* 49:2977–82. doi: 10.1021/es505362x.

Panis, L. I. 2010. New directions: Air pollution epidemiology can benefit from activity-based models. *Atmospheric Environment* 44 (7):1003–4. doi: 10.1016/j.atmosenv.2009.10.047.

Park, Y. M., and M.-P. Kwan. 2017. Individual exposure estimates may be erroneous when spatiotemporal variability of air pollution and human mobility are ignored. *Health & Place* 43:85–94. doi: 10.1016/j.healthplace.2016.10.002.

Piedrahita, R., Y. Xiang, N. Masson, J. Ortega, A. Collier, Y. Jiang, K. Li, et al. 2014. The next generation of low-cost personal air quality sensors for quantitative exposure monitoring. *Atmospheric Measurement Techniques* 7 (10):3325–36. doi: 10.5194/amt-7-3325-2014.

Rai, A. C., P. Kumar, F. Pilla, A. N. Skouloudis, S. Di Sabatino, C. Ratti, A. Yasar, and D. Rickerby. 2017. End-user perspective of low-cost sensors for outdoor air pollution monitoring. *Science of the Total Environment* 607–608:691–705. doi: 10.1016/j.scitotenv.2017.06.266.

Steinle, S., S. Reis, and C. E. Sabel. 2013. Quantifying human exposure to air pollution—Moving from static monitoring to spatio-temporally resolved personal exposure assessment. *Science of the Total Environment* 443:184–93. doi: 10.1016/j.scitotenv.2012.10.098.

Wang, J., and M.-P. Kwan. 2018. An analytical framework for integrating the spatiotemporal dynamics of environmental context and individual mobility in exposure assessment: A study on the relationship between food environment exposures and body weight. *International Journal of Environmental Research* and *Public Health* 15 (9):2022. doi: 10.3390/ijerph15092022.

Wang, J., M.-P. Kwan, and Y. Chai. 2018. An innovative context-based crystal-growth activity space method for environmental exposure assessment: A study using GIS and GPS trajectory data collected in Chicago. *International Journal of Environmental Research and Public Health* 15 (4):703. doi: 10.3390/ijerph15040703.

Wang, Q., M.-P. Kwan, K. Zhou, J. Fan, Y. Wang, and D. Zhan. 2019. Impacts of residential energy consumption on the health burden of household air pollution: Evidence from 135 countries. *Energy Policy* 128:284–95. doi: 10.1016/j.enpol.2018.12.037.

Wang, Y. P., and A. Murie. 2011. The new affordable and social housing provision system in China: Implications for comparative housing studies. *International Journal of Housing Policy* 11 (3):237–54. doi: 10.1080/14616718.2011.599130.

Yoo, E. H., C. Rudra, M. Glasgow, and L. Mu. 2015. Geospatial estimation of individual exposure to air pollutants: Moving from static monitoring to activity-based dynamic exposure assessment. *Annals of the Association of American Geographers* 105 (5):915–26. doi: 10.1080/00045608.2015.1054253.

Zhao, P., M.-P. Kwan, and S. Zhou. 2018. The uncertain geographic context problem in the analysis of the relationships between obesity and the built environment in Guangzhou. *International Journal of Environmental Research and Public Health* 15 (2):308. doi: 10.3390/ijerph15020308.

Zou, B., J. Wilson, F. Zhan, and Y. Zeng. 2009. Air pollution exposure assessment methods utilized in epidemiological studies. *Journal of Environmental Monitoring* 11 (3):475–90. doi: 10.1039/b813889c.

Individual Vacant House Detection in Very-High-Resolution Remote Sensing Images

Shengyuan Zou and Le Wang

The formation and demolition of vacant houses are the most visible sign of city shrinking and revitalization. Timely detection of vacant houses has become an inevitable task to aid the "Smart City" initiative. Two pressing problems exist for vacant houses, however: (1) No publicly accessible information is available at the individual house level and (2) the decennial census survey does not catch up with the rapidly changing status of vacant houses. To this end, remote sensing provides a low-cost avenue for detecting vacant houses. Traditionally, remote sensing was accredited for its success in deriving biophysical parameters of human settlements, such as the presence and physical size of buildings. It is still a challenge, though, to infer the functions of buildings, such as land-use types and occupancy status. In this study, we aim to detect individual vacant houses with very-high-resolution remote sensing images through a smart machine learning method. Our proposed method entails three steps: ground-truth data collection, classification, and feature selection. As a result, a new building change detection method was developed to collect ground-truth vacant house data from multitemporal images. Important features for classification of houses were identified. Subsequently, we carried out a classification of vacant houses and yielded promising results. Furthermore, the results indicate that both the area of the vacant house parcels and the healthy conditions of the surrounding vegetation contribute most to the detection accuracy. Our work shows the potential of using remote sensing to detect individual vacant houses at a large spatial extent. *Key Words: machine learning, remote sensing, smart city, vacant house.*

空置房屋的形成和拆除是城市萎缩和复兴的最明显标志。及时发现空置房屋已成为支持"智慧城市"倡议的一项重要任务。但空置房屋面临两个紧迫问题：（1）没有可公开获取的个人住房信息；（2）十年一次的人口普查跟不上空置住房迅速变化的形势。为此，遥感技术为发现空置房屋提供了一个低成本的途径。遥感技术在传统上用于获取人类住区生物物理参数，例如建筑物的存在和实际规模，在这方面的成功也受到各方认可。然而，推断建筑物的功能，如土地使用类型和入住情况仍面临着挑战。在本研究中，我们将通过智能机器学习方法，利用高分辨率遥感图像检测个人空置房屋。我们提出的方法需要三个步骤：基本事实数据收集、分类和特征选择。这是一种全新建筑物变化检测方法，用于从多时相图像中收集真实的空置房屋数据。首先要确定房屋分类的重要特征，对空置房屋进行分类。这方面取得了可喜的成果。此外，研究现实，空置地块的面积和周围植被的健康状况，对检测精度的贡献最大。我们的研究证明，以遥感技术在大空间范围内探测个人空置房屋具有重大潜力。关键词:*机器学习、遥感、智慧城市、空置房屋。*

La aparición de casas desocupadas y su demolición son uno de los signos más visibles del encogimiento y revitalización de las ciudades. La detección temprana de casas desocupadas se ha convertido en tarea inevitable para ayudar a la iniciativa de la "Ciudad Inteligente". Sin embargo, hay dos problemas apremiantes en relación con este fenómeno: (1) No existe información disponible para el público a nivel de casas individuales; y (2) el estudio del censo decenal no se mantiene al día con el estatus en cambio rápido de las casas desocupadas. A este respecto, la percepción remota brinda una alternativa de bajo costo para detectar casas desocupadas. Tradicionalmente, a la percepción remota se le reconoció el éxito de derivar parámetros biofísicos de los asentamientos humanos, tales como la presencia y tamaño físico de los edificios. No obstante, todavía es todo un reto inferir las funciones de los edificios, tales como el tipo de uso del suelo y estatus de ocupación. En este estudio, apuntamos a detectar casas desocupadas individuales por medio de imágenes de percepción remota de muy alta resolución usando un método de aprendizaje inteligente con máquinas. El método que proponemos supone tres pasos: recolección de datos reales en el terreno, clasificación y selección de rasgos. Como resultado, se desarrolló un nuevo método de detección de cambios en los edificios para recoger datos reales en el terreno sobre unidades desocupadas a partir de imágenes multitemporales. Se identificaron los rasgos considerados importantes para la clasificación de las casas. Enseguida, efectuamos una clasificación de las casas desocupadas con base en lo cual produjimos resultados prometedores. Lo que es más, los resultados indican que tanto el área del lote de las casas desocupadas como las condiciones sanitarias de la vegetación circundante aportan la mayor contribución en términos de la exactitud de la detección. Nuestro trabajo muestra el potencial de usar percepción remota para detectar individualmente casas desocupadas en extensiones espaciales grandes. *Palabras clave: aprendizaje con máquinas, casas desocupadas, ciudad inteligente, percepción remota.*

Vacant houses, which generally refers to residential houses that have been abandoned for several years, have become a prominent problem in the United States. As a market phenomenon, vacant houses are the end result of various social and economic forces (Accordino and Johnson 2000; Masson 2014). After World War II, the suburbanization of industry and population triggered abandonment of houses in the central cities (U.S. General Accounting Office 1978). Moreover, international competition and changing technology in the mid-1970s resulted in manufacturing job and population reduction in the inner-city areas, especially in the industrial cities of the Northeast and Midwest, thus exaggerated the abandonment (White 1986; Dewar and Thomas 2012). At the same time, the increasing use of high-risk mortgages in minority-race neighborhoods further aggravated the housing vacancy (Dewar and Thomas 2012). As a result, there were total 16,842,710 vacant properties across the United States in 2016 based on the American Community Survey (ACS) 5-Year Estimates, which caused severe problems. Associated with a weak housing market, limited access to health and education resources, and high crime rates, the concentration of vacant houses lowers the quality of life for remaining residents from economic, social, and security perspectives (Schweitzer, Kim, and Mackin 1999; Cohen 2001; Silverman, Yin, and Patterson 2013; Huuhka 2016). It should be noted that segregated African Americans in great poverty usually live in neighborhoods where vacant houses cluster. Therefore, better housing management in these neighborhoods is needed to ameliorate urban inequality and assist the revitalization of central cities in the United States (Silverman, Yin, and Patterson 2015; Lynch and Mosbah 2017), which calls for studies on vacant houses.

There are three existing public vacant house data sources at present:

1. *Census decennial survey data.* The smallest unit in the decennial census survey is the census block.
2. *ACS annual estimate data.* ACS data are confined to the census block group level with a very large margin of estimation error. Using these first two sources of vacant house data, various studies focused on the correlation with socioeconomic variables at an aggregate level. Housing vacancy rate has been used as a considerable factor in geographic analysis of disease (Green et al. 2003), rate of crime (Krivo and Peterson 1996), housing markets (Rosen and Smith 1983; Wheaton 1990), small-area population estimation (Silvan-Cardenas et al. 2010), and vegetation dynamics monitoring (Troy et al. 2007; Wang et al. 2016; Jia et al. 2019; Wang et al. 2019).
3. *Parcel data.* Public access to digital parcel data and updating frequency vary considerably from one city to another (Deng and Ma 2015) and thus could not catch up with the need for data quality for various research purposes and urban management in shrinking cities.

Even though individual-level vacant house data are essential but deficient, previous studies rarely focused on addressing this problem. In this regard, remote sensing provides an economic way of exploring vacant houses, due to its synoptic view and frequent updates. Long-term abandonment results in poorly maintained vegetation neighboring vacant houses and thus causes overgrowth or a dead state of vegetation (Ryznar and Wagner 2001; Hoalst-Pullen, Patterson, and Gatrell 2011; Pearsall and Christman 2012). Also, in the previous study on the relationship between socioeconomic characteristics and residential property vacancy, the estimated median housing value, which could be reflected in physical appearance of houses, is significantly correlated with the vacancy rate at the census tract level (Silverman, Yin, and Patterson 2013). Thus, unusual surrounding vegetation and physical condition indicate that housing vacancy is visible and mapping from the sky is possible. With the rapid development of sensors, very-high-resolution (VHR) images provide a great deal of spectral information, which makes it feasible to automatically explore vacant houses from remote sensing (Deng and Ma 2015).

Studying vacant houses through remote sensing images has barely been explored by scholars (Ryznar and Wagner 2001; Chen et al. 2015; Deng and Ma 2015; Post et al. 2015; Du et al. 2018). Nighttime imagery has been employed to estimate the vacancy rate at the city and metropolitan area levels (Yao and Li 2011; Chen et al. 2015). Using 1.0-km spatial resolution nighttime light images (Defense Meteorological Satellite Program-Operational Linescan System; DMSP-OLS), vacancy rate could only be derived at a very coarse spatial scale (city or metropolitan area scale) instead of at a fine spatial scale. Deng and Ma (2015) tried to predict residential vacancy at the individual level in Binghamton, New York, through remote sensing images and geographic information system (GIS) transportation variables. The only feature extracted by Deng and Ma from remote sensing images is mean value of the Normalized Difference Vegetation Index (NDVI) in each parcel, which represents the surrounding vegetation circumstance (Deng and Ma 2015). Whereas VHR images provide abundant spectral, spatial, geometric, and textural information, only one feature, mean NDVI, is not sufficient to express residential vacancy in the image.

It is difficult to generate a method to detect vacant houses from VHR images without solving two inevitable problems. The first is a lack of ground-truth data. In remote sensing studies, ground-truth data are necessary for the interpretation and analysis of what is being sensed (Hoffer 1972). Specifically, ground-truth data are needed as training and test samples in a machine learning approach to detect vacant houses. As we mentioned, though, there are no free public individual vacant house data in many cases. Second, there is a lack of studies to determine important features. VHR remote sensing images contain abundant spectral, spatial, geometric,

and textural information, but which category of information or which feature is effective in detection of vacant houses is unknown. Furthermore, obtaining effective preknowledge about features or patterns is difficult when distinguishing vacant houses even through visual interpretation. With these two existing problems, detecting vacant houses with remotely sensed images is a difficult challenge at present. The feasibility of detecting vacant houses in remote sensing images is still undetermined. Thus, we want to explore the possibilities while solving these two problems and to provide clues for further work.

The overall objective of this study is to estimate the feasibility of detecting vacant houses using VHR images. Corresponding to two existing problems, there are two specific objectives of this study: (1) automatically obtain vacant house ground-truth data as training and test samples through building change detection from multitemporal images and (2) determine important features in a machine learning classification.

Study Area and Data

Our study area is located in the city of Buffalo (42.8864° N, 78.8784° W) in Erie County, on the shores of Lake Erie in upstate New York (Figure 1).

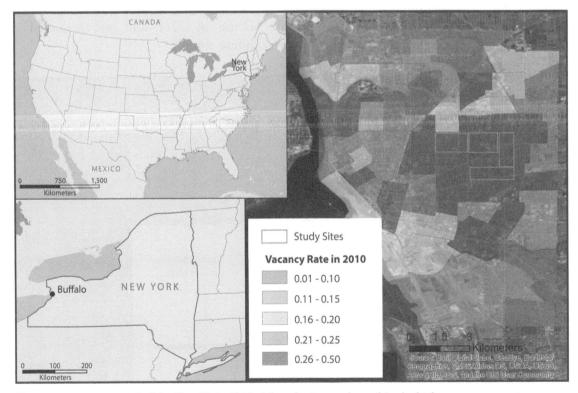

Figure 1. Housing vacancy rate in Buffalo, New York, 2010. All study sites are located in the high-vacancy-rate regions.

Buffalo suffered a decline from the 1970s to the 2000s, with a loss of over 70 percent of manufacturing jobs and 50 percent of population. In 2000, there were 8,684 abandoned properties in Buffalo. The city conservatively estimated its direct five-year costs from a vacant residential property at $20,060 for maintenance (Masson 2014) and announced a "5 in 5" demolition plan in 2007, a plan to demolish 5,000 vacant properties in five years (Brown 2007); the median cost of demolition of a vacant house was $16,989 in 2013 (Masson 2014). Thus, the vacancy rate in Buffalo reached a peak value of 22.8 percent in 2006 and then decreased. That peak rate also indicates that Buffalo was one of the cities with the highest vacancy rates in the United States in 2006 (Masson 2014). Specifically, four study sites are delineated as our study area in the city of Buffalo (Figure 1). All of them are in the suburban region in a high-vacancy-rate residential area (over 25 percent in 2010). Their area is from $0.72\,\text{km}^2$ to $1.16\,\text{km}^2$, with a total area of $3.98\,\text{km}^2$. Each of these four study sites contains a comparable number of houses, from 827 to 1,152 residential properties. The total number of residential houses in these four regions was 3,936 in 2011.

Our remotely sensed data are a digital orthoimage product from the New York State High Resolution Statewide Digital Orthoimagery Program, New York State Clearinghouse (see http://gis.ny.gov/gateway/mg/index.html; Figure 2). For Erie County, the orthoimagery has been updated every three years since 2002. In this study, orthoimages with 1-ft (30.48 cm) spatial resolution and four bands (natural color and near infrared) in 2011 and 2014 were employed. The projection system is the North American Datum of 1983, New York State Plane Coordinate System (measure in feet). Radiometric correction and registration were accomplished before the next image processing step started. Parcel data in 2012 were used to segment VHR images. With this predefined land lot boundary, houses and their corresponding surrounding contexts were integrated together to present the built environment (Figure 3). This saved considerable time and effort in image segmentation and building boundary delineation.

Method

Ground-Truth Data Collection (Building Change Detection)

According to the demolition plan, change trajectory of a building provides a more visible way to determine whether a house was vacant or not in the past. In this case, if a house appeared in the 2011 image and then disappeared in the 2014 image, it was determined to be a vacant house in 2011 and was collected as a vacant house sample.

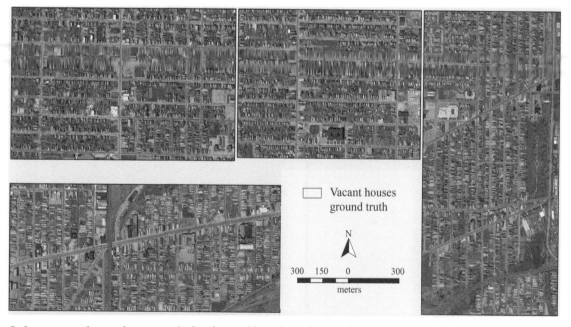

Figure 2. Orthoimages in four study sites overlaid with parcel boundary of vacant houses.

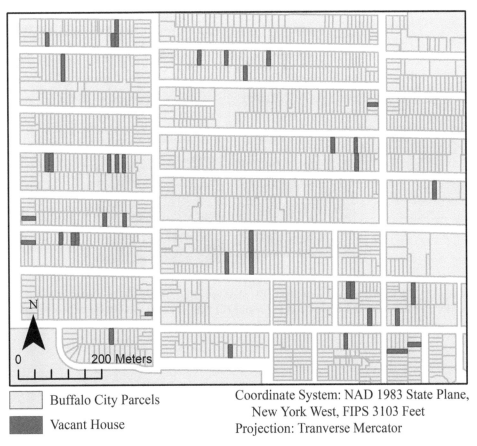

N

0   200 Meters

____ Buffalo City Parcels

▓ Vacant House

Coordinate System: NAD 1983 State Plane,
New York West, FIPS 3103 Feet
Projection: Tranverse Mercator

Figure 3. Parcel map and derived ground-truth vacant houses in Study Site 1. NAD = North American Datum; FIPS = Federal Information Processing Standard.

MBI and Shadow Extraction and Refinement. Among many building change detection methods, which is a developed topic in remote sensing (Maktav, Erbek, and Jürgens 2005), a morphological building index (MBI) proposed by Huang and Zhang (2011) achieved satisfactory accuracy rates in change detection (Huang, Zhang, and Zhu 2014). As an indicator of building presence, MBI has some weaknesses in detection: partially identified roof due to roof heterogeneity, dark or shadowed roof, and incorrectly detected roof due to bright vegetation and bare soil (Huang and Zhang 2011). Large dark shadows in urban areas are visible and detectable in VHR images and thus could perform as a supplementary building indicator in change detection (Sirmacek and Unsalan 2008; Huang and Zhang 2012). After calculating the MBI for each pixel, candidate buildings are extracted by setting a threshold on the MBI image. These candidate buildings are then refined by removing small areas as noise and eroding the MBI image using a 3×3 structure element to remove road-like narrow objects. To extract shadows from high-spatial-resolution images, threshold is an ideal method because of the spectral content of the images (Das and Aery 2013). Bimodal histogram splitting is a practical method to find a proper threshold to separate shadow and nonshadow areas in single-band, high-resolution image (Dare 2005). Candidate shadow is the joint shadow area in the red band and green band. Vegetation shadow and objects with low reflectance could be wrongly included in the candidate building shadow, however. Thus, candidate shadow is refined by removing the false building shadow. NDVI value and a 3×3 structure element are used to filter out dark vegetation and small dark objects. An example of the image processing results is shown in Figure 4.

Threshold. Shadow feature images and MBI feature images are segmented into parcels using the parcel boundaries in vector format. The number of shadow feature pixels, $S^{t1}(i)$, and the number of MBI feature pixels, $MBI^{t1}(i)$, at time $t1$ are calculated for

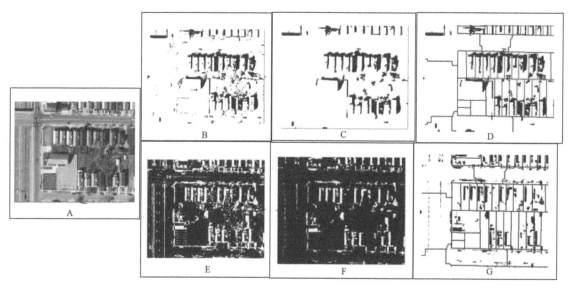

Figure 4. (A) Original high-spatial-resolution image. (B) Shadow extraction result from the red and green bands. (C) Shadow refinement result. (D) Segmentation and shadow shifting result. (E) MBI threshold result. (F) MBI refinement result. (G) MBI feature map segmentation result. MBI = morphological building index.

each parcel i. To detect a change in building presence, a threshold-based change detection method combining refined shadow and refined MBI is applied at the decision level. Two variables used to take the threshold are the amount of MBI feature and shadow feature pixels aggregating in each parcel.

The shadow and MBI criteria for a parcel with building demolition are shown in Equation 1. In contrast, Equation (2) are the criteria for unchanged buildings.

$$\left.\begin{array}{c} MBI^{t1}(i) \geq T(MBI)^{t1} \\ \text{or} \\ S^{t1}(i) \geq T(S)^{t1} \end{array}\right\} B^{t1}(i) = 1, \qquad (1)$$

$$\left.\begin{array}{c} MBI^{t2}(i) < T(MBI)^{t2} \\ \text{or} \\ S^{t2}(i) < T(S)^{t2} \end{array}\right\} B^{t2}(i) = 0. \qquad (2)$$

Four thresholds are set to $T(S)^{t1}$, $T(S)^{t2}$, $T(MBI)^{t1}$, and $T(MBI)^{t2}$ for shadow and MBI pixel numbers at times $t1$ and $t2$. $B^{t1}(i) = 1$ or 0 indicates building presence or absence at time $t1$ in parcel i. Thus, when

$$B^{t1}(i) - B^{t2}(i) = 1, \qquad (3)$$

the parcel i is detected as a parcel with a building demolished during the time between $t1$ and $t2$. Otherwise, parcel i is detected as a parcel without a building demolished between $t1$ and $t2$.

Feature Extraction

After collecting samples, we extract features for each vacant house sample from the VHR image in 2011. Features are distributed into three categories: spectral, geometric, and textural (Zheng, Wu, and Liu 2010; see Table 1). We aimed to include all features in the initial feature set in classification and then select features based on their performance. Among sixty-seven features, a majority (fifty-eight) are from remote sensing, spectral, and textural information. The remaining nine geometric features are from parcel boundary, which is relatively stable to the land use information within parcels. Traditionally, spectral signatures are the most basic feature for detection and classification starting from the pixel-based stage. In the object-based stage, spectral information is still important (Hu and Wang 2013), especially considering its excellent performance in differentiating vegetation and background (Blaschke 2010). Geometric features are widely used in object-based classification. Area, shape index, and density are important in differentiating land-use classes. Texture features reflect the texture information for the roof and vegetation, which might be two key indicators in vacant house detection. Texture can reflect the detailed pattern of the surface and help estimate slight changes in land cover.

Table 1. All sixty-seven features

Feature category	Feature name	Feature index
Spectral features (10)	Mean of B	17
	Mean of G	18
	Mean of R	20
	Mean of NIR	19
	SD of B	21
	SD of G	22
	SD of R	24
	SD of NIR	23
	Mean of brightness	16
	Mean of NDVI	59
Geometric features (9)	Area (m^2)	1
	Length (m)	43
	Width (m)	58
	Length/width	34
	Compactness	6
	Rectangular fit	48
	Shape index	15
	Density	53
	Main direction	25
Texture features for B, G, NIR, R (12 × 4 = 48)	GLCM homogeneity	64–67
	GLCM contrast	60–63
	GLCM dissimilarity	39–42
	GLCM entropy	49–52
	GLCM Ang.2nd moment	54–57
	GLCM mean	2–5
	GLCM SD	11–14
	GLCM correlation	35–38
	GLDV Ang.2nd moment	26–29
	GLDV entropy	44–47
	GLDV mean	30–33
	GLDV contrast	7–10

Notes: B = blue; G = green; R = red; NIR = near-infrared; NDVI = Normalized Difference Vegetation Index; GLCM = gray-level co-occurrence matrix; GLDV = gray-level difference vector. All of the textural features have one value for each band, which is calculated at all directions. Features are based on the aggregated pixel value in each parcel.

Feature Selection and Classification in Random Forest

Feature estimation and classification are achieved together in a random forest classifier (Breiman 2001) and implemented using a fivefold cross-validation. Out-of-bag (OOB) error is calculated for each feature and used to evaluate feature irreplaceability in classification. Exploring the redundant and effective features in the feature set, the top ten important features are selected as input features for the second time classification. Then, a multicollinearity test is performed to examine the multicollinearity issue among ten selected features and to refine the selected feature set. Based on their variance inflation

Table 2. Quality comparison for building demolition detection using MBI and shadow, and using MBI only

Method	Correctness	Completeness	Quality
MBI + shadow	90.0%	67.5%	62.8%
MBI	38.0%	43.2%	25.3%

Notes: MBI = morphological building index.

factor (VIF), which is used as the indicator of multi-collinearity, and their actual meaning, four features are selected as the final feature set for the third time classification. Accuracies of the second and third classifications are compared with the first one to estimate the efficiency of feature selection.

Results

Building Change Detection Result

Our proposed building demolition detection method was tested and compared with results using the MBI only (Table 2). One hundred and ten demolished houses were detected with 90 percent accuracy and were used as vacant house samples. In addition, 670 unchanged houses were randomly selected from the study area and used as occupied house samples.

Classification Result

Sixty-seven features were extracted from these 780 parcel objects from the VHR image in 2011 and, together with their feature set, were randomly and evenly divided into five parts. Each part includes twenty-two vacant houses and 134 occupied houses. For each time, one of five parts containing 20 percent of the samples was selected as the training set, and the remaining 80 percent of the samples, comprised the test data set. We then repeated the training and test course five times. First, random forest classification with fivefold cross-validation involved a total of sixty-seven features. We built 200 decision trees and selected five features for each tree. Combining the fivefold test result, thirty-four vacant houses were correctly detected among real vacant houses, and seventy-six were missed (Table 3). Of forty detected vacant houses, thirty-four were selected correctly and six were falsely detected. Vacant houses were detected with high accuracy (85.0 percent), which means that we were able to detect vacant houses correctly, although we were not able to cover most of the vacant house samples in our detection. Due to the large amount of occupied houses, overall accuracy is mainly attributed to the high accuracy in occupied houses. Thus, 99.1 percent of occupied houses could be classified correctly. The accuracy of vacant house detection was low, however, at 30.9 percent, which means that we missed 69.1 percent of vacant houses in the detection. The kappa coefficient was 0.41, representing a significantly better performance than random classification.

Feature Importance Estimation and Feature Selection

OOB error is shown in Figure 5. Features were ranked based on their OOB error and stratified. The ten most important features were selected to construct a refined feature set (Table 4). Student's t test was conducted to test whether the feature values of vacant houses and occupied houses were significantly different (from different normal distribution with

different mean value). The result of the t test shows that nine of ten important features had significantly different means in vacant houses and occupied houses within a 99 percent confidence interval. The remaining important feature, gray-level co-occurrence matrix (GLCM; Albregtsen 2008) entropy of near-infrared (NIR), which is a textural feature ranking ninth, also had significantly different means within a 90 percent confidence interval. All of them showed a statistically significant difference between vacant houses and occupied houses.

Of the ten important features, three are related to the area of the parcel. Interestingly, we found that smaller residential houses are more likely to be vacant. They usually have low prices because they were built en masse with small parcels and low costs. These smaller houses in areas of smaller parcels

Table 3. Confusion matrix for classification results with total of sixty-seven features

Classification	Vacant	Occupied	Total	User's accuracy
Vacant	34	6	40	85.0%
Occupied	76	664	740	89.7%
Total	110	670	780	
Producer's accuracy	30.9%	99.1%		Overall accuracy: 89.5%
				Kappa coefficient: 0.41

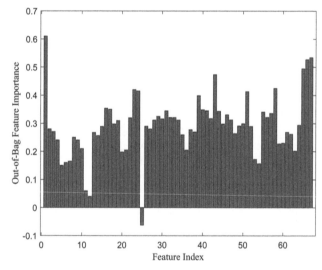

Figure 5. Out-of-bag feature importance among a total of sixty-seven features. Feature index refers to Table 1.

Table 4. Top ten important features in random forest classification and t test results

Rank	Feature name	Mean in vacant	Mean in occupied	Statistical significance test
1	Area	3,378	4,599	***
2	GLCM Homogeneity of NIR	0.124	0.157	***
3	GLCM Homogeneity of R	0.124	0.156	***
4	SD of NIR	31.69	36.46	***
5	Mean of NIR	119.5	114.3	***
6	Width	31.36	37.67	***
7	Length	111.2	127.9	***
8	SD of R	32.62	37.39	***
9	GLCM Entropy of NIR	8.085	8.131	*
10	Mean of R	105.4	100.3	***

Notes: GLCM = gray-level co-occurrence matrix; NIR = near-infrared; R = red.
***$p < 0.001$. **$p < 0.01$. *$p < 0.05$.

tended to be sold to low-income residents, who were engaged in manufacturing jobs that have disappeared in the last century, thus making them more likely to have abandoned their homes. In addition, spectral and textural features in the NIR band strongly suggest that vacancy status is related to the surrounding vegetation condition. Vacant house parcels have a higher value and lower standard deviation in the NIR band. The textural features in NIR bands show that textures appear less complex in the vacant house parcels than in the occupied house parcels. Vegetation has relatively high reflectance in NIR in urban areas. This finding is consistent with and expands on previous work (Ryznar and Wagner 2001; Hoalst-Pullen, Patterson, and Gatrell 2011; Pearsall and Christman 2012; Deng and Ma 2015). The lower standard deviation in the NIR bands and less complexity in GLCM indicate less active vegetation in the vacant house parcel. Features in the R band imply the relationship between concrete roads and vacancy status. Concrete roads generally have high reflectance in the R band (Herold, Gardner,

and Roberts 2003) and could be eroded and destroyed by overgrowing vegetation. Compared with the first time, the accuracy of the second measurement increased slightly, although fifty-seven of sixty-seven features were removed from the feature set. As listed in Table 5, high overall accuracy was maintained at 89.6 percent when producer's accuracy improved from 30.9 percent to 42.7 percent. The kappa coefficient was improved to 0.48, which indicates that classification with selected features performed better than that with all features.

A multicolliearity test was implemented for these ten features to refine them. Table 6 shows the correlation coefficients between selected features. The VIF was 82.9, showing significant multicollinearity issue in this feature set. Thus, we divided the ten features into four groups based on the correlation. If the correlation coefficient between two features was over 0.8, these two features were assigned to one group. Using this criterion, four groups were formed as follows: (1) area, length, and width; (2) mean of NIR and mean of R; (3). standard deviation of NIR

Table 5. Confusion matrix for classification results with ten selected features

Classification	Vacant	Occupied	Total	User's accuracy
Vacant	47	18	65	72.3%
Occupied	63	652	715	91.2%
Total	110	670	780	
Producer's accuracy	42.7%	97.3%		Overall accuracy: 89.6% Kappa coefficient: 0.48

Table 6. Correlation coefficient matrix of ten selected features

	Area	Length	Width	Mean of NIR	Mean of R	SD of NIR	SD of R	GLCM entropy of NIR	GLCM homogeneity of NIR	GLCM homogeneity of R
Area	1.00	0.68	0.75	0.24	−0.14	−0.11	−0.29	0.11	−0.01	−0.06
Length	0.68	1.00	0.12	0.21	−0.06	−0.21	−0.33	0.16	−0.12	−0.16
Width	0.75	0.12	1.00	0.18	−0.14	0.01	−0.14	0.03	0.08	0.04
Mean of NIR	0.24	0.21	0.18	1.00	0.76	−0.43	−0.24	0.56	−0.59	−0.56
Mean of R	−0.14	−0.06	−0.14	0.76	1.00	−0.39	0.04	0.54	−0.59	−0.51
SD of NIR	−0.11	−0.21	0.01	−0.43	−0.39	1.00	0.76	−0.47	0.76	0.73
SD of R	−0.29	−0.33	−0.14	−0.24	0.04	0.76	1.00	−0.29	0.51	0.60
GLCM entropy of NIR	0.11	0.16	0.03	0.56	0.54	−0.47	−0.29	1.00	−0.88	−0.88
GLCM homogeneity of NIR	−0.01	−0.12	0.08	−0.59	−0.59	0.76	0.51	−0.88	1.00	0.98
GLCM homogeneity of R	−0.06	−0.16	0.04	−0.56	−0.51	0.73	0.60	−0.88	0.98	1.00

Notes: NIR = near-infrared; R = red; GLCM = gray-level co-occurrence matrix.

Table 7. Confusion matrix for classification results with four features

Classification	Vacant	Occupied	Total	User's accuracy
Vacant	43	22	65	66.2%
Occupied	67	648	715	90.6%
Total	110	670	780	
Producer's accuracy	39.1%	96.7%	Overall accuracy: 88.6%	
			Kappa coefficient: 0.43	

and standard deviation of R; and (4) GLCM entropy of NIR, GLCM homogeneity of NIR, and GLCM homogeneity of R. It should be noted that for each correlation group, features have similar actual meanings: features about the parcel size in the first group; mean of two spectral bands, R and NIR, in the second group; standard deviation of two spectral bands, R and NIR, in the third group; and textural features of R and NIR in the fourth group. We selected one feature from each group to be the representative feature and the final feature set included four features. The representative feature would have low correlation coefficients compared to features in other groups. The final feature set included area, mean of R, standard deviation of R, and GLCM entropy of NIR. The VIF was 1.67, indicating no significant multicollinearity problem in this feature set. With the final feature set, we classified the houses again and their result is shown in Table 7. Compared with results using ten features, the classification using the final feature set had poorer performance but was still better than the classification using sixty-seven features. With the least number of features, the third time classification had good overall accuracy (88.6 percent) but declining accuracy for vacant houses, at 39.1 percent for producer's accuracy and 66.2 percent for user's accuracy. The kappa coefficient was 0.43, which was better than the first classification (0.41) but lower than the second classification (0.48).

Discussion

To our knowledge, this is the first attempt to detect vacant houses from remote sensing imagery. Classification and accompanying feature analysis represent an initial attempt to identify the important features in vacant house detection. The important

features obtained by this approach are valuable when referring to the previous survey and inference.

The overall accuracy in all three classifications was over 88 percent, which is a good classification result. Because we had 670 occupied house samples and only 110 vacant house samples, the major contribution of this result is for occupied houses, because both producer's accuracy and user's accuracy of occupied houses were high. Regarding vacant houses, the user's accuracy was 85.0 percent using sixty-seven features, 72.3 percent using ten features, and 66.2 percent using four features. This indicates that the vacant houses we detected were more likely correct. We missed many vacant houses in the detection, however; the producer's accuracy was 30.9 percent using sixty-seven features, 42.7 percent using ten features, and 39.1 percent using four features. The low producer's accuracies indicate that using VHR images and the proposed method was unable to completely detect vacant houses. Referring to the definition of vacant houses, we only included the demolished vacant houses as samples, because usually they are long-term abandoned, vacant houses, resulting in bias in the sample set. In addition, we only included single-family residential properties by filtering out other properties before implementing the proposed method. Thus, the proposed method was only tested on single-family, long-term abandoned, vacant properties in this study.

In another finding, more complicated and comprehensive features were able to be used in classification. In this study, as a beginning stage, we used sixty-seven remote sensing features. Based on our result, spectral and textural features in R and NIR bands were more important than NDVI in feature selection. In general, it can be observed that vegetation indexes, such as the NDVI, do not have a standard universal value, with research often showing different results (Bannari et al. 1995). In further work, more complicated features such as time-series spectral and textural features of vegetation and geometric features of houses' footprints could also be introduced to improve the classification performance.

The proposed method for detecting house vacancy status in this article required only VHR images and parcel boundaries. Considering the popularizing of VHR imagery data as the only required updated data source, this method is practical when generalizing to other study areas with high vacancy rates.

Furthermore, the proposed method is almost fully automatic and easily implemented for smart city needs. Demolition has been a common strategy in many cities to deal with the vacancy problem since 1978. Having the ability to capture ground-truth vacant houses, the proposed method has broad potential to be promoted throughout the United States. Other data sources could be involved to help determine occupancy status, e.g., measurements of racial equity, poverty rate, and estimated house value (Silverman, Yin, and Patterson 2013; Deng and Ma 2015). One potential data source is Google Street View (GSV) images. As an emerging data source, GSV images provide another method to view the street-level community, including the housing environment.

Vacant house detection using remote sensing images also has broader applications. With individual vacant house data support, we can more intelligently manage urban living environments. There are two major avenues for further applications.

1. *Small-area population estimation.* A method developed by Deng, Wu, and Wang (2010) was used to estimate small-area population in regions with a constant vacancy rate, like Austin, Texas. When assigning population into residential properties in a high-vacancy-rate area, however, housing vacancy is a significant problem. The vacant house detection method introduced in this article provides a solution for this problem. With this updated detailed vacancy status input, estimated population at a large scale could be assigned accurately to census blocks or individual houses. This would be a breakthrough with respect to generalizing developed small-area population methods in shrinking cities.

2. *House price estimation.* Vacant houses are an important factor in the housing market, not only by negatively affecting neighborhood house prices but also by determining the expected length of time for sale (Han 2014). Individual residential vacancy status was rarely addressed, however, and thus needs to be remedied in house price modeling (Glaeser and Nathanson 2017). In addition, the change in vacant houses is essential in understanding the mechanisms at work in the housing market. With a strong theoretical relationship between vacancy and price (Wheaton 1990), competitive supply in the housing market explains the existence of longer run "structural" vacancy. Detailed information on vacant houses makes experiments accessible, whereas the manual acquisition of detailed residential status is extremely time consuming, tedious, and subjective.

Conclusion

Vacant houses play an important role in urban shrinking and revitalization approaches. As a significant sign of population loss, vacant houses have negative effects on a neighborhood in economic, security, and urban management terms. This is the first attempt to detect vacancy status at the individual house level using remote sensing data. In this study, we proposed a detection method for individual vacant houses with remote sensing images. A new building change detection method was developed to collect ground-truth data. A machine learning algorithm, random forest, was used to classify house occupancy status. Important features were selected to refine the classification and were proven to be effective. Specifically, area of parcels and spectral and textural information in R and NIR bands were indispensable when distinguishing vacant houses from occupied houses in VHR images. Furthermore, our result implies that lower house prices, abnormal surrounding vegetation, and concrete roads are related to housing vacancy status. Remote sensing could not detect the occupancy status directly but could detect the secondary indicators of long-term vacancy to help manage housing environments in urban areas.

Acknowledgements

We gratefully acknowledge the Editor, David R. Butler, and three anonymous reviewers for their valuable insights and constructive suggestions made to improve this article.

References

Accordino, J., and G. T. Johnson. 2000. Addressing the vacant and abandoned property problem. *Journal of Urban Affairs* 22 (3):301–15. doi: 10.1111/0735-2166.00058.

Albregtsen, F. 2008. *Statistical texture measures computed from gray level coocurrence matrices*. Image Processing Laboratory, Department of Informatics, University of Oslo.

Bannari, A., D. Morin, F. Bonn, and A. R. Huete. 1995. A review of vegetation indices. *Remote Sensing Reviews* 13 (1–2):95–120. doi: 10.1080/02757259509532298.

Blaschke, T. 2010. Object based image analysis for remote sensing. *ISPRS Journal of Photogrammetry and Remote Sensing* 65 (1):2–16. doi: 10.1016/j.isprsjprs.2009.06.004.

Breiman, L. 2001. Random forests. *Machine Learning* 45 (1):5–32. doi: 10.1023/A:1010933404324.

Brown, B. W. 2007. *Mayor Brown's "5 in 5" demolition plan*. City of Buffalo, Buffalo, NY.

Chen, Z., B. Yu, Y. Hu, C. Huang, K. Shi, and J. Wu. 2015. Estimating house vacancy rate in metropolitan areas using NPP-VIIRS nighttime light composite data. *IEEE Journal of Selected Topics in Applied Earth Observations and Remote Sensing* 8 (5):2188–97. doi: 10.1109/JSTARS.2015.2418201.

Cohen, J. R. 2001. Abandoned housing: Exploring lessons from Baltimore. *Housing Policy Debate* 12 (3):415–48. doi: 10.1080/10511482.2001.9521413.

Dare, P. M. 2005. Shadow analysis in high-resolution satellite imagery of urban areas. *Photogrammetric Engineering & Remote Sensing* 71 (2):169–77. doi: 10.14358/PERS.71.2.169.

Das, S., and A. Aery. 2013. A review: Shadow detection and shadow removal from images. *International Journal of Engineering Trends and Technology (IJETT)* 4 (5):1764–67.

Deng, C., and J. Ma. 2015. Viewing urban decay from the sky: A multi-scale analysis of residential vacancy in a shrinking U.S. city. *Landscape and Urban Planning* 141:88–99. doi: 10.1016/j.landurbplan.2015.05.002.

Deng, C., C. Wu, and L. Wang. 2010. Improving the housing-unit method for small-area population estimation using remote-sensing and GIS information. *International Journal of Remote Sensing* 31 (21):5673–88. doi: 10.1080/01431161.2010.496806.

Dewar, M., and J. M. Thomas, eds. 2012. *The city after abandonment*. Philadelphia, PA: University of Pennsylvania Press.

Du, M., L. Wang, S. Zou, and C. Shi. 2018. Modeling the census tract level housing vacancy rate with the Jilin1-03 satellite and other geospatial data. *Remote Sensing* 10 (12):1920. doi: 10.3390/rs10121920.

Glaeser, E. L., and C. G. Nathanson. 2017. An extrapolative model of house price dynamics. *Journal of Financial Economics* 126 (1):147–70. doi: 10.1016/j.jfineco.2017.06.012.

Green, C., R. D. Hanna, T. K. Young, and J. F. Blanchard. 2003. Geographic analysis of diabetes prevalence in an urban area. *Social Science & Medicine* 57 (3):551–60. doi: 10.1016/S0277-9536(02)00380-5.

Han, H.-S. 2014. The impact of abandoned properties on nearby property values. *Housing Policy Debate* 24 (2):311–34. doi: 10.1080/10511482.2013.832350.

Herold, M., M. E. Gardner, and D. A. Roberts. 2003. Spectral resolution requirements for mapping urban areas. *IEEE Transactions on Geoscience and Remote Sensing* 41 (9):1907–19. doi: 10.1109/TGRS.2003.815238.

Hoalst-Pullen, N., M. W. Patterson, and J. D. Gatrell. 2011. Empty spaces: Neighbourhood change and the greening of Detroit, 1975–2005. *Geocarto International* 26 (6):417–34. doi: 10.1080/10106049.2011.585439.

Hoffer, R. M. 1972. The importance of ground truth data in remote sensing. The Laboratory for Applications of Remote Sensing, Purdue University. https://ntrs.nasa.gov/archive/nasa/casi.ntrs.nasa.gov/19730007768.pdf

Hu, S., and L. Wang. 2013. Automated urban land-use classification with remote sensing. *International Journal of Remote Sensing* 34 (3): 790–803. doi: 10.1080/01431161.2012.714510.

Huang, X., and L. Zhang. 2011. A multidirectional and multiscale morphological index for automatic building extraction from multispectral GeoEye-1 imagery. *Photogrammetric Engineering & Remote Sensing* 77 (7):721–32. doi: 10.14358/PERS.77.7.721.

Huang, X., and L. Zhang. 2012. Morphological building/shadow index for building extraction from high-resolution imagery over urban areas. *IEEE Journal of Selected Topics in Applied Earth Observations and Remote Sensing* 5 (1):161–72. doi: 10.1109/JSTARS.2011.2168195.

Huang, X., L. Zhang, and T. Zhu. 2014. Building change detection from multitemporal high-resolution remotely sensed images based on a morphological building index. *IEEE Journal of Selected Topics in Applied Earth Observations and Remote Sensing* 7 (1):105–15.

Huuhka, S. 2016. Vacant residential buildings as potential reserves: A geographical and statistical study. *Building Research & Information* 44 (8):816–39. doi: 10.1080/09613218.2016.1107316.

Jia, M., Z. Wang, C. Wang, D. Mao, and Y. Zhang. 2019. A new vegetation index to detect periodically submerged Mangrove forest using single-tide sentinel-2 imagery. *Remote Sensing* 11 (17): 2043. doi: 10.3390/rs11172043.

Krivo, L. J., and R. D. Peterson. 1996. Extremely disadvantaged neighborhoods and urban crime. *Social Forces* 75 (2):619–48. doi: 10.2307/2580416.

Lynch, A. J., and S. M. Mosbah. 2017. Improving local measures of sustainability: A study of built-environment indicators in the United States. *Cities* 60:301–13. doi: 10.1016/j.cities.2016.09.011.

Maktav, D., F. S. Erbek, and C. Jürgens. 2005. Remote sensing of urban areas. *International Journal of Remote Sensing* 26 (4):655–69. doi: 10.1080/01431160512331316469.

Masson, M. J. 2014. Vacant and abandoned housing in Buffalo. Buffalo Commons Program, Cornell University ILR School. https://digitalcommons.ilr.cornell.edu/cgi/viewcontent.cgireferer=https://scholar.google.com/&httpsredir=1&article=1263&context=buffalocommons

Pearsall, H., and Z. Christman. 2012. Tree-lined lanes or vacant lots? Evaluating non-stationarity between urban greenness and socio-economic conditions in Philadelphia, Pennsylvania, USA at multiple scales. *Applied Geography* 35 (1–2):257–64. doi: 10.1016/j.apgeog.2012.07.006.

Post, C., B. Ritter, E. Akturk, A. Breedlove, R. Buchanan, C. Che, J. Fravel, et al. 2015. Analysis of factors contributing to abandoned residential developments using remote sensing and geographic information systems (GIS). *Urban Ecosystems* 18 (3):701–13. doi: 10.1007/s11252-014-0424-6.

Rosen, K. T., and L. B. Smith. 1983. The price-adjustment process for rental housing and the natural vacancy rate. *The American Economic Review* 73 (4):779–86.

Ryznar, R. M., and T. W. Wagner. 2001. Using remotely sensed imagery to detect urban change: Viewing Detroit from space. *Journal of the American Planning Association* 67 (3):327–36. doi: 10.1080/01944360108976239.

Schweitzer, J. H., J. W. Kim, and J. R. Mackin. 1999. The impact of the built environment on crime and fear of crime in urban neighborhoods. *Journal of Urban Technology* 6 (3):59–73. doi: 10.1080/10630739983588.

Silvan-Cardenas, J. L., L. Wang, P. Rogerson, C. Wu, T. Feng, and B. D. Kamphaus. 2010. Assessing fine-spatial-resolution remote sensing for small-area population estimation. *International Journal of Remote Sensing* 31 (21):5605–34. doi: 10.1080/01431161.2010.496800.

Silverman, R. M., L. Yin, and K. L. Patterson. 2013. Dawn of the dead city: An exploratory analysis of vacant addresses in Buffalo, NY 2008–2010. *Journal of Urban Affairs* 35 (2):131–52. doi: 10.1111/j.1467-9906.2012.00627.x.

Silverman, R. M., L. Yin, and K. L. Patterson. 2015. Municipal property acquisition patterns in a shrinking city: Evidence for the persistence of an urban growth paradigm in Buffalo, NY. *Cogent Social Sciences* 1 (1):1012973. doi: 10.1080/23311886.2015.1012973.

Sirmacek, B., and C. Unsalan. 2008. Building detection from aerial images using invariant color features and shadow information. Paper presented at the 2008 23rd International Symposium on Computer and Information Sciences, Suleyman Demirel Cultural Center, Istanbul Technical University, Istanbul, Turkey.

Troy, A. R., J. M. Grove, J. P. M. O'Neil-Dunne, S. T. A. Pickett, and M. L. Cadenasso. 2007. Predicting opportunities for greening and patterns of vegetation on private urban lands. *Environmental Management* 40 (3):394–412. doi: 10.1007/s00267-006-0112-2.

U.S. General Accounting Office. 1978. *Housing abandonment: A national problem needing new approaches.* Washington, DC: U.S. General Accounting Office.

Wang, L., M. Jia, D. Yin, and J. Tian. 2019. A review of remote sensing for mangrove forests: 1956–2018. *Remote Sensing of Environment* 231: 111223. doi: 10.1016/j.rse.2019.111223.

Wang, L., C. Shi, C. Diao, W. Ji, and D. Yin. 2016. A survey of methods incorporating spatial information in image classification and spectral unmixing. *International Journal of Remote Sensing* 37 (16): 3870–910. doi: 10.1080/01431161.2016.1204032.

Wheaton, W. C. 1990. Vacancy, search, and prices in a housing market matching model. *Journal of Political Economy* 98 (6):1270–92. doi: 10.1086/261734.

White, M. J. 1986. Property taxes and urban housing abandonment. *Journal of Urban Economics* 20 (3):312–30. doi: 10.1016/0094-1190(86)90022-7.

Yao, Y., and Y. Li. 2011. House vacancy at urban areas in China with nocturnal light data of DMSP-OLS. In *2011 IEEE International Conference on Spatial Data Mining and Geographical Knowledge Services (ICSDM),* 457–62. Fuzhou, China: IEEE.

Zheng, Y., F. D. Wu, and Y. F. Liu. 2010. A feature analysis approach for object-oriented classification. *Geography and Geo-Information Science* 26:19–23.

The Missing Parts from Social Media–Enabled Smart Cities: Who, Where, When, and What?

Yihong Yuan, ⓘ Yongmei Lu, ⓘ T. Edwin Chow, ⓘ Chao Ye, Abdullatif Alyaqout, and Yu Liu ⓘ

Social networking sites (SNS), such as Facebook and Twitter, have attracted users worldwide by providing a means to communicate and share opinions and experiences of daily lives. When empowered by pervasive location acquisition technologies, location-based social media (LBSM) has become a potential resource for smart city applications to characterize social perceptions of place and model human activities. There is a lack of systematic examination of the representativeness of LBSM data, though. If LBSM data are applied to decision making in smart city services, such as emergency response or transportation, it is essential to understand their limitations to implement better policies or management practices. This study formalizes the sampling biases of LBSM data from various perspectives, including sociodemographic, spatiotemporal, and semantic. This article examines LBSM data representativeness issues using empirical cases and discusses the impacts on smart city applications. The results provide insights for understanding the limitations of LBSM data for smart city applications and for developing mitigation approaches. *Key Words: data quality, location-based social media, sampling biases, smart city.*

诸如脸书和推特等社群网站（SNS），已通过提供沟通和分享日常生活意见与经验的工具，吸引全世界的使用者。以位置为基础的社交媒体（LBSM），因无所不在的位置取得科技成为可能，并成为智慧城市应用来描绘地方的社会观感并模式化人类活动的潜在资源。但我们却缺乏LBSM数据再现的系统性分析。若将LBSM数据应用于诸如紧急应变或交通等智慧城市服务的决策中，则我必须理解其实施更佳的政策或管理实践之限制。本研究对各种观点的LBSM数据抽样误差进行正式化，包括社会人口、时空和语义。本文运用经验案例，检视LBSM数据再现的议题，并探讨其对智慧城市应用的影响。研究结果对于理解LBSM数据之于智慧城市应用的限制和发展缓解方法提出洞见。关键词：数据质量，以位置为基础的社交媒体，抽样误差，智慧城市。

Los sitios de socialización en red (SNS), tales como Facebook y Twitter, atraen usuarios de todo el mundo, al proveer un medio de comunicarse y compartir opiniones y experiencias de las vidas cotidianas. Al ser empoderados con las ubicuas tecnologías de adquisición locacional, los medios sociales basados en localización (LBSM) se han convertido en un recurso potencial para que las aplicaciones de la ciudad inteligente caractericen las percepciones sociales del lugar y modelen las actividades humanas. Hay, no obstante, carencia evidente de un examen sistemático sobre la representatividad de los datos de los LBSM. Si los datos de los LBSM se aplican en la toma de decisiones de los servicios en las ciudades inteligentes, tales como la respuesta a las emergencias o al transporte, es esencial entender sus limitaciones para implementar mejor políticas o prácticas de manejo. Este estudio formaliza los sesgos de la muestra de los datos de los LBSM desde varias perspectivas, incluyendo las sociodemográficas, las espaciotemporales y las semánticas. Este artículo examina aspectos relacionados con la representatividad de los datos de los LBSM usando casos empíricos y discute los impactos sobre las aplicaciones de la ciudad inteligente. Los resultados suministran perspicacias para entender las limitaciones de los datos de los LBSM para las aplicaciones de la ciudad inteligente y para desarrollar enfoques mitigantes. *Palabras clave: calidad de los datos, ciudad inteligente, medios de comunicación sociales basados en localización, sesgos de muestra.*

The White House launched the Smart Cities Initiative in 2015 to support cities, federal agencies, universities, and the private sector in developing new technologies that make cities more inhabitable and equitable (The White House 2016). The following years witnessed a number of federal agencies, private companies, and nonprofits join the table and provide financial support for smart city development. There are several key technologies for smart city applications: (1) sensor-enabled physical infrastructure that provides real-time monitoring of urban resources; (2) communication infrastructure that connects the deployed sensors (e.g., the Internet of Things [IoT]); and (3) big data generated by the various sensors and the associated new theories and applications (Hancke, Silva, and Hancke 2013). These technologies are often inseparable. For instance, the massive amount of data generated by sensors has contributed to the rise of big data (Batty 2013), which in turn expanded the definition of sensing technologies beyond just physical sensors (e.g., Bluetooth sensors). The increase of social networking sites (SNS) where people can share their social lives has introduced new opportunities to monitor individuals' activities and the perception of their surroundings. Researchers have defined location-based social media (LBSM) as "social network sites that include location information" (Roick and Heuser 2013, 764). LBSM has been widely used as potential resources to characterize social perceptions of places and to model human activities in various applications. Innovative concepts such as human sensing and social sensing were introduced into sensing technologies to refer to human observations of both physical and social geographies (Calabrese, Ferrari, and Blondel 2015; Liu et al. 2015).

Like other types of big (geo) data, though, LBSM data have various data quality issues, such as accuracy, precision, completeness, and representativeness (Shi et al. 2018; Yuan, Wei, and Lu 2018). Different SNS tend to attract certain population groups and support the sharing of particular content, making them limited in data representation (Golub and Jackson 2010). In other words, biased sampling (e.g., demographically, spatially, temporally, and semantically) naturally leads to data representativeness issues. If LBSM data are applied to decision making in smart city services, understanding the sampling biases of such data is critical for implementing better policies or management practices. This study examines the representativeness issues of LBSM data caused by sampling biases from sociodemographic, spatiotemporal, and semantic perspectives. The terms *data representativeness issues* and *sampling biases* are used interchangeably in the rest of this article. The main objective is providing a framework to examine LBSM-enabled smart city services and their limitations. We discuss the representativeness of LBSM data and their impacts on smart city applications by incorporating empirical analyses. The results provide valuable inputs for understanding how LBSM sampling biases could manifest themselves in smart city applications.

Challenges for Social Media–Enabled Smart Cities

The implementation of a smart city requires the integration of three essential components: advanced information and communication technologies (ICTs), open governance, and resident-centered services; the third component is often overlooked in real-world smart city services. In other words, there is a tendency to overemphasize the merit of technology in smart city services, even as the core purposes and functions of city operations are ignored (Kitchin 2015). The central goal of a city is to ensure the life quality of its residents through the management, preparation, and delivery of resources and services. To this end, ICTs form the technical backbone for smart cities, the service aspect serves as the ultimate goal, and the open governance aspect provides the means to achieve the goal. A smart city application or service that is built only on the merit of technologies without paying attention to people would risk disconnecting the service from its users. The disconnection challenges go beyond identifying what services are needed; they include where, when, and by whom a service is needed.

In academia, the discussions about smart cities reflect a broad spectrum of views. Although a technocratic approach is not uncommon (e.g., Maeda 2012), researchers recognize the inherent comprehensive characteristics of smart cities (Perera et al. 2014). Harrison et al. (2010) emphasized that smart cities should successfully connect different infrastructures of a city—the physical infrastructure, the information technology infrastructure, the social infrastructure, and the business infrastructure. Mohanty, Choppali, and Kougianos (2016) argued that a smart city is a system of systems where IoT and big data improve a city's operation and help it fulfill its objective of improving life quality. After discussing how big data technologies can support different smart city applications, Al Nuaimi et al. (2015) explored a number of open issues, including the role of social media for smart city applications and its ramifications, how differing levels of access to information affect an individual's power and political position, and the effectiveness and quality of smart city applications.

To ensure efficient services, smart city applications need to be built on real-time measurements and massive data collection. LBSM data, as a complement to traditional sensors, are particularly useful due to their uniqueness in recording human experiences and behaviors at fine spatiotemporal resolutions (Doran et al. 2016). Taking advantage of LBSM data, urban studies have examined spatiotemporal dynamics of cities while seeking insights into the social, cultural, and political aspects of urban life (Hochman and Manovich 2013; Licoppe 2016; Cabalquinto 2018).

Social media data are not universally representative, however. Studies have examined demographic bias of social media data (e.g., Sloan et al. 2015; Yuan, Wei, and Lu 2018). Social and political inequity was found to not only perpetuate the use of social media but also feed back into people's usage of urban space (Boy and Uitermark 2017). In addition to reflecting human experiences, social media might affect human experiences and opinions of space (Evans and Saker 2017). Hence, it is crucial to understand the limitations of LBSM data when applying them to smart city applications.

The Missing Parts from Social Media Data for Smart Cities

Researchers have identified five Ws and one H (who, where, when, what, why, and how) in social media studies (Khosrow-Pour 2018). *Who* refers to the challenges of identifying user groups on social media and evaluating the data quality associated with biased sampling (Longley and Adnan 2016). *Where* and *when* identify the spatiotemporal patterns from social media content, which are the most crucial aspects of LBSM data for geographic information science (GIScience; Zhang et al. 2016). *What* focuses on mining the semantics of user-generated content for urban planning and e-governance (Hu 2018). *How* and *why* focus on the underlying processes within the scheme of social media, such as "How does a social network form on SNS?" and "What are the motivations of SNS users?" Although how and why questions help to understand the theoretical foundations of social media, most social media applications focus on the first four Ws (who, where, when, what; Khosrow-Pour 2018). The rest of this article focuses on the first four Ws to demonstrate the representativeness issues of LBSM for smart city applications.

Who Is Reflected by LBSM Data?

LBSM users are not a random sample in terms of their social, economic, and demographic backgrounds (Golub and Jackson 2010). Pinterest, for instance, particularly attracts women between the ages of twenty-five and thirty-four with average household incomes of $100,000. One crucial challenge in quantifying these biases is obtaining accurate data on social media users because many SNS do not require users to provide personal information. Salganik (2018) discussed common characteristics of big data, such as the lack of demographic information and the representativeness of the data. Previous research either conducted user surveys or harvested user profiles or posts to infer their demographics (Longley and Adnan 2016). A survey-based study by Zickuhr (2013) found that LBSM use in the United States is not equal across age, gender, and race and that young people, women, and ethnic minorities have a greater LBSM presence. Recent studies in computational social science reported similar findings about the biases of digital trace data and the importance of combining such data with traditional survey data (e.g., Foster 2017).

Naturally, such demographic biases might impact the reliability of applying LBSM to urban services. For example, Rizwan et al. (2018) found that check-ins from female users were more spread out in the city, whereas check-ins from male users showed a more clustered pattern in centralized districts. Zhong et al. (2015) identified the connection between user demographic factors (e.g., age, gender, and income) and their points of interest (POIs) check-ins. They constructed a model to effectively predict the demographics of Weibo users based on their check-in time and location.

Our case study shows how the senior population (age sixty-five and older) in China is systematically underrepresented on Weibo (Figure 1). Using a random sample of 230,000 Weibo users who checked in their locations at least once between March and November 2015, Figure 1 displays their distribution using an index of underrepresentativeness, $I_{UR}(i)$,

$$I_{UR}(i) = \log\left(\frac{P_C(i)}{P_W(i)}\right), \quad (1)$$

where $P_C(i)$ and $P_W(i)$ represent the percentage of demographic group i (e.g., seniors) in the census data and Weibo data, respectively. A positive value

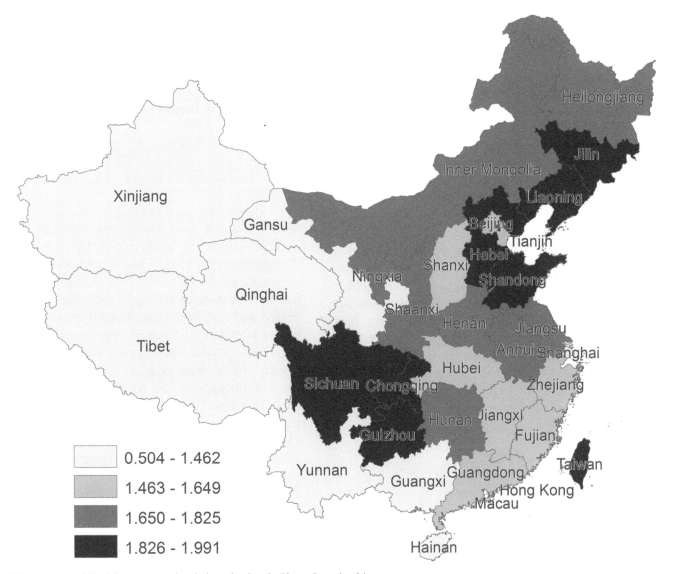

Figure 1. I_{UR} (65+) by province (excluding the South China Sea islands).

indicates that a demographic group is underrepresented in Weibo. The lower the value, the better this demographic group is represented.

This underrepresentation of seniors on Weibo demonstrates a strong regional pattern. For instance, the provinces in northeast China show a significantly clustered pattern of senior underrepresentation (Moran's I z value = 2.20, $p < 0.05$). Another cluster of underrepresentation is in southwest China (e.g., Sichuan, Chongqing, Guizhou). A deeper understanding of the underlying mechanisms for this pattern requires a comprehensive analysis of the factors affecting senior citizen usage of LBSM, such as cultural backgrounds, economic development, resources allocated to support seniors, and so on, which is beyond the scope of

this article. Although this example focused on provincial-level patterns, understanding this unbalanced population representation in social media data is essential for LBSM-enabled smart city applications to effectively serve residents. Future studies should conduct a city- or subcity-level analysis to explore how spatial scales and the modifiable areal unit problem (MAUP) could affect the results.

Despite the challenges in mitigating LBSM demographic biases, it still provides a valuable data source for smart city applications. Salganik (2018) pointed out that even though social scientists are more used to probabilistic random samples from a well-defined population, nonrepresentative data can still provide valuable insights, especially in the exploratory stage of outlier patterns and causations. Therefore, if city officials were to rely on nonrepresentative social media data to engage a broader audience in urban planning and infrastructure renovation, it would be important to identify suitable research questions. For example, it is feasible to answer questions like "Are there abnormal spatial clusters in the city during a musical festival?" based on LBSM data; however, questions like "What is the average number of people impacted by Hurricane Harvey in each county?" requires more representative data and cannot be answered solely based on social media data.

In addition, it is possible to adjust the sampling process through methods like adopting stratified sampling or combining LBSM samples with other public survey data, which would help mitigate the influence of LBSM user sampling biases. Previous studies have applied machine learning algorithms to estimate LBSM users' demographic information, such as age and gender, based on their profile information and the semantics of their posts (Longley and Adnan 2016). Another option is to generate stratified samples across space based on census data (e.g., income).

Where and When Are Things Happening on LBSM?

The public sector has used spatiotemporal information from LBSM data to model human activities in various smart city applications. For example, the Livehoods Project aims to better understand the dynamics of urban dwellers and reimage cities using LBSM data (Cranshaw et al. 2012). Studies have demonstrated, however, that check-in data tend to cluster in certain areas, causing biased profiling of activities across space and through time (Sloan et al. 2015). For example, Bawa-Cavia (2011) identified social hubs (i.e., where social media users are more likely to generate a high density of activities) in London, New York, and Paris. Sun and Paule (2017) also identified that highly rated restaurants are more likely to cluster spatially and receive more ratings on Yelp. Hence, there are inherent biases and representativeness issues in the spatiotemporal data acquired from LBSM.

Austin, Texas (ATX), is among the U.S. cities that actively pursue smart city development (City of Austin 2017). Using a four-month Twitter data set from January to April 2016, we calculated the number of geotagged tweets by census tracts and correlated that with census data to evaluate social media usage in different urban areas (P_{LBSM}). Here we use the residential population data from the 2010 to 2014 American Community Survey conducted by the U.S. Census Bureau (2015).

$$P_{LBSM} = \frac{F_{LBSM}}{\text{Population}}, \qquad (2)$$

where F_{LBSM} is the frequency of geotagged tweets in an area.

As shown in Figure 2, P_{LBSM} is not spatially uniform across the city. Toward the city center (the dark polygons), the amount of LBSM check-ins is disproportionally high. This is potentially due to geographic distribution of POIs in Austin, such as the bars and restaurants on 6th Street, the convention center, and the Texas State Capitol. Another cluster of check-ins is at the Austin Airport (marked by an arrow). It is clear that certain districts in central Austin attract more people to check in; therefore, these locations tend to be oversampled in LBSM data. Therefore, it is essential to quantify such spatial biases and properly adjust the representativeness of LBSM data when developing smart city applications, such as emergency response or transportation, to represent human mobility (Liu et al. 2014).

To analyze the temporal patterns of LBSM, we aggregated that same Twitter data set from Austin by recording the number of geotagged tweets for each hour. To validate the human mobility pattern, we used Bluetooth data collected by ATX along major streets and freeways and aggregated them for each hour during the same period. This data set captures the presence of Bluetooth enabled devices when they pass by a receiver. Although Bluetooth data do not measure physical movement of a well-sampled population perfectly, they capture the "naturally occurring" physical movement better than LBSM does. Due to the lack of ground truth data for human mobility patterns, we used Bluetooth data as a proxy for physical movement in this study. Both data sets are normalized to the range [0, 1] and can be considered an indication of hourly human activity reflected by LBSM and Bluetooth, respectively. There is a clear time lag where the peak activity for Twitter data is a few hours after that of Bluetooth data (Figure 3). This demonstrates the biases of LBSM data in reflecting the temporal patterns of human activities, indicating that there is a discrepancy between LBSM

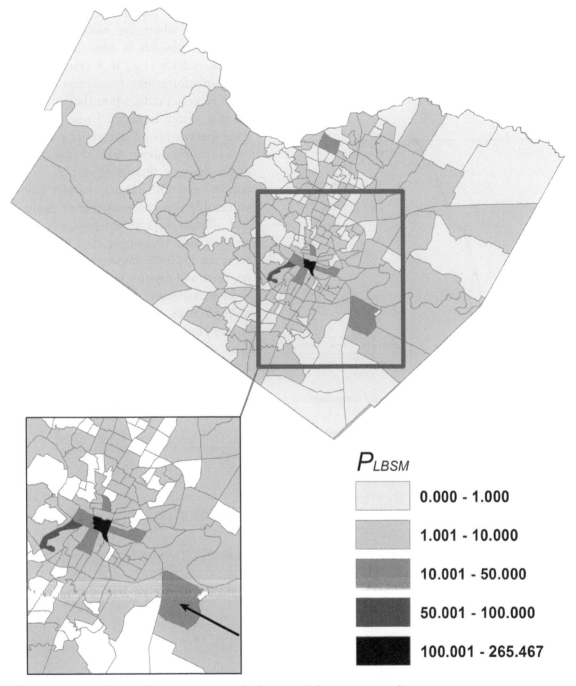

Figure 2. The distribution of P_{LBSM} (the arrow points to the location of the Austin airport).

posts and when the activity referenced by the posts took place. It is likely that LBSM users did not post on LBSM when they were rushing to work, which potentially caused the time lag between when the events occurred and the posting on LBSM of those events. This type of temporal bias has crucial implications for smart city services that respond to real-time mobility patterns of urban dwellers, such as transportation and event planning.

The where and when challenges are beyond simple sampling biases. Locations and time stamps from LBSM data might have different levels of accuracy; space-based geotagged posts with precise x- and y-coordinates are more accurate than place-based posts using a descriptive of or reference to a loosely defined location. For example, Houston can refer to its downtown area, the centroid of the city, or anywhere within the city limit as determined by that social media platform. In the context of multimedia, whether it is a picture or video, the area of interest (AOI) can be captured either on scene (i.e., where it was posted) or off scene (i.e., a certain distance away from the post location). The resulting on- or off-scene capture can be attributed to the time lag

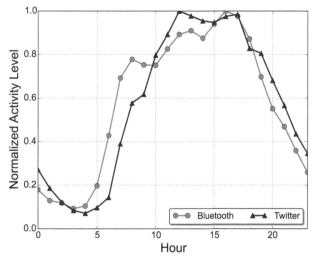

Figure 3. Twitter and Bluetooth hourly activity comparison.

between data capture and posting, which is not uncommon when the multimedia file is large and mobile bandwidth is limited or when the user is moving quickly (e.g., in a vehicle). Moreover, social media platforms like Foursquare allow users to check in to receive points when they are within the vicinity of a certain location. These practices might introduce spatiotemporal uncertainties into LBSM data.

As people share critical information on social media during a disaster (Smith et al. 2017), first responders and GIScientists harvest LBSM to identify vulnerable populations, conduct damage assessments, and allocate appropriate resources for disaster relief and response (Fohringer et al. 2015). Thus, location accuracy is vitally important for effective emergency response and disaster management.

In a case study that harnessed tweets and crowd-sourced data containing water-depth information during Hurricane Harvey in 2017, more than 95 percent of relevant posts contained multimedia (i.e., text-only posts account for ~5 percent). Among all collected tweets, 244 geotagged tweets had a valid location, either represented by a latitude–longitude

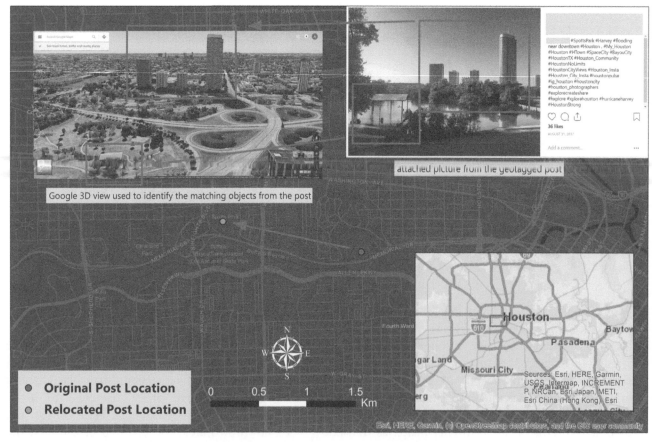

Figure 4. Example of an off-scene post with the inferred original and relocated post locations.

Table 1. Statistical summary (in meters) of on-scene and off-scene spatial error

Post location	Count of posts	Minimum	Maximum	M	SD
On-scene	137 (33)	0 (33.3)	499.5	47.7 (198.1)	106.8 (132.4)
Off-scene	107	512.8	34,175.7	5,218.9	6,618.7

Notes: The numbers in parentheses of on-scene posts exclude the 104 posts with an accurate coordinate from a Global Positioning System–enabled smart phone. See text for more details.

pair or by selecting a place with predefined coordinates (e.g., University of Houston, Downtown Houston, etc.). The geographic features and landmarks captured in the pictures and videos were compared with those found in Google Maps' Street View or 3D views. There were 107 (~44 percent) and 137 (~56 percent) off- and on-scene posts, respectively. In this study, off-scene posts were defined as those located 500 m away from the AOI reference location. By comparing the location of the AOI and the corresponding post location shared on SNS, the distance offset between them was calculated as spatial error (Figure 4). During Hurricane Harvey, the mean spatial errors of geotagged posts related to water depth were 47.7 m (or 198.1 m, excluding 104 posts that had accurate Global Positioning System [GPS]-derived coordinates) and 5,219 m for on- and off-scene posts, respectively (Table 1). This finding supports the cautious use of LBSM and the importance of reporting spatial accuracy for emergency planning and disaster management.

What Do People Talk about on LBSM?

In addition to biases in the who, where, and when aspects, semantic biases from LBSM are worth noting. The content of social media is closely related to the functionalities and characteristics of each SNS platform (Morstatter et al. 2013). Inevitably, various biases exist when conducting sentiment analysis, public opinion collection, and topic extraction from such data sets. Instead of expressing opinions on public matters such as traffic, politics, or urban planning, social media users are more willing to publicly discuss topics related to their personal life (e.g., leisure activities; Lansley and Longley 2016). Although most people, LBSM users included, spend most of their time around a few key locations, it remains difficult to identify the associated land use at these frequently visited locations just by examining the semantics of social media posts (Soliman et al. 2017). Previous findings reported a sentimental

bias in which people are more likely to post on social media under the influence of positive emotions (Mitchell et al. 2013). Hence, if policymakers aim to collect opinions on city services, it is crucial to understand the nature, popularity, and associated sentiment of various topics on social media.

In English, verb–noun phrases (e.g., "attend a wedding") are often used to describe human behavior and activities, but complete verb–noun phrases are not common on the Chinese language Web site Weibo. In a study analyzing the geotagged Weibo posts in Beijing from January 2016 to January 2017 (Figure 5), we used verbs instead of verb–noun phrases to examine activity space. The top ten verbs extracted from Weibo (translated to English) are eat [吃] (195,623), sleep [睡] (45,923), buy [买] (38,449), encourage [加油] (27,252), take pictures [拍(照)] (26,943), have fun [玩] (23,311), work [上班] (19,634), study [学习] (19,323), deliver [送(货)] (16,053), and stroll [逛] (12,888). These verbs correspond to different types of activities, such as employment, leisure, education, and so on. We used fuzzy c-means to cluster the verbs into the following categories: work, study, daily life, leisure, and others. The fuzzy c-means algorithm generates a membership degree representing how well each verb fits into a certain category. The final results in each category include verbs with a membership degree larger than 0.8. Although there might be outliers due to the fuzziness of the algorithm, this method has proven to be effective in categorizing words based on their semantics (Cao, Song, and Bruza 2004). Several verbs in the "others" category, such as prevent [防治] or bribe [贿赂], can potentially be related to public issues; however, these verbs make up less than 1 percent of the entire sample set. Word clouds of two types of activities (work and daily life) are shown in Figure 6. The nonverbs in the word clouds are due to the difference between Chinese and English.

We found that certain types of activities, such as industrial activities (e.g., manufacturing) or political discussions, are rarely discussed on Weibo because social media users prefer to share topics related to

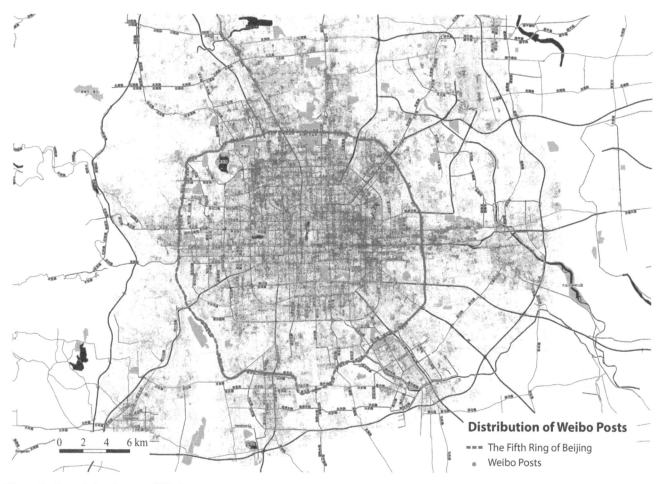

Figure 5. Spatial distribution of Weibo posts.

Figure 6. Word clouds: (A) Work. (B) Daily life.

their daily lives. Another example of thematic bias can be illustrated by a search for keywords like "traffic accident" or "car accident" from Twitter data collected between May 2015 and April 2016 in Austin. The results only yielded 388 records, and 383 of them were from a verified public account, Total Traffic Austin, which is owned by a private company in the business of traffic and weather broadcasting. This demonstrates the lack of discussion of certain issues on SNS, which brings

challenges to the application of LBSM data to smart city services that requires topic extractions.

Discussion and Conclusion

LBSM provides rich data sources for modeling human activities and capturing citizens' perceptions in the age of instant access. This emerging data source also brings challenges for smart city services, however. Understanding these challenges is crucial

Table 2.   Example smart city services and applications based on location-based social media data

Application	Reasoning based on the four Ws
Identify hotspots of city night life	a. Social media particularly attracts young people (Who). b. It is more likely for users to post during leisure activities (What).
Analyze tourist behaviors at transportation hubs (e.g., an airport)	a. Users are more likely to check in at certain locations, such as arriving at an airport or a train station (Where). b. Users are more likely to post to social media when they travel (What).
Model user behaviors during national holidays (e.g., the spring festival in China)	a. Users are more likely to post on social media during holidays (When). b. Users are more likely to travel to new destinations during holidays (Where). c. Users are more likely to conduct leisure activities during holidays (What).

for developing meaningful LBSM-enabled smart city services. This research takes an initial step of formalizing these challenges into a framework of four critical aspects (who, where, when, and what). Nevertheless, there are other challenges that are not fully addressed in this article.

- *Other data quality issues of LBSM:* Although we briefly touched on accuracy and precision, our research focuses on the representativeness of LBSM. In addition to the four Ws, other data quality issues can affect smart city services. First, most LBSM application program interfaces (APIs), including Twitter and Weibo APIs, are only able to obtain around 1 percent of all geo-located posts, raising questions about data completeness. Second, fake check-ins and bots are inevitable issues on LBSM. Without a valid method to address this issue, LBSM-enabled smart city applications could be misguided. Third, the demographic profiles of LBSM users are mostly estimated by algorithms or from self-reported data, and it is a challenge to validate the credibility of such information. Fourth, LBSM data are biased toward overrepresenting "central users" from the perspective of the communication network, where a small group of users generate a disproportionate amount of data. Questions like, "Do we have overrepresentative data from central users in a social network?" are essential to assess LBSM data quality and the scientific rigor of experimental design. Finally, social media platforms are driven by technological advancements and fast-changing culture, so it is important to consider how LBSM data biases evolve. For example, the once-popular

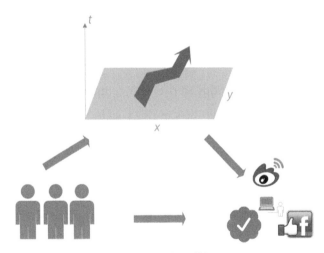

Figure 7. Interactions among the four Ws.

image and video hosting service, Flickr, lost many active users after changing ownership several times, which raises concerns regarding the representativeness of Flickr data.

- *Distinctions between smart city services:* Each smart city service has its own objectives and functionalities, which naturally leads to different data needs. For example, a service that aims to collect public opinions on urban infrastructure will be more sensitive to semantic biases, whereas a service that is designed to respond to certain urban events might be affected more by the spatiotemporal bias embedded in LBSM data. Policymakers should carefully investigate whether and to what degree a smart city service could be influenced by LBSM data biases based on the four Ws discussed in this article. Table 2 lists several sample city services that are particularly well suited to rely on LBSM data. Note that the reasons listed in Table 2 are only hypotheses, which should be carefully tested when developing smart city services.

It is important to note that the four Ws in this study are often inseparable. Specifically, the who aspect (i.e., user group biases) contributes to both spatiotemporal biases (where and when) and semantic biases (what). For example, SNS tend to attract young people, who have their own preferred check-in locations and topics to discuss on social media. Figure 7 expands the social sensing framework discussed in Liu et al. (2015) and illustrates how these four Ws interact with each other.

- Suppose there are n demographic groups, and the number of people in each demographic group is $[U_1, U_2, \ldots U_n]$. The participation rate of each demographic group on LBSM is $[p_1, p_2, \ldots p_n]$. The total LBSM sample S can be calculated as

$$S = \sum_{i=1}^{n} U_i * p_i. \qquad (3)$$

This corresponds to the who component.

- Assume that users in this sample S are conducting m types of activities in real life (e.g., work, leisure, study, etc.), where the number of activities in each type is denoted by $[T_1, T_2, \ldots T_m]$, and the probability of each activity type getting posted on social media is $[q_1, q_2, \ldots q_m]$. As a result, the number of activities reflected on social media can be noted as

$$D = \sum_{j=1}^{m} T_j * q_j \qquad (4)$$

where q_j is highly dependent on the nature of each activity (i.e., the what component), and T_j is determined by sample S from the previous step.

- From a spatiotemporal perspective, activities conducted by users in S are unevenly distributed across space and time. The probability density of type j activity happening in a spatiotemporal unit (x, y, t) can be represented as $v(x, y, t, j)$, where x, y, and t represent latitude, longitude, and time, respectively. The probability of a type j event happening at (x, y, t) appearing on social media is proportional to $v(x, y, t, j)* q_j$. This demonstrates the various factors that might affect LBSM data quality, including the users who are posting in this unit (i.e., the sample S), the specific

location and time of this unit (x, y, t), and the type of activities being conducted in this unit (q_j).

To sum up, user sampling bias is the foundation of social media biases. In the meantime, activities conducted by these users distribute unevenly across space and time. Furthermore, these unevenly distributed activities also have different likelihoods of being posted to social media. Therefore, we should consider the four Ws in a synergistic way when developing smart city services.

Despite a lack of solutions to fully address or quantify these deficiencies of LBSM data, there are several ways to mitigate the potential problem. First, LBSM data can always be supplemented or corroborated by other data sources, such as census data and survey data, to improve the representativeness of LBSM samples. Second, it is important to identify target user groups from SNS data. Although user sample biases are inevitable, researchers can still extract the most representative groups on different SNS sites and design their research objectives according to the user groups available. Third, due to the low spatiotemporal sampling resolution of LBSM data, it is necessary to reevaluate the validity of classic mobility models, measurements, and algorithms when applied to such data sets.

The contribution of this research is twofold. First, we provided a research framework to better understand the relationship between LBSM data and smart city applications through its limitation in reflecting human activities. The results also lay the groundwork for future efforts of applying LBSM data to various smart city services, such as real-time traffic management, early warning systems, and emergency responses. Second, the case studies provided empirical support to quantify the biases of two widely used LBSM data sets (Weibo from China and Twitter from the United States) from four perspectives (who, where, when, and what). The results of this research also provide a reference for policymakers and aid their efforts in applying LBSM data to city services. Future research can focus on extending this framework and identifying the distinctions of biases for different LBSM platforms.

Acknowledgments

Our deep and sincere thanks go to Dr. Ling Bian and Jennifer Cassidento for the tremendous amount of work they put into organizing this special issue. We thank the anonymous reviewers for their constructive comments, which greatly improved the content and clarity of this article. Lei Zhang helped improve the grammar and style of this work.

ORCID

Yihong Yuan (iD) https://orcid.org/0000-0001-6266-9744
Yongmei Lu (iD) https://orcid.org/0000-0003-1994-3458
T. Edwin Chow (iD) https://orcid.org/0000-0002-0386-5902
Yu Liu (iD) http://orcid.org/0000-0002-0016-2902

References

Al Nuaimi, E., H. A. Neyadi, N. Mohamed, and J. Al-Jaroodi. 2015. Applications of big data to smart cities. *Journal of Internet Services and Applications* 6 (1):25.

Batty, M. 2013. Big data, smart cities and city planning. *Dialogues in Human Geography* 3 (3):274–79. doi: 10.1177/2043820613513390.

Bawa-Cavia, A. 2011. Sensing the urban: Using location-based social network data in urban analysis. Paper presented at the First Workshop on Pervasive Urban Applications (PURBA), San Francisco, CA, June 12.

Boy, J. D., and J. Uitermark. 2017. Reassembling the city through Instagram. *Transactions of the Institute of British Geographers* 42 (4):612–24. doi: 10.1111/tran.12185.

Cabalquinto, E. C. 2018. Smartphones as locative media. *Information Communication & Society* 21 (12):1751–54. doi: 10.1080/1369118X.2018.1429481.

Calabrese, F., L. Ferrari, and V. D. Blondel. 2015. Urban sensing using mobile phone network data: A survey of research. *ACM Computing Surveys* 47 (2):1–20. doi: 10.1145/2655691.

Cao, G., D. Song, and P. Bruza. 2004. Fuzzy k-means clustering on a high dimensional semantic space. Paper presented at Advanced Web Technologies and Applications, Berlin, Germany, April 14–17.

City of Austin. 2017. Smart cities strategic roadmap update. Accessed November 20, 2018. http://www.austintexas.gov/edims/document.cfm?id=287471.

Cranshaw, J., R. Schwarts, J. Hong, and N. Sadeh. 2012. The Livehoods Project: Utilizing social media to understand the dynamics of a city. Paper presented at the Sixth International AAAI Conference on Webpages and Social Media, Dublin, Ireland, June 4–7.

Doran, D., K. Severin, S. Gokhale, and A. Dagnino. 2016. Social media enabled human sensing for smart cities. *AI Communications* 29 (1):57–75. doi: 10.3233/AIC-150683.

Evans, L., and M. Saker. 2017. *Location-based social media: Space, time and identity.* Cham, Switzerland: Palgrave Macmillan.

Fohringer, J., D. Dransch, H. Kreibich, and K. Schroter. 2015. Social media as an information source for rapid flood inundation mapping. *Natural Hazards and Earth System Sciences* 15 (12):2725–38. doi: 10.5194/nhess-15-2725-2015.

Foster, I. 2017. *Big data and social science: A practical guide to methods and tools.* Boca Raton, FL: CRC.

Golub, B., and M. O. Jackson. 2010. Naive learning in social networks and the wisdom of crowds. *American Economic Journal of Microeconomics* 2 (1):112–49. doi: 10.1257/mic.2.1.112.

Hancke, G., B. Silva, and G. Hancke, Jr. 2013. The role of advanced sensing in smart cities. *Sensors* 13 (1):393–425. doi: 10.3390/s130100393.

Harrison, C., B. Eckman, R. Hamilton, P. Hartswick, J. Kalagnanam, J. Paraszczak, and P. Williams. 2010. Foundations for smarter cities. *IBM Journal of Research and Development* 54 (4):1–16. doi: 10.1147/JRD.2010.2048257.

Hochman, N., and L. Manovich. 2013. Zooming into an Instagram city: Reading the local through social media. Accessed October 10, 2018. https://firstmonday.org/ojs/index.php/fm/rt/printerFriendly/4711/3698.

Hu, Y. 2018. Geo-text data and data-driven geospatial semantics. *Geography Compass* 12 (11):e12404. doi: 10.1111/gec3.12404.

Khosrow-Pour, M. 2018. *Encyclopedia of information science and technology.* 4th ed. Hershey, PA: IGI Global/Engineering Science Reference.

Kitchin, R. 2015. Making sense of smart cities: Addressing present shortcomings. *Cambridge Journal of Regions, Economy and Society* 8 (1):131–36. doi: 10.1093/cjres/rsu027.

Lansley, G., and P. A. Longley. 2016. The geography of Twitter topics in London. *Computers, Environment, and Urban Systems* 58:85–96. doi: 10.1016/j.compenvurbsys.2016.04.002.

Licoppe, C. 2016. Mobilities and urban encounters in public places in the age of locative media: Seams, folds, and encounters with "pseudonymous strangers." *Mobilities* 11 (1):99–116. doi: 10.1080/17450101.2015.1097035.

Liu, Y., X. Liu, S. Gao, L. Gong, C. Kang, Y. Zhi, G. Chi, and L. Shi. 2015. Social sensing: A new approach to understanding our socioeconomic environments. *Annals of the Association of American Geographers* 105 (3):512–30. doi: 10.1080/00045608.2015.1018773.

Liu, Y., Z. W. Sui, C. G. Kang, and Y. Gao. 2014. Uncovering patterns of inter-urban trip and spatial interaction from social Media check-in data. *PLoS ONE* 9 (1):e86026. doi: 10.1371/journal.pone.0086026.

Longley, P. A., and M. Adnan. 2016. Geo-temporal Twitter demographics. *International Journal of Geographical Information Science* 30 (2):369–89. doi: 10.1080/13658816.2015.1089441.

Maeda, A. 2012. Technology innovations for smart cities. Paper presented at 2012 Symposium on VLSI Circuits (VLSIC), Honolulu, HI, 13–15 June.

Mitchell, L., M. R. Frank, K. D. Harris, P. S. Dodds, and C. M. Danforth. 2013. The geography of happiness: Connecting Twitter sentiment and expression, demographics, and objective characteristics of place. *PLoS ONE* 8 (5):e64417. doi: 10.1371/journal.pone.0064417.

Mohanty, S. P., U. Choppali, and E. Kougianos. 2016. Everything you wanted to know about smart cities: The Internet of things is the backbone. *IEEE Consumer Electronics Magazine* 5 (3):60–70. doi: 10.1109/MCE.2016.2556879.

Morstatter, F., J. Pfeffer, H. Liu, and K. M. Carley. 2013. Is the sample good enough? Comparing data from Twitter's streaming API with Twitter's firehose. Paper presented at the Seventh International AAAI Conference on Weblogs and Social Media (ICWSM-13), Cambridge, MA, July 8–11.

Perera, C., A. Zaslavsky, P. Christen, and D. Georgakopoulos. 2014. Sensing as a service model for smart cities supported by Internet of things. *Transactions on Emerging Telecommunications Technologies* 25 (1):81–93. doi: 10.1002/ett.2704.

Rizwan, M., W. Wan, O. Cervantes, and L. Gwiazdzinski. 2018. Using location-based social media data to observe check-in behavior and gender difference: Bringing Weibo data into play. *ISPRS International Journal of Geo-Information* 7 (5):196. doi: 10.3390/ijgi7050196.

Roick, O., and S. Heuser. 2013. Location based social networks—Definition, current state of the art and research agenda. *Transactions in GIS* 17 (5):763–84.

Salganik, M. J. 2018. *Bit by bit: Social research in the digital age.* Princeton, NJ: Princeton University Press.

Shi, W., A. Zhang, X. Zhou, and M. Zhang. 2018. Challenges and prospects of uncertainties in spatial big data analytics. *Annals of the American Association of Geographers* 108 (6):1513–20. doi: 10.1080/24694452.2017.1421898.

Sloan, L., J. Morgan, P. Burnap, and M. Williams. 2015. Who tweets? Deriving the demographic characteristics of age, occupation and social class from Twitter user meta-data. *PLoS ONE* 10 (3):e0115545. doi: 10.1371/journal.pone.0115545.

Smith, L., Q. Liang, P. James, and W. Lin. 2017. Assessing the utility of social media as a data source for flood risk management using a real-time modelling framework. *Journal of Flood Risk Management* 10 (3):370–80. doi: 10.1111/jfr3.12154.

Soliman, A., K. Soltani, J. Yin, A. Padmanabhan, and S. Wang. 2017. Social sensing of urban land use based on analysis of Twitter users' mobility patterns. *PLoS ONE* 12 (7):e0181657. doi: 10.1371/journal.pone.0181657.

Sun, Y., and J. D. G. Paule. 2017. Spatial analysis of users-generated ratings of Yelp venues. *Open Geospatial Data, Software and Standards* 2 (1):5. doi: 10.1186/s40965-017-0020-9.

U.S. Census Bureau. 2015. American Community Survey. Accessed September 25, 2018. https://www.census.gov/programs-surveys/acs/technical-documentation/table-and-geography-changes/2014/5-year.html.

The White House. 2016. Fact sheet: Announcing over $80 million in new federal investment and a doubling of participating communities in the White House smart cities initiative. Accessed November 11, 2018. https://obamawhitehouse.archives.gov/the-press-office/2016/09/26/fact-sheet-announcing-over-80-million-new-federal-investment-and.

Yuan, Y., G. Wei, and Y. Lu. 2018. Evaluating gender representativeness of location-based social media: A case study of Weibo. *Annals of GIS* 24 (3):163–76. doi: 10.1080/19475683.2018.1471518.

Zhang, W., B. Derudder, J. Wang, W. Shen, and F. Witlox. 2016. Using location-based social media to chart the patterns of people moving between cities: The case of Weibo-users in the Yangtze River delta. *Journal of Urban Technology* 23 (3):91–111. doi: 10.1080/10630732.2016.1177259.

Zhong, Y., N. J. Yuan, W. Zhong, F. Zhang, and X. Xie. 2015. You are where you go: Inferring demographic attributes from location check-ins. Paper presented at the Eighth ACM International Conference on Web Search and Data Mining, Shanghai, China, February 2–6.

Zickuhr, K. 2013. *Location-based services.* Washington, DC: Pew Research Center's Internet & American Life Project, Pew Research Center.

PART III
Critical Smartness

The Smart City Conundrum for Social Justice: Youth Perspectives on Digital Technologies and Urban Transformations

Michele Masucci, Hamil Pearsall, ⓘ and Alan Wiig ⓘ

This article employs a social justice framing to examine youth perspectives of the smart city. We examine how youth understand the impact of digital technologies on urban transformations and whether their technology skills and digital literacy give them a sense of ownership over the future of their city. Research was conducted within the context of a six-week summer educational program involving seventy-nine youth of color from public high schools in Philadelphia, Pennsylvania. The program mixed digital skill building with urban fieldwork to prototype solutions to long-standing urban problems: the sort of problems that smart city policies also seek to change. Our research points to a conundrum for youth. Although they embraced technological innovations, they indicated that digital technologies failed to serve the public or address pressing concerns they identified as problematic within the city: crime, drugs, and homelessness. Instead, in their view, digital technologies delivered the most benefit to private spaces in the home and workplace. Furthermore, the youth did not envision that emergent technologies would improve their neighborhoods or communities but only their employment prospects. This research suggests that the emergent smart city is reproducing actual as well as perceived urban inequities: Wealthy residential neighborhoods and spaces of the new economy become "smart," but much of the city remains left behind. These patterns create a paradox for youth who invest in digital skills while remaining on the margins of technology-driven, smart urban change. *Key Words: digital divide, Philadelphia, smart city, social justice, youth.*

本文运用社会正义的架构, 检视青年对智能城市的观点。我们检视青年如何理解数码科技对于城市转变的冲击, 以及他们的科技技能和数码智识能否为其带来拥有自身城市的未来之意识。本研究在六週的夏季教育计画脉络中进行, 并包含宾州费城公立高中的七十九位少数族裔青年。该计画混合数码技能发展与城市田野工作, 为长期存在的城市问题提出模范解方——这些问题类别亦为智能城市政策所寻求改变的。我们的研究指出年轻人所面临的难题。仅管他们拥抱科技创新, 却表示数码科技无法服务公众或应对他们指认为城市中的问题之急迫考量: 犯罪、毒品与无家可归。反之, 在他们的观点中, 数码科技对于家或工作场所等私人空间的助益最大。此外, 青年并不展望浮现中的技术能够改进他们的邻里或社区, 而只会增进其就业前景。本研究指出, 浮现中的数码城市正在再生产实际与认知的城市不均: 富裕的住宅邻里与新经济空间成为"智能的", 但城市大半部分却仍被抛诸脑后。这些模式, 为投资数码技能、同时位于技术驱动和智能城市变迁边缘的青年创造了矛盾。关键词: 数码落差, 费城, 智能城市, 社会正义, 青年。

Este artículo emplea un marco de justicia social para examinar las perspectivas que tienen los jóvenes de la ciudad inteligente. Exploramos el modo como los jóvenes entienden el impacto de las tecnologías digitales sobre las transformaciones urbanas, y si sus habilidades tecnológicas y capacidad digital les proporcionan un sentido de propiedad en relación con el futuro de su ciudad. La investigación se llevó a cabo en el contexto de un programa educativo de verano de seis semanas en el que se involucraron setenta y nueve jóvenes de color de las escuelas públicas de Filadelfia, Pensilvania. El programa mezcló la construcción de habilidad digital con trabajo de campo urbano para buscar el prototipo de soluciones a problemas urbanos de larga duración: la clase de problemas que las políticas de ciudad inteligente también buscan cambiar. Nuestra investigación apunta hacia una problemática para jóvenes. Aunque ellos han adoptado innovaciones tecnológicas, indicaron que las tecnologías digitales fueron incapaces de servir al público o abocar apremiantes preocupaciones que ellos identificaron como problemáticas dentro de la ciudad: crimen, drogas y carencia de albergue. En vez de eso, de acuerdo con su puno de vista, las tecnologías digitales entregaban los mayores beneficios a los espacios privados en el hogar y en el lugar de trabajo. Aún más, los jóvenes no columbraban que las tecnologías emergentes mejorarían sus vecindarios o comunidades sino tan solo sus prospectos de empleo. Esta investigación sugiere que la ciudad inteligente que emerge está reproduciendo inequidades urbanas tanto reales como percibidas: Los espacios residenciales afluentes y los espacios de la nueva economía se hacen "inteligentes", pero la mayor parte de la ciudad se queda rezagada. Estos patrones crean una paradoja para los jóvenes que invierten en habilidades digitales mientras permanecen al margen del cambio urbano inteligente orientado por la tecnología. *Palabras clave: ciudad inteligente, divisoria digital, Filadelfia, justicia social, juventud.*

In the United States today, using digital technologies to address urban issues is a thoroughly established objective of governance and policymaking (Townsend 2013). Whether specific visions of the smart city come from multinational technology corporations or emerge in a more generative fashion through participatory prototyping like hackathons (Perng, Kitchin, and MacDonncha 2018) or digitally mediated civic engagement (Hollands 2008), the expectation is that adding digital technologies to the city can and will "solve" urban problems big and small. Yet, how, where, and for whom these "smart" solutions are deployed and what sorts of issues are tackled by these efforts remain objects of scholarly concern (Shelton, Zook, and Wiig 2015). As Kitchin, Cardullo, and Di Feliciantonio (2018) argued, latent within the smart city is the expectation of social justice and a right to the smart city: that these solutions will improve the lives of all citizens. Building on this social justice framing, we argue that for the smart city to achieve its stated intents, the benefit of these technologies must be equitably distributed throughout a city and must address issues that citizens feel are most pressing.

Through an examination of youth perspectives on digital technologies and the geography of the smart city, we question whether this social justice framework underpinning the smart city is being met. We held group conversations and mapping activities with seventy-nine youth of color from Philadelphia on their perceptions of the impacts of digital technologies on the city today and in the near future. Engaging with youth was a means of thinking through digitally driven urban change with those who will live in the realized, actualized smart city. Our research sought to understand how engaged young people make sense of digital technologies in the context of urban change and whether their emerging technology skills gave them a sense of ownership over the smart city.

Our research uncovered a conundrum, however. Although the youth were investing in their technology skills for future job prospects, they saw the majority of these digital-urban changes as essentially private conveniences in the home and workplace.

These youth felt that this transformation did not address what they considered widespread and pressing social ills such as crime and gun violence, drug abuse, and homelessness: problems they saw as endemic in urban public spaces. Consequently, if smart technologies are primarily designed to change private, personal places and fail to generate substantial change in public spaces or to equitably benefit the public, geographers concerned with smart spaces and places (Cardullo and Kitchin 2018) must take seriously the uneven geographical development of these urban and technological transformations and what it means for future generations.

Literature Review and Background: Youth and the Urban–Digital Divide in Philadelphia

Two decades into the twenty-first century, Philadelphia offers a productive case for critically examining the impacts of smart technologies on actually existing cities. After more than five decades of postindustrial decline, Philadelphia's economy has rebounded and rebranded around education, medicine, and tourism (Wiig 2016) that has revitalized the downtown core and surrounding residential neighborhoods. Alongside and contributing to this transformation, the city government has rolled out new modes of smart city citizen engagement such as a 311 information and complaint service (Nam and Pardo 2014) and an IBM-sponsored workforce education smartphone app (Wiig 2016). The ongoing disenfranchisement of entire working-class, industrial-era neighborhoods remains a significant issue, however. Beyond the transformed core of Philadelphia, significant poverty remains. More than 400,000 citizens, or 25 percent of the population, live below the poverty line, in neighborhoods where the vast majority of crime occurs and drug abuse remains a prominent concern (Howell 2018).

What is prevalent across the entire city is the wireless digital connectivity that facilitates smart spaces and places (Wiig 2013). This ubiquitous connectivity allows for the widespread adoption of Internet-enabled smartphones and the more recent and related growth of the Internet of Things and platform services including on-demand ride-share apps such as Lyft and Uber (Srnicek 2017). Although this encompassing digital connectivity can offer a foundation for convenient services, it does not inherently address long-standing urban inequities (Gilbert and Masucci 2011, 2018): The digital divide has urban, geographic implications that spill over into the geography of the smart city.

The social justice implications of the digital divide typically focus on the idea that a lack of technology access or lack of digital skills knowledge consequently differentiates access to services, education, jobs, and capital (Mossberger, Tolbert, and McNeal 2007). Extending this critique of the digital divide by applying a social justice framework to the smart city focuses on how marginalized groups are negatively affected by uneven technology access and how lower levels of access and fewer skills equate with and reinforce lower economic participation and status, less education, and fewer options for navigating opportunities (Hollands 2008). One of the paradoxes in the social justice literature around the digital divide is the limited attention to youth as a group at the margins of social equality yet at the forefront of technological adaptation (Grant and Eynon 2017). Scholars do acknowledge that youth are often seen as "digital natives," meaning that ubiquitous technologies have resulted in their seamless acquisition and adoption of digital skills (Hargittai 2010; Jones et al. 2010), hypothesized to be the backbone of entry into digital society and thus the smart city. Yet, the positionality of youth within the context of social and geographic inequity affects how they use technologies and how they envision the ways in which digital technologies will affect their lives. This understanding of urban social justice and the divides—digital and otherwise—still latent in the smart city structured our research agenda and methods described next.

Methods: Youth-Driven Research into "Solving" Urban and Digital Inequities

We framed our research with seventy-nine youth of color between fourteen and eighteen years old as a conversation about the near-future city these youth will inhabit: what it would look like, the sorts of jobs they would have, the sorts of neighborhoods they will live in, and what role digital technologies play in shaping this future. These youth were recruited through their participation in a six-week summer work-ready program at a large public university in Philadelphia. From its inception in 2004, the program has approached the digital divide as a social and urban issue that shapes the ways in which participants experience their communities as well as their efficacy to promote individual and community change. The delivery of the program places a high priority on the coproduction of knowledge; as a result, participant learning is situated within problem-based and hands-on experiences (Pearsall et al. 2015; Masucci, Organ, and Wiig 2016).

The researchers recognize that the youths' voices and opinions must be part of any conversations about how to address intractable urban problems. With these digital native youth working alongside college student mentors, the program trains participants in applying science, technology, engineering, and mathematics (STEM) through digital design skills (photography, word processing and blogging, cinematography and video editing), computer and mobile app programming, as well as entrepreneurial thinking and urban geography field methods. This learning framework is then applied to prototype community and contextually relevant solutions to long-standing urban problems that the youth often are confronted with as part of their daily lives and everyday journeys.

Specifically, our research entailed group conversations where we spoke for forty-five to sixty minutes with eight groups of eight to twelve youth. First, we outlined our motivations for conducting the research: (1) that it would lead to scholarly presentations and publications and to evaluate the success of the program itself and (2) to give the youth an opportunity to voice their opinions about the state of their city, in particular after participating in a program that is grounded in, and has been funded to cultivate, civic engagement and civic leaders (Peters 2015). These group conversations were conducted in July 2017 and July 2018, with four conversations per year, with forty-one youth in 2017 and thirty-eight in 2018. Between 65 and 70 percent of the youth were female in both years. We explained that the conversations were not being recorded and their thoughts and comments would be anonymized and that, further, we would document each conversation with quotes and summaries of comments on the classroom's whiteboards that could, at the end of the time, be edited by the youth to clarify a point or remove a point that anyone felt should not be recorded. This open note-taking style was well received by the groups: Some groups referenced earlier comments and multiple youth took pictures with their smartphones to document the conversation themselves because they felt that it was interesting, meaningful, and insightful.

To encourage the youth to think spatially about where they find urban problems and the potential of digital technologies to affect change to these problems, we started each conversation with two mapping exercises. We first asked them to individually take about five minutes to draw a mental map of their neighborhood, with a focus on the urban environment (in 2017) or on technologies (in 2018). This activity encouraged the students to think spatially about local environments and technologies at the neighborhood scale. Second, the youth worked in groups of three or four to identify and locate the wider environmental issues, broadly defined as "the environment" to include social, economic, and ecological or public space issues, facing Philadelphia (2017) or the areas of the city that were changing and the drivers of these changes (2018). From this second mapping exercise, we asked a series of open-ended questions about the future of Philadelphia, the sort of changes new technologies were bringing to the city, where those changes would be located, and whether they felt included or excluded from these changes.

Out of these open-ended questions, we began the group conversation, writing answers and comments on the whiteboard. We analyzed the data by systematically reviewing the maps generated through the group exercise and the responses recorded on the whiteboard to draw out key themes related to youth perceptions and understandings of Philadelphia. These included social and environmental issues; the role of technology in creating, exacerbating, or ameliorating these issues; how the youth see their developing technology skills affecting the future of the city; and how different types of spaces are shaped by technologies. The findings of the group conversations are summarized and theorized in the next section.

Youth Perspectives on "Smart" Technologies and Urban Change

In our group conversations, we worked together to conceptualize how digital connectivity is changing Philadelphia through the eyes of the youth who will be the occupants and citizens of this near-future smart city. Our findings point to a smart city conundrum. The youth saw much of the technological innovation that comprises the core infrastructure of the smart city as corporate, consumer-oriented improvements that have done little to transform the daily lives of many Philadelphians, including themselves, despite the fact that they attested to the ways in which technologies (e.g., smartphones) have changed their lives. Although critical smart city studies highlight the role that private firms and public–private partnerships play in advancing smart city technologies (Viitanen and Kingston 2014; Rossi 2016), the youths' experiences suggested that these technologies were further bifurcated across public–private lines. Additionally, the youth discussed the ways in which technologies shaped their lives—primarily to make everyday tasks more convenient—and noted that they saw little potential for technologies to have a transformative impact on civic issues (e.g., eliminating homelessness). We organize our results along two axes to capture these important ways in which the youth discussed smart technologies and urban change: public–private and convenient–transformative (see Figure 1). Along the public–private axis, the youth discussed technological innovations in public spaces that primarily involved upgrades to public transit. Of particular benefit were electricity and USB outlets on buses allowing a smartphone to be charged and monitors on buses that alerted passengers about travel time to the next stop. The youth were frequent users of these limited, yet accessible technologies serving the public. They also mentioned many consumer-driven technologies that exclusively served private spaces, from smart refrigerators to smart home speakers like Amazon's Alexa. Although few of the youth had direct experiences with many of these technologies, they had heard of them through friends, family, and the media.

Along the convenient–transformative axis, the youth reflected on the ways in which these technologies affected the city, their neighborhoods, and their own lives. On the one hand, they noted that many technologies made their lives easier, such as the charging ports on buses, improving the efficiency of their daily tasks. On the other hand, they also commented that digital technologies often failed to provide a transformative societal impact. The youth across all eight groups were remarkably consistent in identifying what they saw as the key problems facing Philadelphia: homelessness, crime and violence, and

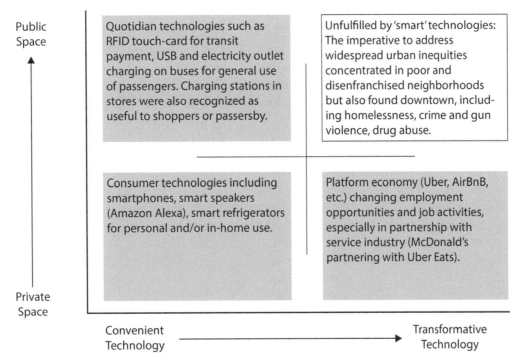

Figure 1. The geography of smart urban technologies, from private to public and convenient to transformative.

drug abuse. They expressed a strong skepticism that the types of technologies deployed across the city in either the public or private realm would meaningfully address these critical issues of social justice. In what follows, we elaborate on each of the quadrants in Figure 1 to more directly address the extent to which young people see digital technologies as a means to improving their city.

Private–Convenient

The youth readily summed up many examples of recent technological innovations for corporate consumer goods, from smartphones to self-driving cars. Interestingly, several groups quickly arrived at the conclusion that the types of technologies that received the most attention and hype in the media (e.g., the 2017 release of the Tesla Model 3) were primarily about serving the interests and needs of the middle class and above. In several groups, youth also reflected on how these consumer goods were largely designed to make life more efficient and more convenient for certain individuals, while failing to address broader and arguably more pressing city needs (e.g., homelessness). In 2017, two separate groups referenced Elon Musk, which suggests how his highly visible persona has shaped and defined what technology is and who it aims to serve.

Their observations about how digital technologies largely affected private spaces also reflected their comments about how the city was changing and what those changes meant for challenging or reinforcing sociospatial inequities. In 2018, multiple groups discussed gentrification in Philadelphia, which they characterized as certain neighborhoods having more new people and more new housing construction. Residential development anchored change in the city, and many of the new technologies in the city have been designed for private use inside these new homes. Several groups commented on the presence of security cameras in certain neighborhoods to deter crime and reduce drug dealing and drug use. Other groups also shared their observation that the influx of people and new residential development also pushed lower income families to the margins of the city, further limiting their ability to take advantage of the new services and technologies that the residential development generated, in addition to making it harder for these displaced families to get to jobs in the city center or elsewhere.

Public–Convenient

The youths' experiences of technologies that served the public were almost exclusively about transportation in Philadelphia. City buses were recently equipped with smartphone charging stations (Aguilar 2017), which many youth used on a daily basis. Additionally, the public transportation system had recently been modernized to use RFID-enabled debit cards rather than tokens (SEPTA 2018). Many groups also mentioned the Indego bike share program, which launched in 2015 (Indego 2017). Although this was technically a privately funded transit initiative, many youth saw the bike share as a new mode of transportation that is heavily equipped with technologies for locking the bikes and tracking usage. Very few of the youth had ever accessed the bike system, and several youth pointed out that the bike docks were primarily located in the downtown and its adjacent neighborhoods, with limited access to those who live farther away. Their comments highlighted how this new, highly visible smart city transit feature largely targeted a different (arguably wealthy and professional) demographic, a point of view taken by urban geographers as well (Stehlin 2019).

Several groups keenly commented that new transportation features are in place because there are more people living in the city, an observation supported by recent U.S. Census data (U.S. Census 2017). Such updates to public transportation are not necessarily seen by the youth as equitable, however. For instance, the youth noted that despite such upgrades to the transportation system that improved their experiences, many other public spaces in the city remained underserved and undermaintained. Multiple youth lamented the poor state of the streets due to potholes that were so large that they would damage vehicles. Further, although there were some changes and improvements across the city, the youth primarily observed them in terms of aesthetic changes (e.g., updates to building facades). Additionally, several youth commented that older adults struggled to understand the recent updates to the transportation system (e.g., the RFID-enabled payment card). For them, these digital technologies were about serving a young population, and they expressed some concerns about what it would mean to age in a city that required residents to keep up with the latest innovations in digital technologies to perform routine tasks.

Private–Transformative

Youth commented less frequently on technologies that served the private–transformative nexus, yet their insights are suggestive of the ways in which the platform economy will likely transform the everyday urban experience. Youth described how Uber and Lyft's ride-share services are dramatically altering mobility across the city, for both people and goods. Their observations about these services are interesting because few of the youth had direct experiences with them. All the same, they were familiar with the arguments for the ride-share industry improving transportation services in areas underserved both by public transit and traditional taxis and against the ride-share model, because it provides variable service to individuals and drivers might choose to not accept a ride request from certain people, in particular people of color like the youth. Regardless, the youth saw that ride-share services and the employment of an on-demand, private driver were reshaping mobility in the city.

The youth also expressed some concerns about how the platform economy could shape their future employment prospects. The mention of Uber Eats food delivery, in particular the service's partnership with the McDonald's fast food chain and the delivery of fast food to the home, led to a discussion about touchscreens for the ordering of meals and workforce automation in fast food, based on jobs that these youth or their friends had or that they experienced while eating at the restaurant. Regarding these sorts of digital–urban changes, the youth felt little control over what they regarded as the rapid technological shifts in mobility and employment. Whether useful or a hindrance, these services were quickly becoming normalized into their daily lives.

Public–Transformative

Many narratives of the promise of the smart city evoke this final category of public–transformative, where technologies have the potential to create a utopian city (Söderström, Paasche, and Klauser 2014). The youth were consistent and insistent, however, that the technologies deployed in cities do not and cannot change the critical social issues facing Philadelphia today, including homelessness, crime and violence, and drug abuse. It is worth noting that a 2016 Pew Trust survey of "poor and non-poor" Philadelphians found the same concerns reported citywide, in order of concern: crime, drugs, and safety; education and schools; jobs, economy, economic development, and lack of economic opportunities; and poverty and homelessness (Howell 2018). These concerns can be seen as the core social justice issues facing Philadelphia, and yet smart technologies do not address these issues.

The majority of groups conveyed a nuanced understanding of the sociospatial inequities in different neighborhoods and how digital technologies reinforced existing patterns of social polarization and exclusion. They noted considerable investment in the gentrified Center City neighborhood, with new skyscrapers and residential construction, whereas other neighborhoods with persistently high poverty rates experienced little investment and remained characterized by crime, drugs, trash, and vacant lots or abandoned buildings. They noted neighborhoods that were changing, anchored by the presence of universities and aligned real estate investments, as well as neighborhoods, such as Kensington, that experienced highly variable change benefiting certain populations more than others. They marked formerly industrial neighborhoods along the Delaware River, noting legacy pollution that lingers in the soil, water, and air. Their maps and subsequent conversation conveyed the unevenness of smart city innovations introduced to a city that has struggled to recover following extensive deindustrialization over the past fifty years. These digital technologies have brought quality-of-life improvements to certain wealthy neighborhoods, while doing little to deal with the neighborhoods that have suffered the most from long-term poverty and unemployment.

The youth also asserted that these technologies might never substantively change the lives of certain populations even if they had routine access to them. There was broad consensus that many of these technologies made their lives more convenient or comfortable, but they did not see these technologies (even their smartphones) as fundamentally transformative to how they live their lives. Arguably intractable issues, such as homelessness, crime, and violence, might require social and political interventions rather than technological responses, yet the youth gestured at the concerns that technological innovations certainly were not targeting these societal issues. Alternatively, in several groups, youth indicated that they considered some technologies, such as social media, as a driver of crime (e.g., cyberbullying). The youth did not see a technological fix for poverty or violence, a discouraging ending point for a conversation about the city's "smart" future.

Technology Skills and the Future of the City

Our discussion about whether the youth thought that their technology skills would give them a sense of ownership over their future smart city revealed a nuanced understanding of the fundamental role that technologies play in underpinning the development of the city, as well as their ability to affect change. These youth who participated in the six-week program embraced technology skills as part of their future, and many of them expressed an interest in becoming engineers, software developers, and professionals in other technology-intensive fields. The youth saw their skills as a way to improve their job placement prospects; several groups discussed how they sought to acquire technology skills as a backup plan in case their primary career goal did not work out. The youth were cognizant of the strong emphasis that schools place on STEM and the growing employment opportunities afforded by such expertise. Few of the youth were passionate about the role that their technology skills would play in their careers. One group discussed how they felt that technology in schools interfered with their learning (e.g., a focus on learning how to use Microsoft PowerPoint vs. understanding course content itself). Youth acknowledged the need to keep up with technology to stay relevant in multiple professions, however. For instance, many youth planned to seek jobs in health sciences or law and anticipated the ways in which technology would change the workforce (e.g., automation, robots assisting with surgeries). Some youth expressed a desire to seek careers that would not be threatened by automation, but they were not sure what sort of career that would be.

The youth saw their technology skills as a baseline expectation for being part of a digitally engaged citizenry, but they did not see a way that these skills would give them a greater voice in the development of the city as a whole or even their own neighborhoods. They saw public policy and governance as the primary means to address core social issues like gun violence. One group discussed how they wanted the voting age to be lowered so that they could vote as teenagers. Although the youth were optimistic that their technology training would improve their lives as individuals through high-paying jobs, few expressed any notion that these jobs would allow them to bring meaningful social change to their neighborhoods. Several youth indicated that they wanted to contribute to "big ideas" that would affect change but saw a technology-driven career path as separate from this civic effort.

Conclusion

The relative invisibility of youth in the more established digital divide literature (Grant and Eynon 2017), let alone their absence in the limited literature on smart citizens (Shelton and Lodato 2019), illustrates how youth sit at the margins of the smart city. This status reinforces and is reinforced by the role that digital technology plays in their everyday experiences. The types of experiences cued up for youth to gain and demonstrate digital skill prowess—like hackathons, computer programming trainings, digital social game play, and use (and development) of smartphone applications—all forecast a lifetime of being on the "inside" of the technologies integral to their daily lives. Youth, as those citizens who will inherit the near-future smart city, also should have greater ownership over its production. A number of factors, however, like the burden they often face to provide a pathway for older family members to use technologies, the limited financial capacity of youth on the margins to keep up with the changes in technologies, and the technology industry's well-established scheduled obsolescence approach to generating a continued demand for products and services (Reid 2018), all serve to maintain the (low) status of youth rather than provide a platform for navigating opportunities in the smart city.

That the argument assuming digital access equates with social and economic advancement (Mossberger, Tolbert, and McNeal 2007) continues to hold sway is in and of itself a social justice concern, as it divorces the concept of digital access from other means of production and capital and implies that digital technologies have the inherent ability to empower youth. Our work suggests that youth are well aware of this paradox and struggle to simultaneously keep up with technological change and seek resources to advance their opportunities through education and expanding social networks. Yet, for many youth, this quest is an elusive one, fraught with difficult decisions about how to direct time toward gaining access to the smart city; navigating these technologies in their personal life, at school, and in jobs; supporting family members and others to do the same; and advancing their educational and economic opportunities. For geographers concerned with social justice, these findings trouble our understandings of the impact of digital technologies and prod us to work toward and advocate for smart spaces and places that serve the public and actively redress long-standing urban inequities.

Acknowledgments

We acknowledge the staff, in particular Jean Akingeneye, the student mentors, and the youth participants of the Building Information Technology Skills Program at Temple University, Philadelphia, Pennsylvania. Without their support and insights, this research would not have been possible.

Funding

The Building Information Technology Skills Program (BITS) was initiated in 2004 with generous support from the National Science Foundation (Award #0423242) and was supported during the study period (2017-2018) with funding from the Philadelphia Youth Network.

ORCID

Hamil Pearsall ⓘ https://orcid.org/0000-0003-2287-7586
Alan Wiig ⓘ http://orcid.org/0000-0002-6224-2633

References

Aguilar, R. 2017. SEPTA unveils new bus model with USB charging ports. NBC 10 Philadelphia. Accessed November 19, 2018. http://www.nbcphiladelphia.com/news/local/SEPTA-New-Buses-USB-Port-Other-Features-414660063.html.

Cardullo, P., and R. Kitchin. 2018. Smart urbanism and smart citizenship: The neoliberal logic of "citizen-focused" smart cities in Europe. *Environment and Planning C: Politics and Space*. Advance online publication. doi:0263774X18806508.

Gilbert, M., and M. Masucci. 2011. *Information and communication technology geographies: Strategies for bridging the digital divide*. Vancouver: University of British Columbia Praxis (e)Press.

Gilbert, M., and M. Masucci. 2018. Defining the geographic and policy dynamics of the digital divide. In *Handbook of the changing world language map*, ed. S. Brunn and R. Kehrein, 1–19. Cham, Switzerland: Springer.

Grant, L., and R. Eynon. 2017. Digital divides and social justice in technology-enhanced learning. In *Technology enhanced learning: Research themes*, ed. E. Duval, M. Sharples, and R. Sutherland, 157–68. Cham, Switzerland: Springer.

Hargittai, E. 2010. Digital na(t)ives? Variation in internet skills and uses among members of the "net generation." *Sociological Inquiry* 80 (1):92–113.

Hollands, R. 2008. Will the real smart city please stand up? Intelligent, progressive or entrepreneurial? *City* 12 (3):303–20.

Howell, O. 2018. *Philadelphia's poor: Experienced from below the poverty line*. Philadelphia: Pew Trusts. Accessed November 20, 2018. https://www.pewtrusts.org/-/media/assets/2018/09/phillypovertyreport2018.pdf.

Indego. 2017. About. Accessed November 19, 2018. https://www.rideindego.com/about/.

Jones, C., R. Ramanau, S. Cross, and G. Healing. 2010. Net generation or digital natives: Is there a distinct new generation entering university? *Computers & Education* 54 (3):722–32.

Kitchin, R., P. Cardullo, and C. Di Feliciantonio. 2018. Citizenship, justice and the right to the smart city. SocArXiv. doi: 10.31235/osf.io/b8aq5.

Masucci, M., D. Organ, and A. Wiig. 2016. Libraries at the crossroads of the digital content divide: Pathways for information continuity in a youth-led geospatial technology program. *Journal of Map & Geography Libraries* 12 (3):295–317.

Mossberger, K., C. Tolbert, and R. McNeal. 2007. *Digital citizenship: The Internet, society, and participation*. Cambridge, MA: MIT Press.

Nam, T., and T. Pardo. 2014. The changing face of a city government: A case study of Philly311. *Government Information Quarterly* 31:S1–S9.

Pearsall, H., T. Hawthorne, D. Block, B. Walker, and M. Masucci. 2015. Exploring youth socio-spatial perceptions of higher education landscapes through sketch maps. *Journal of Geography in Higher Education* 39 (1):111–30.

Perng, S. Y., R. Kitchin, and D. MacDonncha. 2018. Hackathons, entrepreneurship and the making of smart cities. *Geoforum* 97:189–97.

Peters, M. 2015. Philadelphia's urban apps & maps studios helps youth develop tech solutions for urban communities. Accessed November 19, 2018. https://knightfoundation.org/articles/philadelphias-urban-apps-maps-studios-helps-youth-develop-tech-solutions-urban-communities.

Reid, A. 2018. *The smartphone paradox: Our ruinous dependency in the device age*. Cham, Switzerland: Springer.

Rossi, U. 2016. The variegated economics and the potential politics of the smart city. *Territory, Politics, Governance* 4 (3):337–53.

SEPTA. 2018. SEPTA key FAQ. Accessed March 16, 2019. http://www.septa.org/key/faq.html.

Shelton, T., and T. Lodato. 2019. Actually existing smart citizens: Expertise and (non)participation in the making of the smart city. *City* 23 (1):35–52.

Shelton, T., M. Zook, and A. Wiig. 2015. The actually existing smart city. *Cambridge Journal of Regions, Economy and Society* 8 (1):13–25.

Söderström, O., T. Paasche, and F. Klauser. 2014. Smart cities as corporate storytelling. *City* 18 (3):307–20.

Srnicek, N. 2017. *Platform capitalism*. London: Polity.

Stehlin, J. 2019. *Cyclescapes of the unequal city: Bicycle infrastructure and uneven development*. Minneapolis: University of Minnesota Press.

Townsend, A. 2013. *Smart cities: Big data, civic hackers, and the quest for a new utopia*. New York: W.W. Norton & Company.

U.S. Census. 2017. U.S. Census Bureau QuickFacts: Philadelphia City, Pennsylvania; Philadelphia County, Pennsylvania. Accessed November 19, 2018. https://www.census.gov/quickfacts/fact/table/philadelphiacitypennsylvania,philadelphiacountypennsylvania/PST045217.

Viitanen, J., and R. Kingston. 2014. Smart cities and green growth: Outsourcing democratic and environmental resilience to the global technology sector. *Environment and Planning A: Economy and Space* 46 (4):803–19.

Wiig, A. 2013. Everyday landmarks of networked urbanism: Cellular antenna sites and the infrastructure of mobile communication in Philadelphia. *Journal of Urban Technology* 20 (3):21–37.

Wiig, A. 2016. The empty rhetoric of the smart city: From digital inclusion to economic promotion in Philadelphia. *Urban Geography* 37 (4):535–53.

"Smart" Discourses, the Limits of Representation, and New Regimes of Spatial Data

Craig Dalton, Clancy Wilmott, Emma Fraser, and Jim Thatcher

As "smart" urbanism becomes more influential, spaces and places are increasingly represented through numeric and categorical data that have been gathered by sensors, devices, and people. Such systems purportedly provide access to always visible, measurable, and knowable spaces, facilitating ever more rational management and planning. Smart city spaces are thus governed through the algorithmic administration and categorization of difference and structured through particular discourses of smartness, both of which shape the production of space and place on a local and general level. Valorization of data and its analysis naturalizes constructions of space, place, and individual that elide the political and surveillant forms of technocractic governance on which they are built. This article argues that it is through processes of measurement, calculation, and classification that "smart" emerges along distinct axes of power and knowledge. Using examples drawn from the British Home Office's repurposing of charity outreach maps for homeless population deportation and the more recent EU EXIT document checking application for European citizens and family members living in the United Kingdom, we demonstrate the significance of Gunnar Olsson's thought for understanding the ideological and material power of smartness via his work on the very limits of representation. The discussion further opens a bridge toward a more relational consideration of the construction of space, place, and individual through the thinking of Doreen Massey. *Key Words: data, Massey, Olsson, place, smart cities, space, spatial data*

由于"智慧型"城市化建设的影响力日益加强，在描述空间和地点时，人们越来越倾向于使用传感器、相关设备和人工收集的数字和分类数据。据称，这类收集系统可进入一直可见、可测量、可感知的空间，帮助进行更合理的管理与规划。因此，智慧型城市空间通过算法执行和差异分类进行管理，以特定的智慧型语言体系作为结构，这样的管理方式和结构决定了空间和地点在局部及广义层面的生成方式。通过数据的验证和分析，使空间、地点和个体的构建趋向于自然化，无需使用"技术治国"背景下的政治和监管形式。本文认为，正是通过测量、计算和分类流程，权力和知识这两条轴线开始呈现"智慧型"特征。通过两个例子：其一，英国内政部修改其慈善宣传地图，以解决无家可归者被驱逐出境的问题；其二，近期提出了一项申请，要求检查英国境内的欧洲公民及其家庭成员的脱欧证明文件，以及Gunnar Olsson的详尽陈述，我们证明了他的思想对于理解智慧型意识形态和客观力量的重要性。本文的讨论通过 Doreen Massey 的思想，进一步构建一座桥梁，让人们可以对空间、地点和个人的构建进行更理性的思考。关键词：数据、Massey、Olsson、地点、智慧型城市、空间、空间数据。

A medida que el urbanismo "inteligente" se hace más influyente, los espacios y los lugares crecientemente se representan por medio de datos numéricos y categóricos que se han obtenido por sensores, aparatos y gente. Esos sistemas supuestamente dan acceso a espacios siempre visibles, medibles y conocibles, que facilitan un manejo y planificación cada vez más racional. Los espacios de la ciudad inteligente son entonces gobernados a través de la administración y categorización algorítmica de la diferencia, y estructurados con base en discursos particulares de agudeza, los cuales configuran la producción de espacio y lugar a nivel local y general. La valorización de los datos y su análisis naturaliza la construcción del espacio, el lugar y el individuo que suprime las formas políticas y vigilantes de la gobernanza tecnocrática sobre las cuales están construidos. Este artículo sostiene que es a través de procesos de medición, cálculo y clasificación como surge lo "inteligente" junto a ejes perceptibles de poder y conocimiento. Mediante el uso de ejemplos extraídos de los mapas que reconvierten el alcance de la caridad respecto de la deportación de la población carente de vivienda de la Oficina Británicas del Hogar, y del más reciente documento EXIT de la UE sobre el control de comprobación para los ciudadanos europeos y su familia residentes en el Reino Unido, demostramos el significado del pensamiento de Gunnar Olsson para la comprensión del poder ideológico y material de la inteligencia, a través de su trabajo sobre los propios límites de la representación. La discusión, además, tiende un puente hacia una consideración más relacional de la construcción de espacio, lugar e individuo a partir de las ideas de Doreen Massey. *Palabras clave: ciudades inteligentes, datos, datos espaciales, espacio, lugar, Massey, Olsson.*

A leading homelessness charity passed key information about migrant rough sleepers to Home Office enforcement teams and may well have done so without their consent.

—Taylor (2018a)

The app [is] analogous "to border guards knocking on every door in the U.K. and forcing EU nationals to show documentation."

—Brewster (2018)

Smart geographical rhetorics—whether concerning cities, devices, spaces, or places—are promising more rational, calculated geographical policies and procedures, with critical implications for the production of space and place (Anthopoulos, Janssen, and Weerakkody 2019). Existing research has revealed that carbon-neutral towers and other positive uses of "smart" technologies are not the only outcomes of smart initiatives (Söderström, Paasche, and Klauser 2014). In practice, data-based decision making facilitates both empowering and disempowering experiences, ranging from self-driven data navigation using crowdsourced apps (Hind and Gekker 2014) to the identification and deportation of homeless people in cities across the United Kingdom (Jackson 2015). In practice, the language and framing of the smart city frequently obscure a landscape of governmental and corporate imperatives in which individuals, spaces, places, and things experience the uneven promises and pitfalls of "smart" technology, often in terms of representation (Rose and Willis 2018) and generating inequality (Shelton, Zook, and Wiig 2015).

Smart city initiatives reflect geographic scalar shifts, as the power to measure, calculate, and classify becomes more centralized, and the power of signification becomes resituated at divergent, often conflicting scales of space and place. Scalar relations have interested geographers for decades (Marston, Jones, and Woodward 2005; Moore 2008), yet, as smart cities promulgate, measurements, calculations, and classifications produced for specific purposes are reappropriated, resulting in unexpected conflicts between stakeholders across multiple scales.

The definition of the smart city often encompasses only those places defined specifically by "smart" initiatives (Milton Keynes, Madrid, or Songdo International Business in Seoul are key examples). In this article, we examine smart urbanism more broadly via the relations between data and body, representation and self, individual and society, and representational structures of smartness. These relations underpin the constant negotiation between the structural and governmental logics in smart city discourses and the lived, embodied, and situated experiences of place (whether in cities or beyond; Dowling, McGuirk, and Maalsen 2018). We examine these tensions through the impacts and experiences resulting from two nominally "smart" cases—a London charity that shared personal information about homeless people with a government agency to identify, locate, and deport noncitizens and two of the author's own experiences with the newly released EU EXIT identity checking application—as examples of the future of data-driven smart technologies in regulating bodies, spaces, and places. Through these cases, we explore how societal valorization of data and associated analytic capabilities naturalize constructions of space and the individual—constructions that elide the political and surveillant aspects of technocratic governance that undergird smart city discourses. Such discourses, as Datta (2018), Cordullo and Kitchin (2018), and Maalsen (2019) explained, reinforce values of smart citizenship and overlook neoliberal and colonial logics of governance. This, we argue, requires a critique focused on the relations between power and knowledge and space and place. To undertake this critique, we first outline current research on the technosocial nature of measurement, calculation, and classification in smart discourses; we then consider the limits of representation through rationalism and, in particular, cartographic reasoning in terms of the interweaving of individual (placed) experience and global governmental logics, before presenting two key cases that exemplify the slippages between the operation of smartness, lived experience, and the representation of space and place.

Highlighting the epistemic limits and ontological breaks that occur between representation and individual within these systems, we emphasize the works (and critiques) of Gunnar Olsson regarding the translation of abstract classifications and calculations into material, bodily, and social subjugations and of Doreen Massey regarding the distance between these abstractions and lived experience, including the relationality between space and place, against the totalizing logic of abstract global technology and "real" life (Massey 2005). These critiques hinge on linguistic and indexical correspondence, the space of representation and speech that structures encounters with place and power, primarily through the use of representation to secure and stabilize spatial processes. As Massey (2005) wrote of a territorial geographical imagination: "Words such as 'real,' 'everyday,' 'lived,' 'grounded,' are constantly deployed and bound together; they intend to invoke security, and implicitly—as a structural necessity of the discourse—they counterpose themselves to a wider 'space' which must be abstract, ungrounded, universal, even threatening" (184–85).

The writings of Olsson offer a structuralist and linguistic analysis of the way in which power works through the indexicality, symbolism, and semiotics of representation itself with a distinctly spatial (rather than temporal) lens. Olsson's work, however, has been critiqued as a self-indulgent reaffirmation of masculinist geographical knowledge and dehistoricized subjectivity (Sparke 1994) at the cost of situated, communal, and sensitive acknowledgment of the specificity of space and place, as well as a failure to engage with the theoretical work undertaken by feminist geographers such as Rose (1993). To this end, we aim to complement these representational analyses with the arguments made by Massey (2005) about the way in which representational power comes to bear as it is entangled with the lived, material, and conformational; that is, language, but not without bodies or emplaced experiences and relations. In particular, Massey's rejection of a place–space dualism, or "technology-led" understandings of "information as disembodied and of globalization as some kind of other realm, always somewhere else" (185), are of particular use in examining smart cities and their discourses.

By using the representational and spatial critique of Massey (2004, 2005) to complement Olsson's structuralist, epistemic claims to the basis of linguistic ontology—and thereby power through calculation—an experience of space is described in which self and other are situated simultaneously within and beyond and governed through language and the limits of representation. Specifically, although data governance or governmental logics might be understood via Foucault's (2002b) writings on disciplinary societies, we are specifically interested in outlining the role that linguistic ontologies play in both promising (and failing to achieve) total and absolute representation of urban people, processes, and places. This is a cautious distinction: As Foucault (2002a) argued, and Belyea (1992) reminded us, "Texts must be examined independently of the persons who write them" (3). Yet, our argument is precisely that the meaning and power of data are at once embodied by the data themselves and by the person or authority who seeks to use it and that this will shift as the same data are used in different contexts. By leveraging Olsson's structuralist (and often humanist) thinking around signification as thought and action, we demonstrate how, as equivalences are made between data and the lived world, the politics of generating and repurposing social and spatial data occurs at multiple scales. Furthermore, as differences defined by data and analysis become rescaled, signifiers (and therefore social implications) shift in the move from one context to another. Both the cases discussed—the Home Office's acquisition of homeless people's personal information and the use of the mobile phone EU EXIT app to verify the status of (already settled) European Union (EU) nationals and their families in the United Kingdom—are examples of such shifts, where one set of data is repurposed in a different spatial context, albeit in the same place. The contextual shifts in social and spatial data are crucial when considering the wider implications of smart city governance—particularly when working between the affective experiences of inequality and representational expressions of "smartness" (Shelton and Lodato 2018). Thus, Olsson's framework provides a means to bridge critical discourse and lived materialities without discounting the possibilities of smart technologies and cities. We demonstrate how, as differences become datafied, they are rescaled and, in the process, signifiers slip.

The Unstable Foundations of Smart Cities

Cities are frequently the subjects of geographical discourse on smartness (Silva, Khan, and Han 2018). Smart city initiatives come in many different forms, ranging from smartphone apps for reporting potholes (Shelton, Zook, and Wiig 2015) to the wholesale development of new neighborhoods (Scola 2018), but in each case, as Olsson suggested, the discourses around them move beyond the sayable to the representable and into the power of representation to fix and deny being in both space and place. Scholars have been quick to note the discursive underpinnings and associated material dimensions of such smart initiatives. Söderström, Paasche, and Klauser (2014) characterized smart cities as a form of corporate storytelling, a sales pitch that places the company at an "obligatory passage point" (Callon, as cited in Söderström, Paasche, and Klauser 2014) on the way to a systems-based, planned technological urban utopia. In practice, delivering on such rhetoric is extremely difficult (Wiig 2016). Even if they work as designed and intended, smart city initiatives often involve data-focused, technical solutions to urban problems that lack consideration or engagement with social processes and political conditions (Luque-Ayala and Marvin 2015). Nevertheless, those political dimensions are vital to understanding how the braiding of identity at an ontological level between individuals and societies (Olsson 1979) that emerge with and through smart cities and associated technologies is highly contingent.

Narratives of smart cities cut across scales from the individual to the global; however, at each step, greater quantification and analysis lead inexorably toward an apparently better, more organized, and more efficient self, city, nation, and planet (Wilson 2015). Fundamentally, this occurs through the generation, capture, and analysis of data. In this context, classification and measurement is the process of quantification by which something is rendered into data, calculated into being meaningful, and used to enact change. That something, somewhere, or someone—within narratives of smartness—could stretch across multiple scales and might refer to persons, objects, or flows.

Entangled with the notion of an abstract global, the technological emphasis of smartness tends to juxtapose proximity and materiality, much like global and techno-focused imaginaries of the past, in which spaces of power and control were (conveniently) abstracted from a supposedly more distinct sense of place (Massey 2005, 183), as if individuals, objects, or sites can be simultaneously fixed, but untouched, by governmental logics and their representations. Within semiotic theory and, specifically, theories of signification (Olsson 1980; de Saussure 1987), the politics of conceptualizing objects as coherent, namable, and durable things (and so, measurable, classifiable, and calculable) is critical to understanding how representation matters in the contemporary datacentric context of smart cities. Crucially, as Olsson (1980) argued from de Saussure, the formation of the sign does not simply link a word with a thing but instead defines the thing itself at an ontological level, whether it is subject or body, place or space, neighborhood or nation-state. If urban spaces are already proliferated with signs, then smart city discourses reshape and rewrite processes of signification into digitally oriented interoperable and calculable systems of representation (cf. Wilmott 2016). Thus, urban processes of quantification hinge on the development of coherent ontological structures through classification (Leszczynski 2009). This is achieved through the standardization of systems of classification and measurement and the relationship between both. In short, the question of what should be measured is conditional on how it should be measured (Beer 2016; Gabrys 2016). For instance, within the smart city, the question of how a body should be rendered as part of a population is codependent on how it is being measured: biologically (i.e., temperature, weight, blood type), socially (race, gender, sexuality, class, age), culturally (ethnicity, religion, taste), economically (income, housing, expenditure), or politically (nationality, suffrage, residence). As the cases in this article discuss, this enables the repurposing of data to be calculated toward a number of analytical ends, some of which might be unanticipated.

Classification measurement secures spatiotemporal correspondence: between objects in time and space or between the points of measurement themselves as an act of differentiation or similarity. In doing so, both classifications and measurements are political actions: first, defining what sociospatial objects are ontologically being measured (spaces, places, bodies, neighborhoods) through difference and indexicality; second, methodologically determining what processes should be measured (i.e., what kinds of data—heat, particles, location, nationality); and third, the production of (inter)nationalized baselines or starting points from which measurement occurs (i.e., meridians, datums, data points, time zones). Thus, acts of quantification, of reducing and representing one thing as another, are fundamental to the processes of the modern state (Scott 1998; Foucault 2008). As such, signification is not unique to smart city processes but is intrinsic to their functioning; it also flows beyond the smart city, affecting a range of places and populations. As Massey (2005) noted of the line, such representation is an attempt to create ontological security and is therefore the basis on which all smart city discourses rest. It assumes a discrete correspondence between the real world and the data produced; both the act of "saying that something is something else and being believed" (Olsson 2000, 1237).

As measuring involves producing and framing the world, setting the limits of what can be epistemologically known and ontologically secured, calculation is the means by which these quantified representations are leveraged in service to the social ambitions of smart city initiatives but also smart technologies and various models of smart urbanism. Through the application of measurements, the city becomes a computer, a now-smart city that "frames the messiness of urban life as programmable and subject to rational order" (Mattern 2017). If measurement is roughly the creation and capture of data, and calculation its processing, the (re)classification is again the end goal of these two processes; it is the rendering of all things as populations (Mulcahire 2018a) according to smart city agendas. This occurs through a refiguring of what counts (and is counted) and measures (and is measured) through the host of sensors and sensing devices that have proliferated within smart city environments (Gabrys 2014). Thus, classifications themselves become calculable, in that they need to be calculated against each other to draw inferences between previously separate categories. In the representational promise of a fixable, classifiable, calculable world, smart city discourses in particular purport to have achieved that long-sought goal of the modern state: controllable, knowable populations. To become populations in the smart city, however, categories must be algorithmically sortable outputs from digital calculations. In short, digitally oriented measurements inform digital

calculations, which are then used to reclassify things, people, places, and sensors into a promised better, more efficient, digital–material world, the space of the smart city itself. From border security (Amoore and Hall 2009) to pop musician safety (Canon 2018), these classifications and measurements are then calculated into conclusions, their power suffusing everyday life and being directly felt through the transformation of probability into certainty and, then, data into materiality (Amoore and Piotukh 2015; Mackenzie 2018b).

Smart city technologies (from facial recognition to data sharing) make equivalences and produce differences. Whether the sets of directions that individuals receive when moving through space or the mortgage rates that are offered depending on the specifics of place and individual, the manipulation and control of bodies in space rests on these calculation and classification processes (Thatcher 2013; Dalton and Thatcher 2015). As the given examples demonstrate, these logics also apply to wider contexts—national borders, collective citizenship, global security—stretching the rhetoric and practice of smart cities across a far wider field than the underlying knowledge production. As Shelton, Zook, and Wiig (2015) pointed out, "This new kind of expertise tends to be embodied in far off places and organizations which must be brought in from outside" (17). Under these conditions, local knowledge is only valued and included to the extent that it can be collected and fit into the data models and technological imperatives of smart city initiatives. Furthermore, this classificatory sorting sits at the nexus of power and knowledge formations—the power to state "something is something else" but also to make it exist as such (Olsson 2000; Foucault 2002b). As Gabrys (2016) wrote, however, "Processes of producing data are also processes of making sense" (41). Smart city systems are fundamentally concerned with securing a specific epistemological and ontological understanding of space and place (Kitchin, Lauriault, and McArdle 2015).

The computable, calculative solutions offered through smart urban technologies belie a double transformation that occurs with respect to lived experience. Processes of calculation rest on an epistemic myth of the world as perfectly knowable and representable, what Greenfield (2017), Wyly (2014), boyd and Crawford (2012), and others have referred to as a resurgent, unreformed logical positivism.

Smart solutionism belies a second transformation of the real: Algorithmically calculated probability becomes the highest form of knowledge and certainty (Beer 2018). At the extreme, this form of knowledge production eschews situated, theoretical, and historical understandings in favor of what Gregg (2015) called the "data spectacle" wherein data require only "the indication of potential to achieve veracity" (40). In doing so, calculation also smooths over the difference established in the first instance of measurement. In short, smart city discourses do not prioritize accurate measurement or calculation but rather engage in a "fantasy of command and control" (Gregg 2015, 37) that serves Söderström, Paasche, and Klauser's (2014) corporate sales pitches through which society is transformed into a superficially calculable whole. This process homogenizes collective and individual experience, bringing many places under a single mechanism of calculation, producing a spatial logic that shapes, but also denies, the diversity of lived material encounter.

The Limits of Cartesian Reason

Issues of individuals versus governance, ontological complications, and epistemological divides have long been the focus of Olsson. Alongside Farinelli, his critique of "cartographic(al) reason" is specifically useful in the context of smart cities for connecting ontological and epistemological disjunctures with expressly geographical scales and lines (Farinelli 1998; Olsson 2007). Whereas Foucault's work understands control through discursive epistemes, in Olsson, ontological concerns arise around material encounters and individual experiences, and epistemological framings specifically reflect what is and can be known through representation(s), including those that structure a sense of space and place. For Olsson (and Farinelli 1998), cartographical reason thus denotes the moment when the act of inscription ceases to be purely descriptive and instead becomes a platform for decision making and action—in short, reasoning. We argue that for smart cities, as well as smart city initiatives and discourses of smartness, structures of reasoning emerge in the crucial distinction between measuring urban processes and the deployment of calculation and classification as models for future building, planning, and policy.

The distance between measurement and reasoning stretches across spatiotemporal scales, from long-term urban planning to near-real-time, data-driven reasoning involving individuals. Massey (2005) recognized the territorial capacities of meaning production through reason and abstraction, in the context of space–place and global–local dualisms. Massey's push for a relational spatial politics insists on a mutual constitution of local and global, asking: "Where would you draw the line around the lived reality of your daily life?" (Massey 2005, 185), gesturing toward a critique of reductive inferences that name the local place as more "real" and denying the interconnectedness of the individual with national and global governmental spaces and structures.

For Olsson, deploying structures of reasoning relies on making logical inferences between object and subject, the finger and the eye (Olsson 1991). This is where the ontological and epistemological interweave at the level of denotation, through body and symbol. Applying data knowledges into material structures (and vice versa) engages this form of linguistic indexicality. This notion underpins much of Olsson's work on thought and action. He described how language and identity are intertwined and how the ability to point at an object and give it a name is at once separative (between self and other) and connective (between self and language and between objects; Olsson 1980). Language is a "prison house," forming and delimiting our thinking, yet it is impossible to think beyond it: We think and talk in it, even when talking about it. This is a "chiasm," where language and action (epistemology and ontology) occupy a kind of Mobius strip in their paradoxical separation and connection (Olsson 1993). Yet, Massey (2005) made a crucial distinction between representation and space: Space is open, brimming with possibility and always in formation and process; representation, on the other hand, fixes spatial processes into stasis, leaching out material vibrancy in pursuit of indexical, descriptive, or numeric qualities. In between, Massey (2004) insisted on a lived and situated sense of place and space that includes the past and the feeling of emplacement. In this sense, the way in which data operate within and produce space is necessarily reductive, omitting the emplacement of individuals in specific sites but also resisting open spatial possibilities to further fixed and calculable space(s).

It is in translating between space and representation that we find both Olsson's and Massey's anxieties about the limitations and politics of representation. For instance, for Olsson (1982) "/" is a fundamental sign for understanding the ability of the state to draw inferences between previously separate categories, or "state capitalism" (Olsson 1982). Signs like "/" and "-" hold the crux of power, to "mislead by mixing ontologies" (Olsson 1982, 29), suggesting links even where there might be none. In doing so, they hide how language is being deployed to deliberately draw relations between ideas in the pursuit of political agendas, as we demonstrate in the next section. Here, where Olsson would argue that there are inferences between a/b or a = b, we can see a new set of inferences emerging through political initiatives, documented in critical accounts of post-eighteenth-century governmentality and, contemporarily, smart city measurement: a/1, a = 1. The ability to connect one concept, a human being, a space, a place, to a numeric value becomes embedded in the linguistic play of city planning (now including smart cities), which as Massey (2005) argued, has real material consequences, and "is a dangerous basis for politics" (185).

The UK Home Office

The UK Home Office is a particularly interesting governmental structure for understanding the politics of (spatial) representation that underlies smartness and the slippages between lived experience and abstract data. The two cases that follow demonstrate the ways in which the logic of smart cities, including many of the technologies and attendant data, are routinely, seamlessly adapted for territorial control of space in both local and global contexts. Both cases deal with an individual's right to occupy particular places and spaces, mobilized through questionable rhetorics of citizen and noncitizen, deployed in specific sites, through the use of smart technologies.

Our first case of data's meaning slipping across scales took place amidst the United Kingdom's recent "hostile environment" policy on international migration (Jackson 2015). The Home Office began to collaborate with charitable organizations in London to target homeless non-UK citizens for deportation. Some collaboration was directly on the ground, but much involved repurposing the charities' data about people who were sleeping on the street to identify and locate foreign nationals. When this partnership became public, it sparked a scandal and governmental investigation (Taylor 2018a).

The data in question were part of the Combined Homelessness And Information Network (CHAIN),

a shared database funded by the London Mayor's Office and administered by St. Mungo's, a major homelessness charity, with access to the data granted by the Greater London Authority (GLA). Each homeless person has a listing, logged by charity outreach workers, with their name, history of homelessness, special needs, gender, age, and, crucially, their regular location and nationality (St. Mungo's 2018; Greater London Authority 2019). In May 2015, the Home Office secretly obtained GLA permission to access the CHAIN database. Those data now served a Home Office program to remove homeless non-UK nationals. If a homeless person declined contact or refused an outreach worker's offer of help to voluntarily return to his or her country of origin, the Home Office would send officers to his or her regular location to detain and deport the person by force (Taylor 2018b). CHAIN geographic data facilitated the detentions by indicating where to find the homeless person in question and, in the aggregate, geographic "hotspots" where potential deportees might be concentrated. "We are trying to build in a timeline on the map so you can see where non-UK nationals have moved to over time, which hopefully will also be able to help you establish priorities by seeing patterns" (Townsend 2017). Deportation rates rose an estimated 41 percent for EU nationals, totaling 698 EU nationals deported by May 2017[1] (Ironmonger 2018).

The second example stretches from 2018 to the time of writing. In the wake of Brexit, the UK Home Office introduced an app-based immigration application system for EU migrants hoping to settle permanently in the UK. EU EXIT is an Android-only app that cross-references multiple data channels, using smartphones to conduct biometric passport or residence card scans and iProov facial recognition technology to verify applicants' identities before checking against preexisting residential data (Brewster 2018). The power of the smart technology discourse is evident in the limited pushback against such a vast data-gathering exercise, which underwent testing with a group of University of Manchester health service workers in November and December 2018 and a public beta test ending in March 2019 (the former including two of the authors of this article). In April 2019, the app was rolled out to all UK-based EU nationals. There is evidence to suggest that this technology will lead to further use of facial recognition technology provided by iProov, as well as radio-frequency identification, near-field communication, and other technology provided by WorldReach and ReadID, which are likely to be used at UK ports and borders after Brexit (GOV.UK 2019; Parliament.UK 2019), indicating the expansion of the logic of smart technocratic governance beyond noncitizen individuals and into far wider spheres.

This still emerging case represents one of thousands of new interactions between spatial and social data under the guise of "smart" rationalizations—where abstractions of place and identity feed into a national network but with profoundly personal consequences. In both cases, the ontology of the data scaled between the hyperlocal coordinates at a single street or block and the centralized, more global administrative hubs of commercial servers,[2] CHAIN, and Home Office. This extreme scalability is characteristic of data use in recent years (Dalton 2018) but also of the omnipresence and naturalization of particular smart discourses, which in turn shape smart initiatives and the spaces and places of smart cities more broadly. In the CHAIN–Home Office example, it was nonsmart resistance (in the form of media discourses and community pushback) that worked against the implied objectivity of the CHAIN database and its representation of homelessness as something appropriate for the Home Office to calculate and act on within the City of London. The local scale of street and place came up against the national scale of CHAIN, creating a slippage between lived experiences of place, wider occupation of smart spaces and places through software, and the governmental logics of measurement, calculation, and classification.

In the EU EXIT example, at stake is the right to occupy one's existing home or place of work but also the very representation of home and space in the smart city: The EU right to remain, presently held by millions of EU nationals in the United Kingdom, has been brought into question before the conclusion of Brexit. Data that previously existed for ease of transit (e.g., biometric residence card data) are repurposed and triangulated via smart technologies into disempowering significations and determinations. Although borders, locations, and everyday practices have not changed, this amalgamation of smart biometric technology and data demands new measurements and calculations (nationality, years and place of residence, even criminality) to determine classification (settled, presettled, rejected) and in turn to support or deny the right to reside in the United Kingdom. This right is first determined by the algorithms, calculations, and biometric scans of

an app, before ever being assessed by Home Office staff. The lived and situated encounters of local people are rendered as dates and documentation to be verified by human eyes only once confirmed that the scanned passport (using NFC technology) and the scanned face (using iProov technology) match the database information (checked using WorldReach). In this context, local people suddenly become noncitizen others; their places are transformed from secure and settled to under threat from the power of smart technology, deployed in service of the governmental logics of measurement, calculation, and classification—a process that is mirrored by the violence seen by Olsson in language and the logics resisted by Massey denying the specificity of place.

Our concern is in the production, reproduction, and reassignment of equivalences between local, situated processes and broader governmental rationalities through analytically created "obligatory passage point[s]" (Callon 1986, as cited in Söderström, Paasche, and Klauser 2014). Like Massey's (2005) relationality, these points support the production of a space of governance and influence specific experiences of place. In London, for example, CHAIN aims to describe and map urban homelessness through the personal data and location of homeless individuals, including their nationality, whereas the EU EXIT app operates at a more diffuse level, measuring the future legal rights of EU citizens throughout the United Kingdom. When these data cease to be purely descriptive and become actioned—by people refusing to rent houses to EU nationals (Westwater 2019), for example, or the Home Office using location data to find and deport vulnerable individuals—measurement, calculation, and classification come into being as political tools that shape not just personal experience but also wider ontological and epistemological claims to the experience and knowledge of particular places as they emerge within the spatially driven reasonings of "smart." Clear lines of division emerge between empowered experts who make and respond to data equivalences and the people whose spaces and places are transformed (Shelton, Zook, and Wiig 2015).

The calculative and classificatory abilities of such interactions build on the critical concerns of both cases at the level of representation and its limits. What the EU EXIT app shows us is the exceptional capacity not only for scalability but for structuring a space of control that operates in many individual places at once—

through the portability of the smartphone app but also through the omnipresence of the surveillant capacities of smart technologies. We argue that it is at these moments of representation or, as Olsson suggested, equivalences and inferences that "smartness" begins political work. Rather than conceptualizing space as multifaceted and vibrant, this technocratic smartness flattens space into a representational data surface, which is easily correlated with other flat data surfaces and might be hijacked or repurposed according to different political and economic agendas.

The original intention for the data, much less its calculation and classification at other scales, neglects important aspects for people who live in these datafied spaces and places, including the jobs, privacy, and human rights of homeless people in London and EU citizens living across the United Kingdom. Furthermore, due to the limitations of the data measurement, calculation, and classification, these processes overlook how such programs and applications might simply redistribute problems. The CHAIN case, for example, might redistribute homelessness internationally and possibly within London, as homeless people become wary of charity workers, but does nothing to resolve the issue of homelessness itself. Conversely, the EU EXIT app repurposes and shares existing data, in concert with private entities, with little attention to identity security, community impact, or the inherent risks of unleashing mobile mass surveillance technology and data with limited forward planning. In Olsson's terms, at the basic level of equivalence and inference, so much geographical context is left out of this "smart" data that its fundamental equation, -/-, will obscure the many social psychological, environmental, and even practical aspects of living in the smart city. Moreover, centralized strategic decisions rely on data that can be compared, contrasted, and classified—a logic of the line (territory, rule, data) over the lived. Whether presenting shortcuts or deporting people, this logic involves enacting control through space and removing bodies from place.

New Openings and Problems

In both cases, data sets are devised and chosen to suit the platform (Android, IProov, EU EXIT) or organization (Home Office) rather than to reflect the heterogeneity of the places and lives they represent. Equivalences and inferences become ideologically embedded in quantitative systems transforming

into bureaucratic systems focused more on reifying their own epistemological status than addressing social needs (Olsson 1974). This calculating approach emphasizes preserving already present political relations of the status quo. Connecting back to Olsson's concerns about language,[3] the problem of signified Cartesian inferences in spatial planning models is precisely this emphasis on producing conditions to ensure its own perpetuity, rather than resolving social problems supposedly targeted by governmental logics.

Signs, for Olsson, are always culturally situated and subjective, demarcating lines of difference and identity: you–me, we–them, individual–society. Through measurement, calculation, and classification, data move between these contexts, creating slippages, even disjunctures along those lines between people and organizations operating on different geographic scales, articulating new political and spatial relations. In arguing that language is a prison house, however, Olsson's thought elides any possibility of reclaiming spaces and lives that have become datafied and rationalized according to external agendas and discourses. Alternatively, Massey (2004) suggested that emplacement, the embodiment of the self as structured through place, is fundamentally meaningful; it is the site at which representation, bodies, and language come together. Far from Olsson's prison, Massey argued that lived and grounded spatial encounters are the site at which meaning must be made in resistance to hegemonic geographies, which we argue include smart discourses and cartographic reason. Massey (2004) cautioned against the exoneration of the local from global spaces and instead demands a critique of local–global politics around individual experience, space, and place:

> These things are utterly everyday and grounded at the same time as they may, when linked together, go around the world. Space is not the outside of place; it is not abstract, it is not somehow "up there" or disembodied. Yet that still leaves a question in its turn: How can that kind of groundedness be made meaningful across distance? (8)

We have argued that such groundedness has been both exploited and obscured by the operations of smart city politics that classify and calculate and increasingly extend beyond smart cities and into wider discourses and bureaucratic functions. Although Olsson's "reason" certainly dictates the operation of smart cities and discourses, place itself presents its own complexity that cannot be reduced neatly to data nodes and points. Within these processes, which places and people are deemed worth defending (Massey 2004) from such processes that rely so heavily on measurement, calculation, and classification? What alternative possibilities are missed? Ultimately, how can we better know the limits of Cartesian reason and recognize the unstable foundations on which smart cities, spaces, and places are built?

Acknowledgments

Clancy Wilmott is now with the Department of Geography, University of California Berkeley.

Notes

1. Data on non-EU nationals were not published.
2. At the time of publication, the authors could not determine where the EU EXIT data are stored or what data are accessible to nongovernment providers, although the privacy policy in the app clearly warns that data may be shared with private-sector organizations. Freedom of information requests about these data have been refused by the Home Office, including a request by Axel Antoni from 2018 (Whatdotheyknow.com 2018).
3. A key differentiating factor from Foucault's work.

References

Amoore, L., and A. Hall. 2009. Taking people apart: Digitized dissection and the body at the border. Environment and Planning D: Society and Space 27 (3):444–64. doi: 10.1068/d1208.

Amoore, L., and V. Piotukh. 2015. Life beyond big data: Governing with little analytics. Economy and Society 44 (3):341–66. doi: 10.1080/03085147.2015.1043793.

Anthopoulos, L., M. Janssen, and V. Weerakkody. 2019. A unified smart city model (USCM) for smart city conceptualization and benchmarking. In Smart cities and smart spaces: Concepts, methodologies, tools, and applications, ed. Information Management Resources Association, 247–64. Hershey, PA: IGI Global.

Beer, D. 2016. Metric power. London: Palgrave Macmillan.

Beer, D. 2018. Data gaze. London: Sage.

Belyea, B. 1992. Images of power: Derrida/Foucault/Harley. Cartographica 29 (2):1–9.

boyd, d., and K. Crawford. 2012. Critical questions for big data. Information, Communication & Society 15 (5):662–79. doi: 10.1080/1369118X.2012.678878.

Brewster, T. 2018. Meet the businesses making millions from the Brexit immigration nightmare. Forbes, November 14. Accessed May 15, 2019. https://www.forbes.com/sites/thomasbrewster/2018/11/14/meet-the-businesses-making-millions-from-the-brexit-immigration-nightmare/#570bb6de4390.

Callon, M. 1986. Eléments pour une sociologie de la traduction: la domestication des coquilles Saint-Jacques et des marins-pêcheurs dans la baie de Saint-Brieuc. L'Année sociologique 36:169–208.

Canon, G. 2018. Surveillance fears grow after Taylor Swift uses face recognition tech on fans. The Guardian, December 13. Accessed December 29, 2018. https://www.theguardian.com/music/2018/dec/13/taylor-swift-facial-recognition-technology-surveillance.

Cardullo, P., and R. Kitchin. 2018. Smart urbanism and smart citizenship: The neoliberal logic of "citizen-focused" smart cities in Europe. Environment and Planning C: Politics and Space 37:813–30.

Dalton, C. 2018. Big data from the ground up: Mobile maps and geographic knowledge. The Professional Geographer 70 (1):157–64. doi: 10.1080/00330124.2017.1326085.

Dalton, C., and J. Thatcher. 2015. Inflated granularity: Spatial "big data" and geodemographics. Big Data & Society 2:1–15. doi: 10.1177/2053951715601144.

Datta, A. 2018. The digital turn in postcolonial urbanism: Smart citizenship in the making of India's 100 smart cities. Transactions of the Institute of British Geographers 43 (3):405–19. doi: 10.1111/tran.12225.

de Saussure, F. 1987. Course in general linguistics. New York: Columbia University Press.

Dowling, R., P. McGuirk, and S. Maalsen. 2018. Realising smart cities: Partnerships and economic development in the emergence and practices of smart in Newcastle, Australia. In Inside smart cities: Place, politics and urban innovation, ed. A. Karvonen, F. Cugurullo, and F. Caprotti, 15–29. London and New York: Routledge.

Farinelli, F. 1998. Did Anaximander ever say (or write) any words on the nature of cartographical reason. Philosophy & Geography 1 (2):135–44. doi: 10.1080/13668799808573640.

Foucault, M. 2002a. Archaeology of knowledge. London and New York: Routledge.

Foucault, M. 2002b. Discipline & punish: The birth of the prison. London and New York: Routledge.

Foucault, M. 2008. Security, territory, population: Lectures at the College de France 1977–1978. New York: Palgrave.

Gabrys, J. 2014. Programming environments: Environmentality and citizen sensing in the smart city. Environment and Planning D: Society and Space 32 (1):30–48. doi: 10.1068/d16812.

Gabrys, J. 2016. Program Earth: Environmental sensing technology and the making of a computational planet. Minneapolis: University of Minnesota Press.

GOV.UK. 2019. EU settlement scheme public beta testing phase report. Accessed May 19, 2019. https://assets.publishing.service.gov.uk/government/uploads/system/uploads/attachment_data/file/799413/EU_Settlement_Scheme_public_beta_testing_phase_report.pdf.

Greater London Authority. 2019. Rough sleeping in London (CHAIN reports). Accessed January 2, 2019. https://data.london.gov.uk/dataset/chain-reports.

Greenfield, A. 2017. *Radical technologies: The design of everyday life.* London: Verso.

Gregg, M. 2015. Inside the data spectacle. *Television & New Media* 16 (1):37–51. doi: 10.1177/1527476414547774.

Hind, S., and A. Gekker. 2014. "Outsmarting traffic, together": Driving as social navigation. *Exchanges: The Warwick Research Journal* 1 (2):1–17.

Ironmonger, J. 2018. EU rough sleepers win damages for illegal deportations. *BBC News*, May 13. Accessed December 20, 2018. https://www.bbc.com/news/uk-44093868.

Jackson, E. 2015. *Young homeless people and urban space: Fixed in mobility.* London and New York: Routledge.

Kitchin, R., T. Lauriault, and G. McArdle. 2015. Knowing and governing cities through urban indicators, city benchmarking, and real-time dashboards. *Regional Studies, Regional Science* 2 (1):6–28. doi: 10.1080/21681376.2014.983149.

Leszczynski, A. 2009. Rematerializing GIScience. *Environment and Planning D: Society and Space* 27 (4):609–15. doi: 10.1068/d1607b.

Luque-Ayala, A., and S. Marvin. 2015. Developing a critical understanding of smart urbanism? *Urban Studies* 52 (12):2105–16. doi: 10.1177/0042098015577319.

Maalsen, S. 2019. Smart housing: The political and market responses of the intersections between housing, new sharing economies and smart cities. *Cities* 84:1–7. doi: 10.1016/j.cities.2018.06.025.

Mackenzie, A. 2018a. *Machine learners.* Cambridge, MA: MIT Press.

Mackenzie, A. 2018b. Personalization and probabilities: Impersonal propensities in online grocery shopping. *Big Data & Society* 5:1–15. doi: 10.1177/2053951718778310.

Marston, S. A., J. P. Jones, III, and K. Woodward. 2005. Human geography without scale. *Transactions of the Institute of British Geographers* 30 (4):416–32. doi: 10.1111/j.1475-5661.2005.00180.x.

Massey, D. 2004. Geographies of responsibility. *Geografiska Annala* 86 (1):5–18. doi: 10.1111/j.0435-3684.2004.00150.x.

Massey, D. 2005. *For space.* London and New York: Routledge.

Mattern, S. 2017. A city is not a computer. *Places Journal.* doi: 10.22269/170207.

Moore, A. 2008. Rethinking scale as a geographical category: From analysis to practice. *Progress in Human Geography* 32 (2):203–25. doi: 10.1177/0309132507087647.

Olsson, G. 1974. Servitude and inequality in spatial planning: Ideology and methodology in conflict. *Antipode* 6 (1):16–21. doi: 10.1111/j.1467-8330.1974.tb00579.x.

Olsson, G. 1979. Social science and human action or on hitting your head against the ceiling of language. In *Philosophy in geography*, ed. S. Gale and G. Olsson, 287–307. Dordrecht, The Netherlands: Springer.

Olsson, G. 1980. *Birds in egg/egg in birds.* London: Pion.

Olsson, G. 1982. -/-. *SubStance* 11 (2):24–33. doi: 10.2307/3684022.

Olsson, G. 1991. *Lines of power/limits of language.* Minneapolis: University of Minnesota Press.

Olsson, G. 1993. Chiasm of thought-and-action. *Environment and Planning D: Society and Space* 11 (3):279–94. doi: 10.1068/d110279.

Olsson, G. 2000. From a = b to a = a. *Environment and Planning A: Economy and Space* 32 (7):1235–44. doi: 10.1068/a3257.

Olsson, G. 2007. *Abysmal: A critique of cartographic reason.* Chicago: University of Chicago Press.

Parliament.UK. 2019. EU exit: Written statement–HCWS1386. Accessed May 20, 2019. https://www.parliament.uk/business/publications/written-questions-answers-statements/written-statement/Commons/2019-03-07/HCWS1386/.

Rose, G. 1993. *Feminism & geography: The limits of geographical knowledge.* Cambridge, UK: Polity.

Rose, G., and A. Willis. 2018. Seeing the smart city on Twitter: Colour and the affective territories of becoming smart. *Environment and Planning D: Society and Space* 37:411–27. doi: 10.1177/0263775818771080.

Scola, N. 2018. Google is building a city of the future in Toronto. *Politico*, July/August. Accessed January 4, 2019. https://www.politico.com/magazine/story/2018/06/29/google-city-technology-toronto-canada-218841.

Scott, J. C. 1998. *Seeing like a state: How certain schemes to improve the human condition have failed.* New Haven, CT: Yale University Press.

Shelton, T., and T. Lodato. 2018. Actually existing smart citizens: Expertise and (non)participation in the making of the smart city. *OSF*, November 2.

Shelton, T., M. Zook, and A. Wiig. 2015. The "actually existing smart city." *Cambridge Journal of Regions, Economy and Society* 2015 (8):13–25. doi: 10.1093/cjres/rsu026.

Silva, B. N., M. Khan, and K. Han. 2018. Towards sustainable smart cities. *Sustainable Cities and Society* 38:697–713. doi: 10.1016/j.scs.2018.01.053.

Söderström, O., T. Paasche, and F. Klauser. 2014. Smart cities as corporate storytelling. *City* 18 (3):307–20. doi: 10.1080/13604813.2014.906716.

Sparke, M. 1994. Escaping the herbarium: A critique of Gunnar Olsson's "chiasm of thought-and-action." *Environment and Planning D: Society and Space* 12 (2):207–20. doi: 10.1068/d120207.

St. Mungo's. 2018. CHAIN—Combined homelessness and information network. Accessed December 20, 2018. https://www.mungos.org/work-with-us/chain/.

Taylor, D. 2018a. Charity may have shared rough sleepers' data without consent, watchdog finds. *The Guardian*, September 21. Accessed December 20, 2018. https://www.theguardian.com/society/2018/sep/21/st-mungos-is-likely-to-have-given-home-office-data-on-rough-sleepers-information-commissioner.

Taylor, D. 2018b. Complaint filed against charity over removal of EU rough sleepers. *The Guardian*, May 14. Accessed December 20, 2018. https://www.theguardian.com/uk-news/2018/may/14/st-mungos-homelessness-charity-complaint-filed-over-unlawful-removal-eu-rough-sleepers.

Thatcher, J. 2013. Avoiding the ghetto through hope and fear: An analysis of immanent technology using ideal types. *GeoJournal* 78 (6):967–80. doi: 10.1007/s10708-013-9491-0.

Townsend, M. 2017. Home office used charity data map to deport rough sleepers. *The Guardian*, August 19. Accessed December 20, 2018. https://www.theguardian.com/uk-news/2017/aug/19/home-office-secret-emails-data-homeless-eu-nationals.

Westwater, H. 2019. Brexit will stop landlords renting to EU nationals under right to rent scheme. *The Big Issue*, April 8. Accessed May 20, 2019. https://www.bigissue.com/latest/brexit-will-stop-landlords-renting-to-eu-nationals-under-right-to-rent-scheme/.

Whatdotheyknow.com. 2018. Settled status app—Data privacy, request by Axel Antoni. Accessed May 19, 2019. https://www.whatdotheyknow.com/request/settled_status_app_data_privacy.

Wiig, A. 2016. The empty rhetoric of the smart city: From digital inclusion to economic promotion in Philadelphia. *Urban Geography* 37 (4):535–53. doi: 10.1080/02723638.2015.1065686.

Wilmott, C. 2016. Small moments in spatial big data: Calculability, authority and interoperability in everyday mobile mapping. *Big Data & Society* 3 (2). doi: 10.1177/2053951716661364.

Wilson, M. 2015. Flashing lights in the quantified self-city-nation. *Regional Studies, Regional Science* 2 (1):39–42. doi: 10.1080/21681376.2014.987542.

Wyly, E. 2014. The new quantitative revolution. *Dialogues in Human Geography* 4 (1):26–38. doi: 10.1177/2043820614525732.

Technology as Ideology in Urban Governance

Luis F. Alvarez León and Jovanna Rosen

This article argues that the turn toward smart cities, emphasizing solutions, services, and infrastructures driven by digital technologies, has reinforced a dominant ideology shaping urban decision making, frameworks, and outcomes. Two core dimensions of this ideology of technology in urban governance interact to consequentially reshape urban processes: (1) the priority of attracting high-technology industries as engines for urban economies and (2) the tendency to reframe urban problems into technological problems, to be addressed by technological solutions. Together, these mechanisms operate in conjunction to privilege technological needs, capacities, and priorities in urban governance, contributing to the widespread exclusion of people and problems beyond the scope of technology. Although not unprecedented, this ideology of technology has acquired renewed potency with neoliberalized urbanism, urban restructuring, and the ongoing information revolution. Furthermore, these changes intensify the ongoing transformation of cities (and space more generally) into digitized spaces tailored for capital accumulation in the context of digital and surveillance capitalism. To illustrate these dynamics, we briefly describe recent events in San Francisco, one of the key sites in the current techno-economic paradigm. *Key Words: digital economy, ideology, smart cities, technology, urban governance.*

本文认为, 人类向智慧城市的转变、追求数字技术驱动解决方案、服务和基础设施的举动, 对城市决策、框架和结果的主导思想起到了强化作用。这种技术意识形态在城市治理中的两个核心维度相互作用, 必然会重塑城市过程: (1) 优先吸引高技术产业作为城市经济的引擎; (2) 未来趋势是将城市问题重新界定为技术问题, 由技术解决方案加以解决。这些机制共同作用, 为城市管理的技术需要、能力和优先事项提供便利, 但会造成技术范围之外的人和问题被广泛忽视。伴随新自由主义城市化、城市重组和正在进行的信息革命, 这种技术意识形态获得了新的效力, 这样的现象在历史长河中也并非前所未有。此外, 城市 (以及更普遍的领域) 正在数字资本主义和监管资本主义环境下向资本积累的数字化空间转型, 这些变化对其也起到了推波助澜的作用。为了阐述这些动态, 我们简要描述了最近在当前技术经济代表城市 —— 旧金山发生的一些事件。关键词: 数字经济、意识形态、智慧城市、技术、城市管理。

En este artículo se argumenta que el giro hacia las ciudades inteligentes, que pone énfasis en las soluciones, los servicios y las infraestructuras jalonadas por tecnologías digitales, ha fortalecido una ideología con la cual se configura la toma de decisiones, las enmarcaciones y los resultados relacionados con lo urbano. Dos dimensiones medulares de esta ideología de la tecnología en la gobernanza urbana interactúan para reconfigurar consecuentemente los procesos urbanos: (1) la prioridad de atraer industrias de alta tecnología como motores de las economías urbanas, y (2) la tendencia a replantear los problemas urbanos como problemas tecnológicos, para ser confrontados con soluciones tecnológicas. Juntos, estos mecanismos operan en conjunción para privilegiar las necesidades, capacidades y prioridades tecnológicas en la gobernanza urbana, contribuyendo de esa manera a la exclusión extendida de gente y problemas que se hallan fuera del alcance de la tecnología. Sin ser única, esta ideología de la tecnología ha alcanzado potencia renovada con el urbanismo neoliberalizado, la reestructuración urbana y la actual revolución de la información. Más todavía, estos cambios intensifican las transformaciones corrientes de las ciudades (y del espacio, en general) en espacios digitales diseñados a la medida para la acumulación de capital dentro del contexto del capitalismo digital y vigilante. Para ilustrar estas dinámicas, brevemente describimos eventos recientes en San Francisco, uno de los sitios claves del actual paradigma tecno-económico. *Palabras clave: ciudades inteligentes, economía digital, gobernanza urbana, ideología, tecnología.*

The information technology (IT) sector has become increasingly intertwined with city governments and influential urban actors, who promote technology to bring prosperity to urban regions and address social ills. Urban leaders have advanced a "smart cities" vision supporting this agenda, reflecting the broader discursive and material deployment of digital technologies toward societal improvement. This vision is supported by a "global discourse network" not just about smart cities but that represents constitutive part of them (Joss et al. 2019, 5). Its building blocks are texts and other expressions produced across the private sector, the scientific knowledge community, and a range of cross-sector actors who promote smart cities logics and expressions (Joss et al. 2019).

In this article we argue that a rising ideology of technology underlies this global discourse network and its material manifestations, which has acquired particular potency in urban governance. We contend that this ideology has come to define and reshape many urban regions through the combined emphasis on IT sector investment and technological solutions to reframe and address urban issues. Furthermore, in reshaping urban governance processes, technology is contributing to urban transformations by turning cities into digital geographies optimized for capital accumulation. The core components of the ideology of technology are identified here, but the specific shape of the urban transformations it enables is informed by the variegated geographies where it takes root. As Datta and Odendaal (2019) remarked, the smart cities narrative is propelled by hopes that are simultaneously ubiquitous and mundanely predictable, and its contradictions are revealed in a diverse array of manifestations.

Focusing on the underlying ideology of these manifestations allows us to frame the various discursive and material elements of smart cities to better understand both their drivers and their effects on urban governance. By the latter we mean the structures, processes, stakeholders, and institutions that determine the urban financial, spatial, and policy outcomes. Ultimately, we aim to show how this particular technological ideology has reshaped urban decision making and related outcomes and how its components privilege technological needs, capacities, and priorities in urban governance, thereby contributing to the exclusion of people and problems beyond the scope of technology. To illustrate these dynamics, we briefly describe recent events in San Francisco, one of the key urban sites in the current techno-economic paradigm.

Smart Cities as a Means to Address Urban Social Problems

The smart cities conceptualization envisions technological tools and digital innovations as reshaping the urban context toward greater efficiency and sustainability. Smart cities, however, are more than technical propositions. Scholars have argued that smart cities represent variant forms of entrepreneurial urbanism, salient since the 1980s (Hollands 2008; Kitchin 2015). This turn, backed by arguments of austerity, effectiveness, cost-saving innovation (Pollio 2016), and anticipation of future crises (White 2016), enabled cities to become smart by inviting private actors to experiment in the urban arena, with transformative impacts. Cities moved from primarily functioning as a place "where new technologies might be born" to "receptacles for technology, the target of its applications" (Glasmeier and Christopherson 2015, 4). Aligned with the broader neoliberal program from which it emerged, the smart city paradigm helped cities become "a new market for urban management … an urban form to be sold, resold, modified or augmented to make money" (Glasmeier and Christopherson 2015, 6). Combined with the state's long-standing role in facilitating innovation by undertaking risky early investment or offering subsidies to create financial incentives, it is unsurprising that local governments and leaders have become key actors enabling smart city logics to take hold (Lazonick and Mazzucato 2013).

Although digital technologies can improve urban problem solving, the frequent alignment between the smart city framework and a more general ideology of technology remains problematic, influencing how we think about, govern, and participate in the city. This has produced a "smartmentality," whereby "cities are made responsible for the achievement of smartness—i.e. adherence to the specific model of a technologically advanced, green and economically attractive city, while 'diverse' cities, those following different development paths, are implicitly reframed as smart-deviant" (Vanolo 2014, 889).

Scholars have documented that a greater reliance on complex technological and informational systems to address social issues, beyond disciplining urban processes, can exclude, marginalize, and hypersurveil vulnerable populations (O'Neil 2017; Eubanks 2018; Mann and Daly 2018; Noble 2018). Placing too much emphasis on and trust in technological tools can additionally black-box decision making, further curtailing the public sphere (Pasquale 2015; Vaidhyanathan 2018). At the urban scale, this can give private corporations disproportionate control over influential urban governance decisions, with little public accountability (Johnson et al. 2017). Technological firms' outsized and undemocratic power over urban affairs, therefore, tightly couples cities' fates to the welfare of the very firms that make them "smart" (McNeill 2015) and paves the way for a transformation of urban space for the accumulation of capital via (increasingly) digitized means.

On the other hand, some researchers have also argued for a shift toward the "Smart City 2.0" (Etezadzadeh 2016), where technological solutions are not foregrounded; instead, endogenously defined social issues are prioritized through citizen participation and empowerment. For example, Trencher (2019) discussed the case of Aizuwakamatsu, a Japanese city whose smart approach prioritized endogenous societal issues over technological solutions: Social considerations drove technological development and customization. Community ownership of each smart solution reinforced the dynamic, which the author argued produced community empowerment.

Although we acknowledge the wide range of smart city types and the possibility for successfully integrating citizen participation, private industry, and government in the service of pressing social challenges, in this article we take on the underlying ideology that informs the imperative to "smartify" urbanism. We see this as a specific expression of the more widespread drive to rely on the convergence of private industry, digital technologies, and market logics to address (or, more narrowly, solve) social problems.

We take a critical position because, despite instances of increased citizen participation, we interpret the rise of technological urban governance solutions and smartfication as manifestations of more generalized trends toward digitization, privatization surveillance, and capital extraction, which in turn feed into new configurations of neoliberal urbanism (Levenda, Mahmoudi, and Sussman 2015; Ho 2017; Cardullo and Kitchin 2018). In this article, we argue that this ideological dimension shapes the smart city project by legitimizing, empowering, and foregrounding digital, networked, and other smart technologies for urban governance, along with the corporations behind them—simultaneously redefining urban political spaces (Barns 2016) and often narrowing the possibilities for understanding and participating in urban life (Tenney and Sieber 2016; Wiig and Wyly 2016; Stehlin 2018).

Rejecting corporate control over urban futures, Sadowski and Bendor (2018) advocated for breaking

discursive dominance over smart city imaginaries to build alternative urban visions. Our article advances this effort by addressing the conceptual underpinnings of tech sector hegemony. By identifying two intertwined dimensions—tech investment attraction to catalyze urban growth and the rise of technological urban solutionism—we show how the tech sector's influence on many urban regions operates at the ideological level to reshape decision making, outcomes, and ultimately the very constitution of cities as digitized environments tailored for capital accumulation.

Digital Technology Industries as Engines of Urban Growth

Facing growing inequality, fiscal constraints, footloose companies, and heightened local revenue competition, cities actively pursue industrial attraction as a growth strategy. Despite its dubious track record and unevenly distributed benefits, this smokestack chasing has only moved toward increasingly competitive, outright bidding to attract technology companies, seen to bring high-paying jobs and young, educated professionals—and a certain prestige. Since the early 2000s, cities responded to the post-dot-com crash tech industry resurgence by deliberately pursuing amenity generation strategies to draw the creative class of young, hip, artistic, and technologically oriented professionals (Florida 2002), with the hopes of becoming the next iteration of tech hubs (MacGillis 2009). The protracted interurban race for the new Amazon headquarters provides only the most recent example of the fierce competition and sincere hope attached to tech firm locational decisions—a competition that Amazon actively promoted through competitive bidding.

This development model is built on a trifecta of technology, talent, and tolerance, with technology foregrounding the latter two (Florida 2000; Gates and Florida 2002). Therefore, we examine this technological idealization and cities' related mobilization around technology to induce urban growth. In the two decades since Florida's arguments first gained traction, digital technologies and their applications have continued to evolve and diversify as the central role of technology in urban governance has become increasingly normalized. The proliferation of urban big data, surveillance, and smart applications has substantively changed how cities work and helped make technology an articulating force in new visions of urban life, leading to the digitization of the urban fabric, remaking it into a site for data collection and monetization or, as one critic put it, a "behavioral data farm" (Rogan 2019). Accordingly, in the trifecta just referenced, the role of technology as an engine of growth is increasingly complemented by the widespread incorporation of technological solutions as core elements of urban governance that can provide fixes for all manner of urban problems. This all-purpose application constitutes the second dimension of technology as an urban governance ideology.

Urban Problems Become Technological Problems

How does an ideology of technology help transform urban problems into technological problems? According to Morozov (2013, 2014), a powerful step is the advancement of *solutionism*, defined as "a pervasive and dangerous ideology [… that views problems] based on just one criterion: whether they are 'solvable' with a nice and clean technological solution at our disposal" (Morozov 2013). Solutionism requires that social settings be redefined as domains with discrete, solvable problems.

Examining the pervasive likening of cities to computers, Mattern (2017) identified a "new ideology on the rise," which redefines cities in terms of information optimization and effectiveness. Mattern (2017) described various influential urban initiatives driven by technology companies but countered that these projects misread cities, failing to understand that "*city-making* is always, simultaneously, an enactment of *city-knowing*—which cannot be reduced to computation."

Here we argue that technological solutionism and its associated redefinition of social (and urban) life, together with technology firm attraction mobilized as an urban economic growth engine, constitute the core manifestations of a more generalized ideology of technology, which has acquired particular potency in urban governance.

Technology as Ideology and Why It Matters

Although the concept of ideology has shifted over time, and engendered much debate, Knight (2006) found an enduring core definition: "a coherent and relatively stable set of beliefs or values" (625). Williams (1985), for example, identified three main uses of the term *ideology* that emerged throughout the nineteenth century. In the first, derived from Napoleon's attack on his political opponents, ideology came to mean "a sense of abstract, impractical or fanatical theory." Second and third senses advanced by Marx and Engels, who saw ideology both as the false consciousness and "the set of ideas which arise from a given set of material interests or, more broadly, from a definite class or group" (Williams 1985). Synthesizing Marx and Engels's views of ideology, Martin (2015) argued that beliefs are ideological when "they are *idealized*, *universalized*, and *detached* expressions of actual social relations" (17–18). These three characteristics—idealization, universalization, and detachment—often manifest when digital technologies are deployed across social domains: Technology's benefits are idealized, its applications are universalized, and it becomes detached from its constitutive social and power relations.

Shearmur and Wachsmuth (2019) described this manifest tendency, noting that "the prevailing 'respectable' attitude towards technology is one of determinism," where "there is no place for asking why new technology is necessary, who benefits from it, or whether it diverts resources from more important endeavors." They see this imperative as another phase in the evolution of the city as a "growth machine," which they term the "innovation machine" and is characterized by the emergence of local coalitions "assembled around urban technology as an unquestioned good" (Shearmur and Wachsmuth 2019, 176–77).

How does an ideology of technology underwrite the smart cities framework? From building new cities on greenfield sites (Shwayri 2013; Cugurullo 2016) to, more commonly, "awkwardly" integrating "smartness" into existing urban configurations (Shelton, Zook, and Wiig 2014, 15), smart cities are sold as both technical and political propositions enveloped by normative discourses. These discourses often constitute utopian visions that can be readily adapted to particular political conditions or structural configurations (Datta 2015; Luque-Ayala and Marvin 2015; Marvin, Luque-Ayala, and McFarlane 2015; Grossi and Pianezzi 2017). The ideological dimension of technology enters here, enrolled to buttress the political work of persuading and repositioning actors and interests, while appearing entirely scientific, objective, and technical. This aligns with neoliberal urbanism's broader discourse and practice by advancing privatized urban visions that often transcend public deliberative realms.

When deploying a smart cities framework, influential urban actors, and predominantly corporations and city governments, idealize conceptions of technology that universalize its applications and detach them from the actual social relations that define urban life. This idealized conception both enables and obscures the actually existing smart city (Shelton, Zook, and Wiig 2014), a concrete manifestation of political and technological orders, with context-specific impacts, composed of regulations, governance practices, and identifiable economic interests. The ideology of technology's double valence allows urban actors to craft political alliances toward two related ends:

1. The attraction on high technology industries.
2. The transformation of urban problems into technological problems, requiring technological solutions.

At this point, technology can be mobilized strategically by urban power holders to advance specific urban futures. By explicitly recognizing the ideology of technology accordingly, we can identify how technology has come to embody a particular set of ideas and ideals, as well as a corresponding urban governance vision, legitimized in public discourse and mobilized beyond democratic debate.

Aligned with the priorities and orientations of neoliberal urbanism, this reframing does more than advance smartness or optimality; at a fundamental level, it enables capital accumulation and political power concentration through ostensibly technical means that transcend public deliberation. Therefore, it is no accident that the development of digital urban tools has increasingly shifted toward building governance mechanisms that, by virtue of their technological complexity and purported efficiency, extend beyond reach for most citizens and consequently elude participation, deliberation, and even understanding—a mode of social organization that Pasquale (2015) labeled the "black box society."

Next, we offer a short vignette to illustrate this ideology and its corresponding political–economic dynamics in urban settings. This describes how the ideology of technology has taken hold and reshaped urban governance and spatial arrangements in a particularly important site of technological production and its incorporation into urban governance: the city

of San Francisco. The next section documents how the technological ideology and its manifestations have become embedded in the city's functioning, consequently producing significant backlash for reshaping urban governance and sociospatial arrangements.

San Francisco: Technology as Growth Engine and Solution

San Francisco is a matured tech cluster that has benefited from this position while struggling to address the challenges of a deeply engrained technological ideology and the urban transformations associated with the tech industry over the past decades.[1] Adjacent to the Silicon Valley, San Francisco has long housed startups and therefore played an important role in regional tech industry development (Saxenian 1996; Stehlin 2016). Over the past decades, major tech firms, including Twitter, Airbnb, Dropbox, Salesforce, and Yelp, have located in San Francisco, and the city currently experiences more rapid tech growth than the rest of the Silicon Valley (McNeill 2015; Stehlin 2016). Silicon Valley and San Francisco have created a mutually reinforcing tech cluster assisted by an educated workforce, strong regional educational institutions and business organizations, supporting firms, and a changing urban culture. The city is now "a veritable playground for the instant wealth of the tech industry," where the inequality produced by the tech industry's economic polarization is extensively documented (Stehlin 2016, 474; see also Guzman and Stern 2015; Walker 2018). San Francisco's emergence as a tech hub in its own right, however, reflects the successful, deliberate technological pursuit along the two dimensions of firm attraction and technological solutionism—an entrenched ideology that has contributed to significant inequality and urban political conflict.

Mayor Ed Lee drove this deliberate tech industry pursuit. As he ascended to mayor-elect, Lee stated, "I want tech companies to start here in San Francisco, and I want them to stay and grow" (Tsotsis 2011). Indeed, with a strong jobs emphasis, Lee became known as "tech's mayor": a steadfast City Hall tech ally leading a governance drive to subsidize and support technological innovation and its supporting industries (Lazonick and Mazzucato 2013). Lee famously advanced the "Twitter tax break," a policy intended to catalyze the redevelopment of the Tenderloin, a neighborhood bordering

City Hall with the city's largest concentration of homeless service providers and homeless residents (Stehlin 2016; "Ed Lee's legacy" 2017). In the late 2000s, the area struggled to cope with high office vacancy rates and was seen as ripe for development to reduce blight (Coté 2014). The city government sought to create an "arts district" to encourage new, trendy businesses and pave the way for future investment. Simultaneously, the city struggled to continue to attract and retain tech firms. Many firms cited San Francisco's payroll tax as a reason for some jobs leaving the city (including Yelp and Trulia) or threatening future headquarters moves away from the it (e.g., Twitter and Zynga; Taylor 2011; Coté 2014).

In 2011, Twitter searched for larger offices, threatening to leave the city. City officials responded with a plan to create a tech hub in the larger Tenderloin neighborhood, leveraging the available office space, the neighborhood's advantageous location, and proximity to transit and the desirable SOMA neighborhood, where Twitter was previous located, to induce the tech giant to stay in the city. The linchpin of this strategy was the city's contested proposal to create a Twitter tax break that lessened payroll taxes, including stock options (over a time period when Twitter went public) for six years for companies that located in the mid-Market or Tenderloin neighborhoods—now rebranded as the Central Market district (Taylor 2011; Coté 2014; Lang 2015; McNeill 2015; Stehlin 2016). The tax break, costing the city $34 million in 2015 alone, was championed by Mayor Lee, whose campaign received significant tech investment and who supported the industry and job creation throughout his policy agenda (Lang 2015; McNeill 2015). This relationship demonstrates the strong alignment between local government and the tech industry—revealing the entrenched ideology that the industry's growth would advance the city's interest, regardless of evidence and advocacy to the contrary—a governance dynamic that has reshaped San Francisco.

The city has continued to prioritize the tech sector's needs, including supporting industries and businesses, as well as the needs of tech employees, who have large salaries to spend (McNeill 2015; Stehlin 2016). Within a state economic context with limited potential for property tax revenue, sales tax offers an important opportunity to bolster economic revenue for cities. For this reason, as young tech employees want to live in the city, the city, in turn, has sought ways to support ways to keep tech firms and tech workers in the city. Google buses, shuttling workers from their homes in San Francisco to the Google campus in Mountain View, have further entrenched this spatial arrangement. This situation benefits both tech firms, looking to attract young, high-skilled workers who do not want to live in suburban Silicon Valley, and the city government, who wants to retain the industry's economic spillovers.

Google buses thus offer a private-sector solution to the public problem of a spatial jobs–housing mismatch, free-riding off of public bus loading zones and therefore disrupting public transit but aligned with the technological solutionism that the entrenched tech ideology promotes. Initially, Mayor Lee proposed allowing tech shuttles free bus stop use, impeding public transportation in the city and subsidizing tech firm operations. After pushback, the city then imposed on them a small fee—less than $8 per stop (Knight 2015; "Ed Lee's legacy" 2017). Google buses have produced years-long public protests, with critics deriding the "techsploitation" brought by tech investment and technological solutionism, driving significant regional inequality and economic polarization, reliant on public investment and subsidies, and exacerbating housing unaffordability (Stehlin 2016; McElroy 2019).

Throughout, Mayor Lee disproportionately protected the tech industry's needs, under the guise of jobs and industrial attraction and retention. Relatedly, Lee was known for his lack of tech industry regulation—including vetoing a law that would have capped at sixty days annually the extent to which landlords can rent units using short-term rentals such as Airbnb at the same time that the city began to experience an affordable housing crisis, with insufficient affordable units in the region available to low- and middle-wage workers, including those working for tech and supporting firms but not in nominally high-skilled roles (Brinklow 2016). He was also criticized for not regulating Uber and Lyft enough, which have further contributed to regional economic polarization ("Ed Lee's legacy" 2017). The city government has also allowed policies to help rebrand San Francisco such as the $110 million sale of naming rights of the Transbay Terminal, a city landmark, to the tech firm Salesforce and the renaming of the San Francisco General Hospital for Facebook's Mark Zuckerberg after a $75 million gift (McElroy 2019).

Table 1 documents some of the key policies enacted by the city of San Francisco to attract

Table 1. San Francisco policies relating to the high-technology attraction (organized chronologically)

Type of policy	Description	Actor	Objective	Date
Tax break	Central Market Street and Tenderloin Area Payroll Expense Tax Exclusion (aka Twitter tax break)	City of San Francisco	Retain Twitter headquarters in San Francisco	2011
Fee	Google and other bus shuttles fee	City of San Francisco	Initially allowed Google buses to use shuttle stops for free; later imposed an $8/stop charge	2014–2016
Veto	Vetoed cap on Airbnb hosting	Mayor Ed Lee	Vetoed measure that would have imposed Airbnb limits of 60 days annually	2016
Naming rights	Naming rights for Transbay Terminal	City of San Francisco	Sold naming rights to San Francisco landmark Transbay Terminal for $110 million to tech firm Salesforce	2018

Notes: Assembled by authors with data from various sources.

high-technology industry. In these ways and beyond, the tech sector has reshaped San Francisco in its image and contributed to growing inequality, offering technological solutions targeted to benefit tech workers. These solutions are detached from the broader systemic issues that the tech sector continues to intensify, from housing market pressures to unequal economic opportunities. Combined with an idealized and generalized belief in the tech industry's benefits, however, such solutions have contributed to a strong ideology of technology. This idealization and solutionism eschews necessary conversations about public investment, economic polarization, and regional housing shortages (Schafran 2013; McNeill 2015; Stehlin 2016). Rather, the entrenched ideology of technology in San Francisco, advanced by tech firms with the support of city leadership, has (re)arranged space to support the tech industry, its urban restructuring objectives, and related capital accumulation.

Conclusion

In this article, we advance a theoretical framework revealing a dominant ideology of technology operating to restructure cities and their governance processes, priorities, and outcomes, "recreating the city, both public and private, as the common workplace of the tech sector" (Stehlin 2016, 475). Combined with the urban inequality it engenders and the public resources it requires, this transformation has produced enduring backlash. We draw evidence from San Francisco's entrenched tech cluster

to illustrate how this ideology operates along the dual dimensions of tech attraction and technological solutionism.

From nineteenth-century Hausmannization to modernist and postmodernist forms of command and control, technology has always been crucial to manage the city. In the current context of digital-surveillance capitalism, though, technology has come to suffuse every aspect of city life and qualitatively transformed the configuration of space (Kitchin and Dodge 2011), particularly in urban settings (Graham and Guy 2002)—a change thoroughly embodied in the smart cities paradigm. Beyond pulling governance processes away from democratic practice (echoing the warnings issued by recent inquiries [Wilson and Swyngedouw 2014] on the rise of the postpolitical), this has the profound effect of channeling urban dynamics into an ever-expanding data stream, which fundamentally turns cities, and space itself, into digitized landscapes optimized for capital accumulation. This process, relying on continuous data collection and monetization, constitutes a contemporary variant of familiar logics of accumulation by dispossession that has aptly been termed *data colonialism* (Thatcher, O'Sullivan, and Mahmoudi 2016).

With the long history of the state subsidizing technological innovation (Lazonick and Mazzucato 2013), it is unsurprising that urban governance structures, government, and leaders have become key actors enabling smart city logics to take hold. It is critically important, however, to uncover the ideological dimension of this dynamic, as it has contributed to accelerate fundamental urban transformations whose benefits are far from widely shared. Most problematically, the dominant ideology of technology shifts governance away from democratic practices, focused on the public good, toward privileging a technologically focused urban future dominated by "smart cities" logics—which, despite their diversity, are underlined by the primacy of the market (Joss et al. 2019) and its inherent focus on capital accumulation, therefore driving socioeconomic polarization. By moving toward a world where cities function as "a smart technological web with tech corporations and city leaders sitting at its center" (Sadowski and Bendor 2018, 13), urban governance processes, priorities, and outcomes become detached from public responsibility and accountability.

Rather, urban governance comes to privilege tech sector growth and reinforcement, helping reconfigure cities as (increasingly digitized) spaces for capital accumulation. Enmeshed in such a dominant ideology, our governance structures and conversations are unlikely to question the tech sector's fundamental limitations as an urban economic engine and the efficacy of technological solutions to address complex urban problems—including the problems and inequalities driven by the tech sector itself. We do not seek to villainize the tech sector entirely; rather, by explicitly framing the tech industry's logics as a dominant ideology, we seek to elucidate the detachment of our technologically driven urban governance structures from its purported and overarching goals and, in doing so, create space for alternative visions of urban life.

Acknowledgments

The authors thank *Annals of the American Association of Geographers* Editor, Prof. Ling Bian and the journal's anonymous reviewers, as well as Prof. Lisa Schweitzer for their generous feedback throughout the development of this article.

Note

1. As Schafran (2013) noted, "In 1999, at the top of the dot-com boom, the Bay Area received twice as much venture capital (5.5%) as the next largest metro area, and almost ten times the US metro average" (9), concentrated within both the Silicon Valley and San Francisco.

References

Barns, S. 2016. Mine your data: Open data, digital strategies and entrepreneurial governance by code. *Urban Geography* 37 (4):554–71. doi: 10.1080/02723638.2016.1139876.

Brinklow, A. 2016. Mayor Ed Lee vetoes Airbnb crackdown. *Curbed San Francisco*, December 9. Accessed September 20, 2019. https://sf.curbed.com/2016/12/9/13898732/airbnb-veto-sf-ed-lee

Cardullo, P., and R. Kitchin. 2018. Smart urbanism and smart citizenship: The neoliberal logic of "citizen-focused" smart cities in Europe. *Environment and Planning C: Politics and Space* 37 (5):813–30. doi: 10.1177/0263774X18806508.

Coté, J. 2014. S. F. hails 'Twitter tax break' as rousing success. *San Francisco Chronicle*, November 3. Accessed July 2, 2019. https://www.sfgate.com/bayarea/article/S-F-controller-hails-Twitter-tax-break-as-5851498.php.

Cugurullo, F. 2016. Urban eco-modernisation and the policy context of new eco-city projects: Where Masdar City fails and why. *Urban Studies* 53 (11):2417–33. doi: 10.1177/0042098015588727.

Datta, A. 2015. New urban utopias of postcolonial India: Entrepreneurial urbanization. *Dialogues in Human Geography* 5 (1):3–22. doi: 10.1177/2043820614565748.

Datta, A., and N. Odendaal. 2019. Smart cities and the banality of power. *Environment and Planning D: Society and Space* 37 (3):387–92. doi: 10.1177/0263775819841765.

Ed Lee's legacy. 2017. *CityLab*, December 12. Accessed June 26, 2019. https://www.citylab.com/equity/2017/12/ed-lees-legacy/548174/.

Etezadzadeh, C. 2016. *Smart city—Future city? Smart city 2.0 as a livable city and future market*. Wiesbaden, Germany: Springer Vieweg.

Eubanks, V. 2018. *Automating inequality: How high-tech tools profile, police, and punish the poor*. New York: St. Martin's.

Florida, R. 2000. Technology, talent, and tolerance. Informationweek.com. Accessed November 21, 2018. https://www.creativeclass.com/rfcgdb/articles/14%20Technology%20Talent%20and%20Tolerance.pdf.

Florida, R. 2002. *The rise of the creative class*. New York: Basic Books.

Gates, G., and R. Florida. 2002. Technology and tolerance: Diversity and high tech growth. Accessed November 21, 2018. https://www.brookings.edu/articles/technology-and-tolerance-diversity-and-high-tech-growth/.

Glasmeier, A., and S. Christopherson. 2015. Thinking about smart cities. *Cambridge Journal of Regions, Economy and Society* 8 (1):3–12. doi: 10.1093/cjres/rsu034.

Graham, S., and S. Guy. 2002. Digital space meets urban place: Sociotechnologies of urban restructuring in downtown San Francisco. *City* 6 (3):369–82. doi: 10.1080/1360481022000037788.

Grossi, G., and D. Pianezzi. 2017. Smart cities: Utopia or neoliberal ideology? *Cities* 69:79–85. doi: 10.1016/j.cities.2017.07.012.

Guzman, J., and S. Stern. 2015. Where is Silicon Valley? *Science* 347 (6222):606–09. doi: 10.1126/science.aaa0201.

Ho, E. 2017. Smart subjects for a smart nation? Governing (smart)mentalities in Singapore. *Urban Studies* 54 (13):3101–18. doi: 10.1177/0042098016664305.

Hollands, R. G. 2008. Will the real smart city please stand up? Intelligent, progressive or entrepreneurial? *City* 12 (3):303–20. doi: 10.1080/13604810802479126.

Johnson, P. A., R. Sieber, T. Scassa, M. Stephens, and P. Robinson. 2017. The cost(s) of geospatial open data. *Transactions in GIS* 21 (3):434–45. doi: 10.1111/tgis.12283.

Joss, S., F. Sengers, D. Schraven, F. Caprotti, and Y. Dayot. 2019. The smart city as global discourse: Storylines and critical junctures across 27 cities. *Journal of Urban Technology* 26 (1):3–34. doi: 10.1080/10630732.2018.1558387.

Kitchin, R. 2015. Making sense of smart cities: Addressing present shortcomings. *Cambridge Journal of Regions, Economy and Society* 8 (1):131–36. doi: 10.1093/cjres/rsu027.

Kitchin, R., and M. Dodge. 2011. *Code/space: Software and everyday life*. Cambridge, MA: MIT Press.

Knight, H. 2015. 5 years of Ed Lee: How San Francisco has changed under the mayor. *San Francisco Chronicle*, October 9. Accessed July 2, 2019. https://www.sfchronicle.com/bayarea/article/5-years-of-Ed-Lee-How-San-Francisco-has-changed-6559790.php?psid=nxj5T.

Knight, K. 2006. Transformations of the concept of ideology in the twentieth century. *American Political Science Review* 100 (4):619–26. doi: 10.1017/S0003055406062502.

Lang, M. 2015. Companies avoid $34M in city taxes thanks to 'Twitter tax break'. *San Francisco Chronicle*, October 19. Accessed September 27, 2019. https://www.sfgate.com/business/article/Companies-avoid-34M-in-city-taxes-thanks-to-6578396.php.

Lazonick, W., and M. Mazzucato. 2013. The risk-reward nexus in the innovation-inequality relationship: Who takes the risks? Who gets the rewards? *Industrial and Corporate Change* 22 (4):1093–128. https://doi.org/10.1093/icc/dtt019.

Levenda, A., D. Mahmoudi, and G. Sussman. 2015. The neoliberal politics of "smart": Electricity consumption, household monitoring, and the enterprise form. *Canadian Journal of Communication* 40 (4):615–36.

Luque-Ayala, A., and S. Marvin. 2015. Developing a critical understanding of smart urbanism? *Urban Studies* 52 (12):2105–16. doi: 10.1177/0042098015577319.

MacGillis, A. 2009. The ruse of the creative class. *The American Prospect*, December 18. Accessed November 21, 2018. http://prospect.org/article/ruse-creative-class-0.

Mann, M., and A. Daly. 2018. (Big) data and the North-in-South: Australia's informational imperialism and digital colonialism. *Television & New Media* 20 (4):379–95. doi: 10.1177/1527476418806091.

Martin, J. L. 2015. What is ideology? *Sociologia, Problemas e Práticas* 2015 (77):9–31. https://doi.org/10.7458/SPP2015776220.

Marvin, S., A. Luque-Ayala, and C. McFarlane. 2015. *Smart urbanism: Utopian vision or false dawn?* London and New York: Routledge.

Mattern, S. 2017. A city is not a computer. *Places Journal* February. Accessed September 26, 2019. https://doi.org/10.22269/170207.

McElroy, E. 2019. Data, dispossession, and Facebook: Techno-imperialism and toponymy in gentrifying San Francisco. *Urban Geography* 40 (6):1–20. doi: 10.1080/02723638.2019.1591143.

McNeill, D. 2015. Global firms and smart technologies: IBM and the reduction of cities. *Transactions of the Institute of British Geographers* 40 (4):562–74. doi: 10.1111/tran.12098.

Morozov, E. 2013. The perils of perfection. *The New York Times*, March 2. Accessed November 20, 2018. https://www.nytimes.com/2013/03/03/opinion/sunday/the-perils-of-perfection.html.

Morozov, E. 2014. *To save everything, click here: The folly of technological solutionism*. New York: PublicAffairs.

Noble, S. U. 2018. *Algorithms of oppression: How search engines reinforce racism*. New York: NYU Press.

O'Neil, C. 2017. *Weapons of math destruction: How big data increases inequality and threatens democracy*. New York: Broadway Books.

Pasquale, F. 2015. *The black box society: The secret algorithms that control money and information*. Cambridge, MA: Harvard University Press.

Pollio, A. 2016. Technologies of austerity urbanism: The "smart city" agenda in Italy (2011–2013). *Urban Geography* 37 (4):514–34. doi: 10.1080/02723638.2015.1118991.

Rogan, K. 2019. The 3 pictures that explain everything about smart cities. *CityLab*, June 27. Accessed July 4, 2019. https://www.citylab.com/design/2019/06/smart-city-photos-technology-marketing-branding-jibberjabber/592123/.

Sadowski, J., and R. Bendor. 2018. Selling smartness: Corporate narratives and the smart city as a sociotechnical imaginary. *Science, Technology, & Human Values* 44 (3):540–63. doi: 10.1177/0162243918806061.

Saxenian, A. 1996. *Regional advantage*. Cambridge, MA: Harvard University Press.

Schafran, A. 2013. Origins of an urban crisis: The restructuring of the San Francisco Bay Area and the geography of foreclosure. *International Journal of Urban and Regional Research* 37 (2):663–88.

Shearmur, R., and D. Wachsmuth. 2019. Urban technology: The rise of "the innovation machine." *Plan Canada* 59 (1):175–79.

Shelton, T., M. Zook, and A. Wiig. 2014. The actually existing smart city. *Cambridge Journal of Regions, Economy and Society* 8 (1):13–25. doi: 10.1093/cjres/rsu026.

Shwayri, S. T. 2013. A model Korean ubiquitous eco-city? The politics of making Songdo. *Journal of Urban Technology* 20 (1):39–55. doi: 10.1080/10630732.2012.735409.

Stehlin, J. 2016. The post-industrial "shop floor": Emerging forms of gentrification in San Francisco's innovation economy. *Antipode* 48 (2):474–93. doi: 10.1111/anti.12199.

Stehlin, J. 2018. Urban platforms, rent, and the digital built environment. *Mediapolis*, October 26. Accessed November 7, 2018. https://www.mediapolisjournal.com/2018/10/urban-platforms-rent-and-the-digital-built-environment/.

Taylor, B. 2011. San Francisco supervisors approve Twitter-inspired payroll tax breaks. *San Francisco CBS Local*, April 5. Accessed September 20, 2019. https://sanfrancisco.cbslocal.com/2011/04/05/san-francisco-to-vote-on-twitter-inspired-payroll-tax-breaks/

Tenney, M., and R. Sieber. 2016. Data-driven participation: Algorithms, cities, citizens, and corporate control. *Urban Planning* 1 (2):101–13. doi: 10.17645/up.v1i2.645.

Thatcher, J., D. O'Sullivan, and D. Mahmoudi. 2016. Data colonialism through accumulation by dispossession: New metaphors for daily data. *Environment and Planning D: Society and Space* 34 (6):990–1006. doi: 10.1177/0263775816633195.

Trencher, G. 2019. Towards the Smart City 2.0: Empirical evidence of using smartness as a tool for tackling social challenges. *Technological Forecasting and Social Change* 142:117–28. doi: 10.1016/j.techfore.2018.07.033.

Tsotsis, A. 2011. SF's Ed Lee on how he'll use his (potential) mayorship to help the tech industry. *TechCrunch*, November 9. Accessed September 27, 2019. http://social.techcrunch.com/2011/11/09/ed-lee-on-how-hell-use-his-potential-mayoral-position-to-help-the-tech-industry/

Vaidhyanathan, S. 2018. *Antisocial media: How Facebook disconnects us and undermines democracy.* New York: Oxford University Press.

Vanolo, A. 2014. Smartmentality: The smart city as disciplinary strategy. *Urban Studies* 51 (5):883–98. doi: 10.1177/0042098013494427.

Walker, R. 2018. *Pictures of a gone city: Tech and the dark side of prosperity in the San Francisco Bay Area.* Oakland, CA: PM Press.

White, J. M. 2016. Anticipatory logics of the smart city's global imaginary. *Urban Geography* 37 (4):572–89. doi: 10.1080/02723638.2016.1139879.

Wiig, A., and E. Wyly. 2016. Introduction: Thinking through the politics of the smart city. *Urban Geography* 37 (4):485–93. doi: 10.1080/02723638.2016.1178479.

Williams, R., Ed. 1985. Ideology. In *Keywords: A vocabulary of culture and society*, rev. ed., 153–57. New York: Oxford University Press. http://www.autodidactproject.org/other/ideo8.html.

Wilson, J., and E. Swyngedouw. 2014. *Post-political and its discontents: Spaces of depoliticisation, spectres of radical politics.* Edinburgh, UK: Edinburgh University Press.

Civic Infrastructure and the Appropriation of the Corporate Smart City

Sung-Yueh Perng and Sophia Maalsen

Concerns have been raised regarding smart city innovations leading to, or consolidating, technocratic urban governance and the tokenization of citizens. Less research, however, has explored how we make sense of ongoing appropriation of the resources, skills, and expertise of corporate smart cities and what this means for future cities. In this article, we examine the summoning of political subjectivity through the practices of retrofitting, repurposing, and reinvigorating. We consider them as civic infrastructure to sensitize the infrastructural acts and conventions that are assembled for exploring inclusive and participatory ways of shaping urban futures. These practices, illustrated by examples in Adelaide, Dublin, and Boston, focus on capabilities not only to write code, access data, or design a prototype but also to devise diverse sociotechnical arrangements and power relations to disobey, question, and dissent from technocratic visions and practices. The article concludes by suggesting further examination of the summoning of political subjectivity from within established institutions to widen dissent and appropriation of the corporate smart city. *Key Words: citizen, infrastructure, political subjectivity, smart city, urban future.*

如今开始有人担心智慧城市创新可能会导致、或固化技术官僚城市治理与公民符号化。但很少有人探究另外一个问题：如何理解企业化智慧城市对资源、技能和专业知识的持续占用，以及这种现象对未来城市意味着什么。在本文中，我们探讨了如何通过改造、改变目的和重新振兴的实践召唤政治主体性。我们将其视之为一种公民基础，以此提高对基本法案和公约的认识，它们的制定目标就是帮助我们以包容和参与的方式塑造城市未来。本文以阿德莱德、都柏林和波士顿为例，证明这些实践的重点不仅在于编写代码、访问数据或设计原型的能力，而且为了设计不同的社会科技安排和权力关系，以对抗、质疑和反对技术官僚的观点和实践。在文章的最后，作者建议进一步思考从既定体制内召唤政治主体性的问题，以此扩大对企业化智慧城市占用的异议。关键词：公民、基础设施、政治主体性、智慧城市、城市未来。

Una cierta preocupación ha surgido sobre las innovaciones de la ciudad inteligente, que pueden conducir a la gobernanza tecnocrática urbana y a la tokenización de los ciudadanos, o a consolidarlas. No obstante, menor ha sido la investigación dedicada a explorar el modo como le sacamos sentido a la actual apropiación de los recursos, habilidades y experticia de las ciudades corporativas inteligentes, y qué significa esto para las ciudades futuras. En este artículo examinamos el llamado de la subjetividad política por medio de las prácticas de actualización, reconversión y revitalización. Las consideramos como infraestructura cívica para sensibilizar los actos infraestructurales y las convenciones que se ensamblan para explorar medios incluyentes y participativos para configurar futuros urbanos. Estas prácticas, ilustradas con ejemplos de Adelaida, Dublín y Boston, se enfocan en las habilidades de no solo escribir código, acceder a los datos o diseñar un prototipo, sino también a concebir diversos arreglos sociotécnicos y relaciones de poder para desobedecer, cuestionar y disentir de las visiones y prácticas tecnocráticas. El artículo concluye sugiriendo más investigación del llamado de la subjetividad política desde el interior de instituciones establecidas para ensanchar la disconformidad y la apropiación de la ciudad corporativa inteligente. *Palabras clave: ciudad inteligente, ciudadano, futuro urbano, infraestructura, subjetividad política.*

In this article we consider infrastructure for smart places by rethinking smart city deployment and appropriation. Smart city developments have gone through various iterations, transitioning from corporate discourses to the deployment of digital infrastructure to improve urban living but have caused considerable concerns (e.g., Marvin, Luque-Ayala, and McFarlane 2016; McLean, Bulkeley, and Crang 2016). Among them, citizen participation is one of the most contentious issues. In early smart city master plans, citizens were often absent, and the city was a "sanitized, orderly and programmable smart polis" (Datta 2015, 53). When present in smart city visions, citizens are subjugated to surveillant infrastructure and monitored constantly to preempt any threats to the safety of the city (Vanolo 2016). Furthermore, the terms of participation are questionable. The "cocreation" and "citizen-centric" approaches to smart city developments, as Cowley, Joss, and Dayot (2018) contended, do not amount to the "public city" because they are shaped by corporate conceptions of citizens as consumers and entrepreneurs who do not prioritize or engage with public interests. Accordingly, citizen participation in smart cities tends to be "postpolitical," focusing on "instrumental rather than normative or political" engagements, creating technological solutions but failing to "challenge or replace the fundamental political rationalities shaping an issue or plan" (Cardullo and Kitchin 2019, 10).

The corporate smart city can be challenged in different ways. On the one hand, Datta (2018) demonstrated the breaches that the "chatur citizen" performs to resist the subjecthood prescribed by the state and corporations. Also, digital civics can be organized to challenge social and political hierarchies, devising radical interventions beyond digital spaces and resisting technocratic understandings of and responses to urban problems (Shelton 2018).

On the other, the ways in which these small-scale interventions are considered in relation to the corporate smart city can also be reframed. They can be considered as humanizing smartness and "using technology to realize progressive ideas, rather than see[ing] the technology as progressive in and of itself" (Hollands 2015, 63; see also de Lange and de Waal 2013). As smart cities develop, we see an emerging number of interventions where smart technologies come face to face with the messiness and complexities of cities and the promise of a smart utopia is recontextualized to the site. As Karvonen, Cugurullo, and Caprotti (2018) noted, corporations have been joined by multiple stakeholders, including "local governments, utility providers, small and medium enterprises, and civil society organizations" (4) in the implementation of smart cities. This means that smart can be reinterpreted, coopted, and appropriated as it materializes in the actually existing city. In Australia, for example, the rollout of smart cities has provided interesting commentary on the corporate smart city, materializing instead through savvy local governance

that remains vendor agnostic, protects their smart assets, and demands that technology products meet their contextual and citizen needs (Dowling, McGuirk, and Maalsen 2018).

In what sense the corporate smart city might be challenged remains a critical issue. The article contributes to this ongoing work by proposing three infrastructural practices to consider whether and how political subjectivity can be invoked to appropriate smart infrastructure and cities for civic purposes. We argue that exploring the practices of "retrofitting," "repurposing," and "reinvigorating" provides a useful way to consider the rupture to smart city innovation and governance and the appropriation of their techniques and infrastructure. Our discussion proceeds by charting the tension between citizenship and the ongoing privatization of infrastructure. It then draws on Isin and Ruppert (2015) and infrastructure studies to suggest how "retrofitting," "repurposing," and "reinvigorating" can build civic infrastructure to appropriate the corporate smart city. Three cases are used to illustrate how political subjectivity might be invoked and the effects that these practices might generate. These three practices reveal possibilities of appropriation and imaginings of a more inclusive smart city.

Infrastructure and Citizen Participation

Although infrastructure is commonly thought of as a political, technological, and discursive technology of state governance (Kooy and Bakker 2008), the increasing privatization of infrastructure has led to concerns over the subjects and exercise of citizenship. The unbundling of infrastructural networks has resulted in "premium networked space" where socially and economically privileged citizens are preferentially provided with accesses to public services (Graham and Marvin 2001). In this transition, the figure of citizens as the customer of state has double exclusionary effects. As infrastructure is turned into a marketplace, "empowering" technologies and services is reserved for the customers who can afford the access (Viitanen and Kingston 2014). Meanwhile, the socially disadvantaged are required to be "fit for market" through building their "calculating subjectivity" and becoming fee-paying, economically responsible citizens, before any access to basic public services is granted (von Schnitzler 2008). Further, although governments and corporations adopt "deliberative rhetoric" to promote the scope of public participation in infrastructure planning, citizens construed as "customers" have limited "decisional influences" on the outcome (Cotton and Devine-Wright 2012).

The citizen subject can be contested, however, in the assembling of alternative discourses, subjectivity, and resources to reconfigure infrastructure. Although infrastructure can extend state and corporate control over people and places, Meehan (2014) and Anand (2015) showed that such control is leaky and its circumvention is inevitable where alternative assemblages of citizen subjects, discourses, norms, and tools constitute an uneven spatiality of power. As Roy (2009) contended, power relations within regimes of "civic governmentality" are far from unidirectional and clear-cut. Instead, they include citizen subjects and state agencies both equally capable of exercising state rule or civic intervention.

Further, the conflicting subjects and exercise of citizenship are inseparable from the situated experiences and knowledges of infrastructure. Davies et al. (2012) problematized discursive rationality as the sole basis of participating in deliberation processes. Instead, they highlighted that the multiple, affective, embodied, and practical rationalities in everyday life are crucial resources for establishing participatory and inclusive engagement processes for infrastructure development. At issue here then is the situatedness of political subjectivity, which, according to Coward (2012), is crucial to understand competing configurations of belonging, exclusion, and ways of enacting togetherness through material engagement with infrastructure and spatially distributed others, both close and distant.

Civic Infrastructure: Challenging the Corporate Smart City

The diverse relations between infrastructure, citizen subjects, and the exercise of citizenship as observed earlier were captured by Isin and Ruppert (2015). They further argued for understanding citizenship through the enactment of legal, performative, and imaginative acts. Rather than perceiving citizens as already formed, they contended that citizenship is a dynamic and contested figure of politics, situated in "a composite of multiple forces, identifications, affiliations, and associations" (Isin and Ruppert 2015, 21). Citizen subjects are brought into being through acts

and conventions of subjectivation, the "summoning" and "taking up" of political subjectivity as an entanglement of different power relations, including disobedience, submission, and subversion. Crucially, these acts, conventions, and entangled power relations produce possibilities to discern, question, and act in dissent to authority in enacting rights.

Subsequent examinations of "smart citizens" focus on questions concerning how political subjectivity can be taken up and what acts of citizenship can be performed. For Vanolo (2016), the political subject is limited because few people possess appropriate technical expertise, access to technology, and thus an influential voice to exercise their rights. For Datta (2018), though, the discursive formation of the "chatur citizen" at odds with Indian smart city policies is a speech act that draws from local governance and urban everyday reality to redefine smart citizens. The different interpretations of the possibilities of "smart citizenship" given ealier, alongside the contradictions in smart cities observed in Karvonen, Cugurullo, and Caprotti (2018), remind us that the taking up of political subjectivity is a process where "composites of multiple subjectivities are likely to emerge from different situations and relations" (Vanolo 2016, 35). This echoes Gabrys's (2016) argument that citizenship is a process of becoming that is constituted by acts of sensing and other digital practices, complicating how the relations between citizens, technologies, cities, and environments are formed.

Isin and Ruppert's (2015) work informs our understanding of civic infrastructure as summoning political subjectivity and creating openings for new infrastructural imaginaries and arrangements to shape alternative smart cities and places. Civic infrastructure includes sociotechnical practices that call on political subjectivity to problematize the corporate smart city and bring into being exercises of judgment and acts of dissent to disrupt technocratic practices and visions of innovation. We highlight three practices that could disrupt corporate smart infrastructure and create openings in different ways: retrofitting, repurposing, and reinvigorating. Retrofitting enacts political subjectivity by indicating the failures of smart infrastructure and mobilizing care and collaboration to repair them. Repurposing builds political subjectivity by discerning and experimenting infrastructural arrangements and effects that transgresses the corporate smart city. Reinvigorating subverts technocratic visions of smart cities by exploring infrastructural arrangements to place the politics of participation at the forefront of innovation. Taken together, these practices resignify participation, innovation, and governance by translating inclusive and participatory visions into practices and protocols for any city and place to become smart. We illustrate these practices with three case studies conducted in three different cities. Although these practices are observed across the cities, we present each practice in relation to a case study for more focused discussion.

Retrofitting

How can infrastructural arrangements for the corporate smart city be retrofitted and in what ways can political subjectivity be summoned for this purpose? This question follows the argument that infrastructure, and the control it implements, can become leaky (Roy 2009; Meehan 2014) and further explores how the insufficiency of corporate infrastructure can be revealed and how openings for alternative arrangements can be explored. The case of Adelaide illustrates how such a possibility might materialize.

Adelaide, South Australia, is a Cisco Lighthouse city and one of Australia's early adopters of smart city approaches. Growing cooperation with the city council has seen Adelaide's smart city initiatives taking a multivendor approach to provide technological solutions that best benefit the city. These initiatives include, among others, increased connectivity ("Adelaide Gig City"), reduced electricity consumption ("Follow Me" responsive LED lighting system) and enhanced citizen innovation (Smart City Studio; Maalsen, Burgoyne, and Tomitsch 2018).

The installation of LoRaWAN, a digital wireless network, shows how retrofitting provides an opportunity to expand smart Adelaide. The network is not only an economical choice for the city but also particularly applicable to projects attuned to community needs:

> LoRaWAN is cost-effective. It's the infrastructure that might enable a range of different scientific projects and experiments, but initiatives that have more of a community focus. (Smart City Studio manager, interview, 17 August 2016)

Indeed, the service-user publicness (Cowley, Joss, and Dayot 2018) and the figure of citizens as state customers (Viitanen and Kingston 2014) are salient in the manager's discussion. However, retrofitting smart infrastructure can be observed as rearranging infrastructure for redefining for whom innovation is pursued. The Studio itself converts a reused office space in a busy part of the State Department of Premier and Cabinet building to attract attention from the public and public officials and invite innovative ideas by showcasing existing smart technologies there. The retrofitting of space attests to the Studio's shift away from consolidating governmental and corporate ownership of innovation. Instead, it attempts to enable initiatives that are collaborative and community building by linking local people with innovative ideas to government resources and infrastructure. The proposal of a community LoRaWAN in a meetup organized by the Studio is an example of the retrofitting of computational networks into the urban fabric. The manager tells a story of one enterprising member of the public who:

> is desperately keen to get access to this [LoRaWAN] because he wants to install sensors in all the possum boxes and the bird boxes in the Adelaide park plans. At the moment there's no mechanism to understand whether or not these boxes are being used and in what way. So, with the rollout that is relatively cost-effective, as opposed to someone coming, walking around with a clipboard, and hoping that the possum is there, and there on the right day. I think that type of data can help to transform the ways in which we support our native animals and the parklands. It's probably quite low range stuff, but it's just transformational. (Smart City Studio manager, interview, 17 August 2016)

Here, although retrofitting LoRaWAN has not been in a directly confrontational relation against corporate smart infrastructure, the practice creates space for looking into certain aspects of urban life that have not been included in Adelaide's smart city agenda. Government and corporate resources have been appropriated for developing ideas and initiatives to transfer ownership to local communities. The reworking of existing infrastructural arrangements in the way just discussed produces retrofitting's ambiguity where the "futurological orientation" of infrastructure rests on the uncertain effectiveness of its current configuration (Howe et al. 2016). The ambiguity requires creativity and social collaboration to continue retrofitting existing technological and government infrastructure to respond to issues that are otherwise unseen (Silver 2014).

Using the case of Adelaide to consider retrofitting illustrates the summoning of political subjectivity by enacting infrastructural arrangements that are open, rather than prescriptive. Retrofitting does so by creating a liminal space in smart city infrastructure, as Zandbergen (2017) also observed in an Internet-of-Things meetup in Amsterdam, to align interests and demands that are not foreseen in the initial setup. Retrofitting smart infrastructure in this way presents the possibility of disobeying existing infrastructural logic and effects as seen in the corporate smart city, adjusting arrangements to meet oncoming demands, and responding to issues that might otherwise remain unseen. That is, retrofitting invokes subversion and demands submission to participate in sustaining its reconfiguration of smart infrastructure through attempts at accessing government and technological infrastructure for community-led innovation. It expands the visions and practices of smart infrastructure beyond consolidating control and ownership of innovation and toward caring for the issues and lives that become obscured in earlier iterations of emerging infrastructure networks.

As Mattern (2018) noted, however, care and caring for urban infrastructure are political, which has implications for who provides and receives care and how care is delivered. This resonates with Cardullo and Kitchin's (2019) concern that many "citizen-focused" smart city projects are conceived in paternalistic ways where the purpose of innovation, the eligibility to innovate, and the terms on which people are engaged in innovation activities are predetermined. We turn to these questions in the two cases that follow.

Repurposing

"Open innovation" and "living labs" are increasingly popular "citizen-focused" innovation approaches adopted for developing smart infrastructure. These approaches aim to produce innovation reflective of citizens' needs. In effect, though, they lead to the intensification of neoliberal governance (Bulkeley and Broto 2013) and reductionist governance practices that foreclose the exploration of issues likely to emerge from taking a "citizen-centric" approach (Joss, Cook, and Dayot 2017). Instead of perpetuating such effects, we consider repurposing and its potential to challenge rather than consolidate existing governance practices. We discuss

Dublin City Council (DCC) Beta to consider how public officials subject themselves to repurpose corporate innovation techniques to experiment and improve governance practices.

DCC Beta "is a live mechanism for imagining, testing, and implementing ways to improve the experience of life in the Capital" (DCC Beta 2018). Since 2012, DCC Beta has conducted many trials to this effect, such as Equinox Cycle Parking (using on-street parking space for temporary bicycle hangars during summer) and Driving Data (understanding car traffic by purposes of travel and cohorts of drivers).

It is the repurposing of design practices, not the success of individual projects, however, that is important. DCC Beta incorporates design practices commonly used at "open innovation" events for citizen engagement, including sticky notes for public brainstorming or props for encouraging participant interactions. Crucially, these corporate design practices are repurposed to experiment with alternative sociotechnical practices of governance. During repurposing, the ways in which municipality employees collaborate with citizens and implement citizen-suggested changes are put under the microscope. DCC Beta develops a "middle-out" approach, which involves DCC on-the-ground staff taking more responsibility in prototype testing. This approach recognizes that simply attracting ideas and suggestions from citizens is insufficient to change governance practices because any suggestion "eventually has to go to probably a member of the local authority wherever in the world they are" (DCC Beta coordinator, interview, 23 November 2015). Accordingly, the municipality has to build capacity so that ideas can be acted on rather than treated as a token of engagement:

> Often that assumes that the staff members that are actually paid by those citizens haven't thought of [the ideas]. Often, they have, but they just don't have a method to actually do it themselves. (DCC Beta 2018)

Iterative design, a common software and service development approach highlighting continuous testing and refinement, has been incorporated into the middle-out approach to address the lack of a method. The approach emphasizes starting with small projects and being transparent about the selection, progress, evaluation, and future iterations. This provides opportunities for overseeing and improving the trials and DCC Beta itself.

Accordingly, with the middle-out approach, DCC Beta repurposes corporate design practices to experiment with how governance practices can implement proposals of change by citizens. DCC Beta does not seek to reduce uncertainty in governance through innovation as observed in many urban living labs. Instead, it exposes issues that arise from transforming citizens from passive consumers to active contributors and in understanding that possible solutions require continuous revision. Therefore, repurposing here no longer concerns policy change for local economic growth and entrepreneurial governance (Lauermann 2018). Rather, the case of DCC Beta illustrates the potentiality of repurposing in reconfiguring infrastructural arrangements to produce effects not initially conceived (Boyce 2016; Rossiter 2017). It does so by "hold[ing] the social and material in suspension" and experimenting with prototypes, not in search for closure or full solutions but for continuous "forking and enabling novel extensions" to reinscribe the right to reassemble infrastructure (Corsín Jiménez 2014, 384).

Accordingly, repurposing ruptures and resignifies smart city governance. The norms of smart city governance can be suspended by discursive and material imaginations around the "chatur citizen" that redefine state and citizen relation (Datta 2018). The case of DCC Beta illustrates how existing city governance might be suspended and reenacted through repurposing corporate innovation techniques to devise infrastructural arrangements to redraw governance practices. Political subjectivity summoned by repurposing creates opportunities for questioning how current urban governance works and fails, a refusal to consolidate the existing political economy of urban innovation, and arrangements for trials and changes within governments. In doing so, reconfigurations of governance practices rupture the citizen-as-consumer and instead create openings by enacting protocols and mechanisms to implement citizens' proposals for shaping future cities.

Reinvigorating

The other questions raised earlier are concerned with the eligibility for and the terms of participation in innovation: Who gets to innovate and how they are involved? Smart city projects are often criticized for the lack of inclusive protocols to engage citizens who are only data points or tokens in innovation and knowledge generation processes (Kitchin, Cardullo, and Di Feliciantonio 2019). The case of Boston's Public Lab illustrates how sensing can be resignified as acts of inclusive witnessing for enacting practices and protocols to engage diverse participants in sensing and innovation activities.

Public Lab started as an activist citizen sensing initiative in the aftermath of the BP oil spill in the Gulf of Mexico in 2010 and has since grown into a community of local and global contributors. It conducts sensing projects with an aim to make visible the previously invisible people, perspectives, and their environmental concerns:

> Public Lab is a community where you can learn how to investigate environmental concerns. Using inexpensive DIY techniques, we seek to change how people see the world in environmental, social, and political terms. (Public Lab 2018)

The use of "inexpensive DIY techniques" addresses the concerns of access and technical expertise required for citizen participation in smart city critiques (cf. Vanolo 2016). Further, the protocols and practices that Public Lab establishes respond to known concerns regarding the discouraging experiences for female and minority participants with open-source or crowdsourcing communities (e.g., Ford and Wajcman 2017). Public Lab establishes various conventions to address the issue, such as increasing female staff and incorporating diversity statements when recruiting new staff to encourage applicants from different backgrounds. These measures produce ripple effects. The diversity within staff "amplifies into the broader organizers group and the much larger community as well" (Public Lab main coordinator, interview, 15 April 2016), and in projects focusing on environmental pollution there is a relatively larger number of participants from diverse backgrounds.

Public Lab also devises collective decision making and careful communication practices to ensure that people "interact with one another in respectful and meaningful ways" and communicate "in a less jargony way or a way that is kinder" (Public Lab main coordinator, interview, 15 April 2016). For this purpose, projects involving only Public Lab members are used to simulate how they would work with external partners by "deal[ing] with what needs to happen internally … [a]nd then you think outwards," building "early agreements in our group around how we would work together, how we would interact with different projects, what our model was going to look like" (Public Lab main coordinator, interview, 15 April 2016).

Practices and protocols that were initially unforeseen have also been established to recognize the importance of otherwise invisible processes of knowledge generation. Researchers in formal institutions are increasingly using the sensing tools that Public Lab builds and also "contribute information [about the tools] once they have figured something out or haven't figured it out" in the wiki pages for relevant tools or projects. The emergent practice of "people … post[ing] about failures or things that didn't go right in the scientific process … teaches people equally as much" (Public Lab main coordinator, interview, 15 April 2016). This sharing practice becomes a protocol within the Public Lab community through the implementation of the CERN Open Hardware License, where participants are expected to report back on their use or modification of the tools developed by Public Lab. This then becomes a convention of "do what you will … but share it back to the community because that is how we learn together" (Public Lab main coordinator, interview, 15 April 2016). Through these practices and protocols, what is recognized and enacted is doing sciences as they are situated in particular places, tools, and people, rather than making these people and processes invisible in the pursuit of innovation.

Accordingly, Public Lab demonstrates the possibilities for sensing "not to be governed quite so much—or in that way" (Gabrys 2016, 190). It illustrates how to reinvigorate sensing with a set of practices and protocols that foreground the diverse people and processes involved in innovation and knowledge generation. Public Lab devises social, organizational, and technical arrangements to establish an infrastructure that sustains the removal of invisible and implicit assumptions that prefigure the absence of diverse knowledge, perspectives, expertise, and experiences. These measures disrupt unequal power relations and social selectiveness in participants' experiences with infrastructural arrangements (cf. Davies et al. 2012). Therefore, reinvigorating as demonstrated in the case of Public Lab energizes the "speculative constructions and additional urban potentialities" (Gabrys 2016, 244) in ways that generate future possibilities and capacities to reroute power relations.

Reinvigorating thus summons political subjectivity that enacts inclusive innovation by establishing appropriate protocols and practices. It exercises careful resignification of sensing by underlining and responding to the epistemically, socially, and politically conflicting relations in the process. Therefore, far from being obedient, reinvigorating creates openings for dissenting from technocratic visions of innovation by exploring infrastructural arrangements that place the politics of participation at the forefront. The arrangements that Public Lab establishes thus act as proclamations and protocols that invoke active, open, and inclusive undertaking of sensing, sciences, and innovation that can better guide future shaping of cities and places.

Conclusion

Smart city infrastructure has rightfully been critiqued for its privileging of technology, corporations, and "experts" at the expense of citizens, but the citizen subjects of smart cities are not passive. In this article, we argue that political subjectivity can be summoned in and through the practices of retrofitting, repurposing, and reinvigorating to establish protocols and practices to include these citizen subjects in shaping future cities. These practices reveal the possibilities of corporate infrastructure to be made accessible, expandable, and changeable, contesting corporate cooptation of the city and appropriating its technologies, resources, and practices. Furthermore, as these practices are undertaken, political subjectivity enacts different power relations that problematize the corporate smart city, as well as devising infrastructural acts and conventions that resignify innovation, governance, and participation. Accordingly, these practices offer a combination of imaginaries, explorations, and protocols for experimenting and establishing open and participatory ways of shaping future cities and places.

Our discussion raises further questions regarding the place and infrastructure for political subjectivity. The selection of a sensing network project, a government initiative, and a "lab" to illustrate infrastructural acts and conventions is intentional. These are unlikely "places" to find civic actions and for political subjectivity to emerge. They deserve greater attention, though. Taking seriously that political subjectivity is situated in entangled forces, affiliations, and associations, these places are not devoid of any opportunity to create openings. The place and infrastructure for political subjects can embrace such complexity.

We recognize, as we have elsewhere (Maalsen and Perng 2017), that the ability to appropriate infrastructure is unevenly experienced and we must continue to critically assess by whom and for whom the smart city is being appropriated. Nevertheless, although we can question whether the examples presented here generated significant change for the city, they are illustrative of the potential of the civic appropriation of corporate infrastructure to build a more inclusive smart city of the future. This is where the value of the framework we have sketched here resides. What would the smart city look like if it was reconfigured by the retrofitting, repurposing, and reinvigorating of civic infrastructure?

Acknowledgments

We thank the reviewers for their thoughtful comments on this article. The research and subsequent work would not be possible without the encouragement from our colleagues who have contributed to and supported the Programmable City project.

Funding

The Programmable City project was funded by an ERC Advanced Investigator Award to Rob Kitchin (ERC-2012-AdG-323636-SOFTCITY). Some of the Australian data were gathered for a scoping project funded by Telstra on smart cities in Australia.

References

Anand, N. 2015. Leaky states: Water audits, ignorance, and the politics of infrastructure. Public Culture 27 (2):305–30. doi: 10.1215/08992363-2841880.

Boyce, A. M. 2016. Outbreaks and the management of "second-order friction": Repurposing materials and data from the health care and food systems for public health surveillance. Science & Technology Studies 29 (1):52–69.

Bulkeley, H., and V. C. Broto. 2013. Government by experiment? Global cities and the governing of climate change. Transactions of the Institute of British Geographers 38 (3):361–75. doi: 10.1111/j.1475-5661.2012.00535.x.

Cardullo, P., and R. Kitchin. 2019. Being a "citizen" in the smart city: Up and down the scaffold of smart citizen participation. Geojournal 84 (1):1–13. doi: 10.1007/s10708-018-9845-8.

Corsín Jiménez, A. 2014. The right to infrastructure: A prototype for open source urbanism. Environment and Planning D: Society and Space 32 (2):342–62. doi: 10.1068/d13077p

Cotton, M., and P. Devine-Wright. 2012. Making electricity networks "visible": Industry actor representations of "publics" and public engagement in infrastructure planning. Public Understanding of Science 21 (1):17–35. doi: 10.1177/0963662510362658.

Coward, M. 2012. Between us in the city: Materiality, subjectivity, and community in the era of global urbanization. Environment and Planning D: Society and Space 30 (3):468–81. doi: 10.1068/d21010.

Cowley, R., S. Joss, and Y. Dayot. 2018. The smart city and its publics: Insights from across six UK cities. Urban Research & Practice 11 (1):53–77. doi: 10.1080/17535069.2017.1293150.

Datta, A. 2015. A 100 smart cities, a 100 utopias. Dialogues in Human Geography 5 (1):49–53. doi: 10.1177/2043820614565750.

Datta, A. 2018. The digital turn in postcolonial urbanism: Smart citizenship in the making of India's 100 smart cities. Transactions of the Institute of British Geographers 43 (3):405–19. doi: 10.1111/tran.12225.

Davies, S. R., C. Selin, G. Gano, and A. G. Pereira. 2012. Citizen engagement and urban change: Three case studies of material deliberation. Cities 29 (6):351–57. doi: 10.1016/j.cities.2011.11.012.

de Lange, M., and M. de Waal. 2013. Owning the city: New media and citizen engagement in urban design. First Monday 18 (11). doi: 10.5210/fm.v18i11.4954.

Dowling, R., P. McGuirk, and S. Maalsen. 2018. Realising smart cities: Partnerships and economic development in the emergence and practices of smart in Newcastle, Australia. In Inside smart cities: Place, politics and urban innovation, ed. A. Karvonen, F. Cugurullo, and F. Caprotti, 15–29. London and New York: Routledge.

Dublin City Council Beta (DCC Beta). 2018. About. Accessed November 11, 2018. https://dccbeta.ie/about.

Ford, H., and J. Wajcman. 2017. "Anyone can edit", not everyone does: Wikipedia's infrastructure and the gender gap. Social Studies of Science 47 (4):511–27. doi: 10.1177/0306312717692172.

Gabrys, J. 2016. Program earth: Environmental sensing technology and the making of a computational planet. Minneapolis: University of Minnesota Press.

Graham, S., and S. Marvin. 2001. Splintering urbanism: Network infrastructures, technological mobilities and the urban condition. London and New York: Routledge.

Hollands, R. G. 2015. Critical interventions into the corporate smart city. Cambridge Journal of Regions, Economy and Society 8 (1):61–77. doi: 10.1093/cjres/rsu011.

Howe, C., J. Lockrem, H. Appel, E. Hackett, D. Boyer, R. Hall, M. Schneider-Mayerson, A. Pope, A. Gupta, E. Rodwell, et al. 2016. Paradoxical infrastructures: Ruins, retrofit, and risk. Science, Technology, & Human Values 41 (3):547–65. doi: 10.1177/0162243915620017.

Isin, E., and E. Ruppert. 2015. Being digital citizens. London: Rowman & Littlefield.

Joss, S., M. Cook, and Y. Dayot. 2017. Smart cities: Towards a new citizenship regime? A discourse analysis of the British smart city standard. *Journal of Urban Technology* 24 (4):29–49. doi: 10.1080/10630732.2017.1336027.

Karvonen, A., F. Cugurullo, and F. Caprotti, eds. 2018. *Inside smart cities: Place, politics and urban innovation.* London and New York: Routledge.

Kitchin, R., P. Cardullo, and C. Di Feliciantonio. 2019. Citizenship, justice, and the right to the smart city. In *The right to the smart city*, ed. R. Kitchin, P. Cardullo, and C. Di Feliciantonio, 1–24. Bingley, UK: Emerald.

Kooy, M., and K. Bakker. 2008. Technologies of government: Constituting subjectivities, spaces, and infrastructures in colonial and contemporary Jakarta. *International Journal of Urban and Regional Research* 32 (2):375–91. doi: 10.1111/j.1468-2427.2008.00791.x.

Lauermann, J. 2018. Municipal statecraft: Revisiting the geographies of the entrepreneurial city. *Progress in Human Geography* 42 (2):205–24. doi: 10.1177/0309132516673240.

Maalsen, S., S. Burgoyne, and M. Tomitsch. 2018. Smart-innovative cities and the innovation economy: A qualitative analysis of local approaches to delivering smart urbanism in Australia. *Journal of Design, Business & Society* 4 (1):63–82. doi: 10.1386/dbs.4.1.63_1.

Maalsen, S., and S.-Y. Perng. 2017. Crafting code: Gender, coding and spatial hybridity in the events of PyLadies Dublin. In *Craft economies*, ed. S. Luckman and N. Thomas, 223–32. London: Bloomsbury.

Marvin, S., A. Luque-Ayala, and C. McFarlane, eds. 2016. *Smart urbanism: Utopian vision or false dawn.* London and New York: Routledge.

Mattern, S. 2018. Maintenance and care. *Places Journal.* Accessed November 21, 2018. https://placesjournal.org/article/maintenance-and-care/.

McLean, A., H. Bulkeley, and M. Crang. 2016. Negotiating the urban smart grid: Socio-technical experimentation in the city of Austin. *Urban Studies* 53 (15):3246–63. doi: 10.1177/0042098015612984.

Meehan, K. M. 2014. Tool-power: Water infrastructure as wellsprings of state power. *Geoforum* 57 (November):215–24. doi: 10.1016/j.geoforum.2013.08.005.

Public Lab. 2018. Home page. Accessed November 11, 2018. https://publiclab.org/.

Rossiter, N. 2017. Imperial infrastructures and Asia BEYOND Asia: Data centres, state formation and the territoriality of logistical media. *The Fibreculture Journal* 29:1–20. doi: 10.15307/fcj.29.220.2017.

Roy, A. 2009. Civic governmentality: The politics of inclusion in Beirut and Mumbai. *Antipode* 41 (1):159–79. doi: 10.1111/j.1467-8330.2008.00660.x.

Shelton, T. 2018. Digital civics. In *Digital geographies*, ed. J. Ash, R. Kitchin, and A. Leszczynski, 250–59. London: Sage.

Silver, J. 2014. Incremental infrastructures: Material improvisation and social collaboration across post-colonial Accra. *Urban Geography* 35 (6):788–804. doi: 10.1080/02723638.2014.933605.

Vanolo, A. 2016. Is there anybody out there? The place and role of citizens in tomorrow's smart cities. *Futures* 82 (Suppl. C):26–36. doi: 10.1016/j.futures.2016.05.010.

Viitanen, J., and R. Kingston. 2014. Smart cities and green growth: Outsourcing democratic and environmental resilience to the global technology sector. *Environment and Planning A: Economy and Space* 46 (4):803–19. doi: 10.1068/a46242.

Von Schnitzler, A. 2008. Citizenship prepaid: Water, calculability, and techno-politics in South Africa. *Journal of Southern African Studies* 34 (4):899–917. doi: 10.1080/03057070802456821.

Zandbergen, D. 2017. "We are sensemakers": The (anti-) politics of smart city co-creation. *Public Culture* 29 (383):539–62. doi: 10.1215/08992363-3869596.

How Smart Cities Became the Urban Norm: Power and Knowledge in New Songdo City

Glen David Kuecker and Kris Hartley🆔

In this article we ask why smart cities have emerged within the international development community as the normative urban logic for confronting systemic global crises. This phenomenon is exemplified by the embrace of smart cities as an implementation tool for UN Habitat's aspirational New Urban Agenda. Our analysis deploys two theoretical approaches in novel combination. First, we reinterpret Foucault's governmentality concept through the lens of Lefebvre's planetary urbanization thesis. This approach reveals global crises and systemic instability as neglected lines of inquiry within the smart city discourse, particularly in scholarship that has viewed technocratic rationality and neoliberalism as primary mechanisms of capitalist reproduction. Our use of Lefebvre positions smart city governmentality within this neglected context. Our second theoretical approach goes beyond critical urban theory's emphasis on social justice to consider its value in explaining rational-technocratic planning vis-à-vis Lefebvre's "critical zone" of full planetary urbanization. We argue that smart cities represent an emergent form of critical zone urbanism. The article begins with a review of governmentality and planetary urbanization that establishes the foundation for the study's case analysis of New Songdo City. We then analyze the Songdo project, its related actors and power brokers, and its evolution from test bed to implementation model that becomes the new urban norm. The conclusion synthesizes elements of the case and novel theoretical approach to highlight distinctions between cities as organically evolving entities and those as products of totalizing technocratic norms. *Key Words: governmentality, planetary urbanization, smart cities.*

我们于本文中, 质问智能城市如何在国际发展社群中浮现作为挑战系统性的全球危机之规范性城市逻辑。此一现象以联合国人居署拥抱智能城市作为梦想的新城市议程的执行工具为例。我们的分析以崭新的方式结合并部署两大理论方法。首先, 我们通过列斐伏尔的全面城市化命题之视角, 重新诠释福柯的治理术之概念。此一方法揭露了全球危机与系统性的不稳定作为智能城市论述中被忽略的探问角度, 特别是在将技术理性和新自由主义视为资本主义再生产的主要机制之学术研究中。我们对列斐伏尔的运用, 将智能城市的治理术置于此一被忽略的脉络之中。我们的第二个理论方法, 超越批判城市理论对社会正义的强调, 从而考量其在理性技术规划相对于列斐伏尔的全面城市化之"关键地带"的解释价值。我们主张, 智能城市呈现出浮现中的关键地带城市主义之形式。本文首先回顾提供新松岛城市研究案例分析基础的治理术与全面城市化。我们接着分析松岛计画、相关行动者与权力中介者, 及其从测试平台至施行模式并成为新兴城市形式的演变。本文结论综合该案例的元素与崭新的理论方法, 以强调城市作为有机演化体和作为全面性技术常规之间的差异。关键词·治理术 全面城市化, 智能城市。

En este artículo nos preguntamos por qué han emergido ciudades inteligentes dentro de la comunidad internacional como lógica normativa urbana para confrontar las crisis sistémicas globales. Este fenómeno se ejemplifica por la acogida de las ciudades inteligentes por la aspirante Nueva Agenda Urbana de Hábitat de las Naciones Unidas, amanera de herramienta de implementación. Nuestro análisis despliega dos enfoques teóricos en novedosa combinación. Primero, reinterpretamos el concepto de gobermentalidad de Foucault a través de la tesis de urbanización planetaria de Lefebvre. Este enfoque revela las crisis globales y la inestabilidad sistémica como líneas olvidadas de indagación dentro del discurso de la ciudad inteligente, en particular en la erudición que ha contemplado la racionalidad tecnocrática y el neoliberalismo como los mecanismos primarios de reproducción capitalista. Nuestro uso de Lefebvre posiciona la gobermentalidad de la ciudad inteligente dentro de este concepto desatendido. Nuestro segundo enfoque teórico va más allá del énfasis en la justicia social de la teoría crítica urbana para considerar su valor en la explicación de la planificación racional-tecnocrática con relación a "la zona crítica" de la urbanización planetaria plena de Lefebvre. Argüimos que las ciudades inteligentes representan una forma emergente de urbanismo de zona crítica. El artículo comienza con una revisión de la gobermentalidad y la urbanización planetaria, con lo que se establece la fundamentación del análisis del caso del estudio, Ciudad Nuevo Songdo. Luego analizamos el Proyecto Songdo, sus actores relacionados y mentores potentados, y su evolución desde el banco de prueba hasta el modelo de implementación que se convierte en la nueva norma urbana. La conclusión sintetiza elementos del caso y el novedoso enfoque teórico para destacar las distinciones entre ciudades como entidades que evolucionan orgánicamente y las que son productos de normas tecnocráticas totalizantes. *Palabras clave: ciudades inteligentes, gobermentalidad, urbanización planetaria.*

The long-evolving discursive turn in social sciences has provided opportunities to enrich urban scholarship with critical theoretical perspectives on power, agency, and class struggle. As an urban norm driven by corporate interests under the endorsement of the international development community, the smart city phenomenon is a fruitful context for interrogating the same issues, and scholarship has responded by applying a neoliberal reading of Foucault's governmentality. Although these studies have enriched understandings about smart cities as a localized development model, there is scant connection to broader forces shaping global systemic resilience in the twenty-first century. Discussions around neoliberalism, although still relevant, are in the twenty-first century insufficient to explain the smart city phenomenon alone. We argue that smart cities in theory and practice cannot be divorced from the global systemic context and that opportunities to connect individual strands of scholarship to problematize this issue remain largely unexploited. This article introduces and applies a theoretical model in which the political economy of smart cities is grounded in the imminence of a perfect storm of global crises (e.g., climate change). The convergence of these and other uncertainties will test the adaptive capacity of economic, social, and political systems at national and global levels.

In recognizing the challenges of the twenty-first century, the global development community has responded with a series of aspirational initiatives aimed at arresting the slide toward systemic collapse. Examples are the Sustainable Development Goals (SDGs), the Paris Agreement, and the Sendai Framework. The implementation model for the SDGs, the New Urban Agenda (NUA), seeks to provide an actionable implementation model by which all actors—from multilateral organizations and national governments to municipalities and communities—can frame their policy models around imminent systemic disruption. At the UN Habitat's World Urban Forum in 2018, at which NUA was the focus, the smart city was presented as a preferred implementation model for enabling the international development community to address global crises. As a "test bed" (Halpern et al. 2013) for smart urbanism, South Korea's New Songdo City (hereafter Songdo) emerged as an exemplary case. We argue that although smart cities are becoming the urban norm, Songdo provided the test bed for transforming the NUA from an aspirational wish list for inclusive and sustainable urbanism into a reality. Despite the growing trendiness of smart cities, however, their capture by elite economic and political interests threatens to undermine their effectiveness and exposes the frailty of development models based on technocratic problem framing.

In interrogating the pivot toward smart cities as the twenty-first-century urban norm, our application of Foucault's governmentality and Lefebvre's planetary urbanization locates scholarly narratives about power and governance in smart cities within urbanization's critical zone: Foucault's governmentality as a project-scale technocratic phenomenon and Lefebvre's planetary urbanization as a systemic crisis whose social and ecological symptoms are drawing increasing attention from the international development community. The following section provides a brief overview of governmentality and planetary urbanization. The analysis and discussion of Songdo highlights how the project has been valorized by the NUA and global development agenda. The article concludes with implications and calls for further research about the implications of the smart city model for organic versus totalizing approaches to urban development.

Governmentality and Planetary Urbanization

The theoretical approach of this article extends Foucault's governmentality beyond its commonly invoked neoliberal and technocratic contexts to the context of Lefebvre's planetary urbanization, which introduces a crucial but underutilized global-systemic variable. As such, this brief review provides a high-level perspective on governmentality and planetary urbanization. We engage Foucault's governmentality due to its flexibility in interrogating meaning around the term *urban* and in explaining the macrostructural transition to a techno-deterministic future defined by smart and test-bed urbanism. Highlighting connections among these analytical elements, Calvillo et al. (2016) asked, "What can these test-beds for urban construction help us see about this broader logic of testing and big data as forms of governmentality—those techniques of measurement, regulation and monitoring—that organise space and manage life?" (156).

Marking Foucault's transition from a structuralist to poststructuralist perspective, the concept of governmentality has origins in *The Order of Things* (Foucault [1966] 1994) and was elaborated throughout Foucault's lectures at the Collège de France from 1970 until his death in 1984. According to Gordon (1991), Foucault defined government as a "conduct of conduct" focused on guiding behavior, with the rationality behind government action interpreted through dispersed agency and legitimacy "capable of making some form of … activity thinkable and practicable both to its practitioners and to those upon whom it is practiced" (3). In one of his final lectures on the topic, Foucault positioned the evolution of governmental rationality within the then-contemporary context of neoliberalism's emergence in Western Europe, finding neoliberalism to be a substantial departure from previous articulations of modern governance and emblematic of a new form of rationality. Scholarship has hitherto connected governmentality with neoliberalism (Gordon, Whimster, and Lash 1987; Kerr 1999; Larner 2000; Ferguson and Gupta 2002; Lemke 2002; Joseph 2013; Brady and Lippert 2016; Zamora and Behrent 2016).

The application of governmentality for this analysis serves to illuminate the exercise and performance of power in the smart city turn. In addressing how power works in daily life, Foucault (1977) distinguished between disciplinary power and security by tracing their historical evolution as instruments of sovereign control. Building on this line of inquiry, Klauser, Paasche, and Söderström (2014) applied governmentality to understand power within smart urbanism and the ability of smart cities to "code" everyday life—a topic that rests at the core of

government rationality. Foucault argued that power evolved from techniques of punishment to those of discipline, a transition exemplified by the Benthanmite panopitcon's mechanism of self-regulatory power that embodies the objectives of modernity. When critics pushed back against what they perceived to be a totalizing and deterministic construct of power, Foucault responded by proposing a more flexible manifestation of governmentality power termed *security*. Klauser, Paasche, and Söderström (2014) argued that smart cities constitute this form of governmentality and "subsume a heterogeneous range of techniques and efforts aimed at governing through code" (870). Methods of technocratic governance, as enabled by the continuing technology revolution, perform governmentality through their role as sources and expressions of power.

Theoretical critiques of power and injustice within urban studies emerge not only from governmentality but also draw on the work of Lefebvre, especially his right to the city thesis as developed by critical urban theorists. At a basic level, critical urban theory is the application of critical theory (derived from the Frankfurt School of sociology) to the analysis of urbanism. Brenner (2009) argued that the subfield emerges from the post-1968 urban studies literature engaging Lefebvre's seminal works on planetary urbanization, the right to the city, and production of space. Its themes are shared by the literature on advocacy planning (Davidoff 1965) and more generally by the discursive turn in social sciences. In Brenner's (2009) assessment, critical urban theory takes aim at the inequities generated by contemporary neoliberal capitalism and its frequent bedfellows, instrumental reasoning and technocratic rationality:

> Critical urban theory rejects inherited disciplinary divisions of labor and statist, technocratic, market driven and market-oriented forms of urban knowledge … [it] emphasizes the politically and ideologically mediated, socially contested and therefore malleable character of urban space … [it] involves the critique of ideology (including social–scientific ideologies) and the critique of power, inequality, injustice and exploitation, at once within and among cities. (198)

Lefebvre's ([1970] 2003) seminal *Urban Revolution* anticipates the current urban historical moment: "Society has been completely urbanized … virtual today, but … real in the future" (1). This proposition resituated questions about power within an urban context, launching a debate about what Castells (1979) framed as the urban question: the extent to which the logic of capitalism has shifted from the modern industrial system to a new form, imagined by Lefebvre, where capitalism is defined by the city rather than the other way around. Lefebvre argued that the urban revolution was a macrostructural transformation driven by a departure from the modern industrial city to a realm he vaguely labeled the critical zone. Known generally as planetary urbanization, Lefebvre's thesis has been taken up by critical urban theorists to develop a language for understanding the practical interconnectedness and theoretical elusiveness of twenty-first-century urbanization. It is thus necessary to avoid reducing critical urban theory to issues only of social justice and to consider planetary urbanization as a global context shaping the smart city turn.

New Songdo City: Analysis and Discussion

Background

Located on reclaimed land near Seoul, Songdo is a master-planned city governed within the jurisdiction of the Incheon Free Economic Zone (IFEZ). The IFEZ advances a national development plan focused on innovation and sustainability as substitutes for South Korea's longtime dependence on industry (Kuecker 2013; Shwayri 2013). Although Songdo receives fiscal support from municipal and national governments, it is a predominantly neoliberal undertaking that embraces corporatized and proprietary models of development (Townsend 2013). For a summary of the mechanics of Songdo's development, see Kshetri, Alcantara, and Park (2014).

The private sector provided much of the capacity for designing, building, and operating the project, and the list of city-building agents is extensive. POSCO, the South Korean construction *chaebol* and one of Songdo's core investors, carried out land reclamation, developed infrastructure, and relocated its own corporate headquarters from Seoul to Songdo. J. P. Morgan made significant investments while providing financing expertise. Cisco Systems played the lead role in developing Songdo's information technology hardware, and architectural firm Kohn, Pedersen, and Fox (KPF) designed several of

Songdo's distinctive buildings. Given this cavalcade of global corporate all-stars, it is notable that a relatively low-profile real estate firm, Gale International, made the biggest financial commitment among all investors. Collectively, these and other firms made Songdo the world's largest private real estate development (Arthur 2012). In 2010, *The Economist* estimated that Songdo would cost US$40 billion once completed (D.T. 2010).

Gale International's original conceptualization of the city was for an amenity-laden international business hub that targets multinational corporations needing regional offices near Hong Kong, Tokyo, and Shanghai; Songdo provides all the amenities of what Glaeser (2012) called a "boutique city." A parallel strategy lured universities (Looser 2012): George Mason University, University of Utah, and the State University of New York, Stony Brook. South Korea's Yonsei University built an international campus in Songdo, and Incheon National University relocated to Songdo early in the city's development phase. On the nonprofit side, the United Nation's Green Climate Fund is located in Songdo, but internal organizational challenges emerged in mid-2018 and have raised questions about the Fund's future.

Songdo as Technocracy

NUA commits the international development community to a construct of inclusive and sustainable urbanism informed by technocratic elements, but the agenda is thin on particulars about implementation. At its 2016 launch in Quito, Ecuador, the creators of the agenda pledged that the next step was to address implementation at UN Habitat's World Urban Forum 9 (WUF9) in 2018. At that event, it was clear that smart cities had emerged as the vehicle for implementing NUA. A test bed for smart city urbanism was already in place with the construction of Songdo and the project was featured in discussions and displays at WUF9.

NUA is a blueprint for urbanization over the next twenty years, a crucial period when interventions for climate change will be collectively decided. NUA's discourse is organized around the twin goals of inclusive and sustainable urbanism and is expressed through a list of 175 aspirational statements (United Nations 2016). These goals reflect a utopian urban imaginary that is rooted in enlightenment-inspired technocratic rationality and instrumental reasoning and is embraced not only by the international development community but also by the scholarly and consultancy communities that validate its directives.

The Songdo project arguably has epistemic roots in the modernist conviction that reason is the supreme arbiter of social, political, and economic affairs (Peters 2017). Songdo is a twenty-first-century embodiment of Cartesian-style instrumental rationality that assumes ends are achieved best through the most efficient, cost-effective means; this mind-set neglects the inconvenient messiness of social and cultural realities. Songdo's instrumental reasoning presumes at an abstract level that sufficient data and analytics deliver desired policy and planning results. According to Kitchin (2014), "The drive towards managing and regulating the city via information and analytic systems promotes a technocratic mode of urban governance which presumes that all aspects of a city can be measured and monitored and treated as technical problems" (9). Kitchin (2014) critiqued technocratic governance for its narrow scope that applies reductionist and functionalist policy strategies to complex cultural and political challenges. Roy (2009) further elaborated on the technocratic delusion by proposing that, in the case of India's cities, urban forms have idioms and their social and cultural particularities constitute the deeper logic of everyday urban life. Songdo reflects a utopian imaginary of instrumental reasoning that largely overlooks these structural societal factors.

The instrumental reasoning that gives Songdo its technocratic credibility is embedded within a neoliberal logic that valorizes the authority of experts with specialized knowledge in the operational triad of urban policy: management, administration, and planning. Expert knowledge legitimizes the power to define problems, whereas the power to produce solutions generates a feedback loop in which power replicates itself through accumulated experience, practice, and knowledge. Technocrats reference themselves as experts and their expertise shapes self-evident problem framing, policy content, and implementation strategies. Furthermore, the application of technical solutions to urban problems can be used to excuse governments from the more challenging tasks of systemic governance reform, wasting resources on initiatives that enrich private providers but fail to generate public value (Hartley 2015b; Vu and Hartley 2018).

The performance of technocratic authority in the framing, building, and branding of the Songdo project—as an embodiment of governmentality—is itself a test bed for the power–knowledge nexus behind NUA (Cohen 2016; Satterthwaite 2016; Caprotti et al. 2017). The problem-solving authority of public policy officials converges with technocracy's multiple forms of expertise to generate a smart city discourse that legitimizes NUA implementation, with Songdo representing a test bed for implementation. Furthermore, instrumental reasoning and technocratic rationality promoted by the global power--knowledge nexus heralds a planetary urbanization (or postindustrial critical zone) shaped by the smart city concept.

Songdo as Test Bed Urbanism

Songdo is an instructive case of test bed urbanism as it illustrates how technocratic reasoning that animates NUA frames the smart city model as a twenty-first-century solution. The path from Habitat III to WUF9 was a seamless transition in the reproduction of expert knowledge and power, recalling images of Gramsci's cultural hegemony (Lears 1985). In such cases, the ideas, practices, and techniques of instrumental reasoning—as deployed by technocrats in test bed settings—are rarely debated or contested. Lears (1985) argued that cultural hegemony describes how elites create "common-sense" among the popular classes; that is, the unreflective acceptance of elite-promoted agendas and ideologies. WUF9 can be interpreted as a moment when smart cities became the consensus approach to NUA implementation, driven by global experts and sold to localities as a silver-bullet solution. The rational, discursive logic of the test bed model makes this emergent common sense possible, and the logic of Songdo in particular excuses the expert elite from unequivocally declaring that smart cities are the solution for implementation; the solution thus appears to emerge endogenously rather than authoritatively.

Viewed through the test bed perspective, Songdo is a "rehearsal of our future" with a "tense relationship between performance and aspiration" (Halpern et al. 2013, 275). Although not representing a viable path to sustainable twenty-first-century urbanism (Kuecker 2015), Songdo constitutes a test bed assemblage of techniques for infrastructure, technology, design, and governance that forms a discursive urban imaginary and "proof of concept" for implementing NUA. The project embodies a particular form of power and knowledge needed to advance smart cities as a viable concept. In proposing the "test bed urbanism" line of inquiry and analysis, Halpern et al. (2013) described Songdo as a departure from the modern urban imaginary driven by enlightenment rationality:

> [Songdo is] reflective of a new form of epistemology that is concerned not with documenting facts in the world, mapping spaces, or making representative models but rather with creating models that are territories. Performative, inductive, and statistical, the experiments enacted in this space transform territory, population, truth, and risk with implications for representative government, subjectivity, and urban form. (274)

Songdo can be considered the moment when knowledge becomes power in facilitating NUA implementation, and the fluid, open, incomplete nature of smart city power is fundamental to how Songdo fulfills that role. The project's test bed urbanism embodies contingency within Lefebvre's planetary urbanization and is the mechanism by which the smart city becomes legitimized as the rational technique of twenty-first-century urbanization. The technocrat's ability to produce knowledge becomes a gesture through which power guides discourse about normative goals. Wainwright (2018) argued that a "politics of knowledge" emerges when knowledge is

> entangled with the exercise of political power, especially at a time when official interpretations of politics are in question. Questions arise about whose knowledge—and whose future matters in public policy, and what kinds of thoughts, ideas, beliefs, exchange of emotions and exercise of skills count as expertise and knowledge in public decision-making, and why. (4–5)

This convergence of policy, technocracy, and planning points to teachable optics within Songdo's test bed function and validates a discourse around urban forms as expressed at events (e.g., WUF9) where discussions and exhibitions perform smart city governmentality. As Scott (1998) argued in *Seeing Like a State* the power to make something legible also carries the power to make something else illegible, and this dynamic is evident in increasingly exclusionary discourses around twenty-first-century urbanism. Governmentality is endemic throughout the

development discourse (Escobar 2012), and issues like informality and the plight of the disadvantaged global majority are marginalized as intractable challenges burdening Davis's (2007) "surplus humanity." Foucault's governmentality is thus a useful lens for understanding how the test bed function is a theater for the performance of technocracy.

Whereas scholars like Klauser, Paasche, and Söderström (2014) prefer to view Songdo's governmentality through the lens of security, others find explanatory value for governmentality in the disciplinary logic of the smart city, one that guides the behavior of citizens and policymakers through a "commonsense" appeal that appears self-regulating but is totalizing in its normative logic; Vanolo (2014) referenced a "discipline mechanism that can be defined as a 'smartmentality'" (889). When the test bed urbanism of Songdo becomes the mechanism for NUA implementation, its discursive power becomes disciplinary and is thus granted the influence to frame knowledge.

> Cities are made responsible for the achievement of smartness—i.e., adherence to the specific model of a technologically advanced, green and economically attractive city, while "diverse" cities, those following different development paths, are implicitly reframed as smart-deviant. (Vanolo 2014, 889)

The concept of governmentality illuminates how smart cities become the disciplinary logic of twenty-first-century urbanism. Songdo is the product of historically rooted, negotiated, and contingent systems of rational thought based on "boundaries of knowledge, ideas, truths, representations and discursive formations" (Wang 2017, S4379). Referencing Foucault, Halpern et al. (2013) argued that Songdo is a heterotopia that forces policymakers and observers not only to reflect on how their choices shape the world but also to consider "alternative realities" through mirror images of attractive and unattractive, ideal and dystopian. It is the heterotopian quality of Songdo that makes it an urban imaginary conveniently deployable by planners to advance a realizable utopian twenty-first-century urbanism.

Conclusion

In a 2014 interview, Michael Berkowitz, president of the Rockefeller Foundation's 100 Resilient Cities campaign, stated that cities "pose the greatest promise. ... We can live much more efficiently in cities, we can live much more harmoniously in cities" (Clancy 2014). This faith in the capacity of cities to solve the great challenges of the twenty-first century shapes the mentality of those with power over urban policy, and Songdo embodies these aspirations. The valorization of Songdo, however, has implications for the discourse and practice of smart cities. Projects such as Songdo are at once progressive and regressive; they look forward in applying modern technology and backward in drawing from the totalizing ideal of master planning that defined mid-twentieth-century urbanism. The appeal of "new town" smart cities such as Songdo, from both planning and policy perspectives, is the blank slate; nevertheless, the complex sociocultural milieu that animates urban life is impossible to master plan and might even be compromised by the types of top-down efforts on which smart city projects rely (Hartley 2015a). By contrast, the retrofitting of existing urban environments presents unique challenges, particularly whether smart city initiatives can be harmonized with a preexisting urban milieu that gives a place the immeasurable air of authenticity. Case-comparative research is needed in this realm.

The power–knowledge nexus that legitimizes Songdo's technocratic authority is a fundamental characteristic of the smart city idiom. It enlists corporations, transnational institutions, governments, nongovernmental organizations, research institutes, and academia to shape public policy by controlling its discourse; this process is actualized by the application of theoretical concepts such as collaborative, network, and joined-up governance. The tight relationship among corporations, organizations, and state actors—enabled largely by neoliberal reforms—generates power and profit for smart city product and service providers. Many of these companies are principals in the promotion of smart urbanism because they have expertise in drafting tenders, connections with sources of transnational capital, and access to the authority of the global consultancy complex that comprises the smart city technocracy. Songdo's test bed function within the implementation of the NUA empowers private corporations to define and drive smart urbanism, as they have capacity advantages over the public sector in technology, innovation, and finance. Much of what the scholarship has claimed about the value of democratizing neoliberal governance applies in the case of smart cities (e.g., collaboration, accountability, and

transparency), but further research is needed to apply models of policy change in offering a discipline-specific explanation for how smart cities are becoming the new urban norm.

We contend that within the planetary urbanization context, the case of Songdo illustrates a form of Lefebvrian critical zone urbanism. We also postulate, though, that humanity might have departed from the age of the critical zone where the postindustrial city no longer defines and explains the urban experience. Public policy researchers need to frame explorations of the smart city within a postcritical zone context that constitutes a departure, or even rupture, from the modern historical period. Likewise, critical urban theory needs to determine whether the global community, in confronting the prospect of systemic collapse (Kuecker 2014), has already entered a postcritical zone period beyond neoliberal and technocratic governmentality and whether that transition can be interpreted through the Songdo case. As Wang (2017) argued, "There is the assertion in the smart city discourse that smartness stands for being good, healthy, and technologically advanced, therefore, the 'smart city' is intended as the ultimate goal for urban development projects" (S4380). Nevertheless, although the profit-making potential of the smart city is high, its technocratic urbanism is confined by the historically documented folly of modernity's rationalist agenda, in which the pursuit of a perfect human condition through reason falls short in tragically ironic ways—as Berman's (1988) discussion of Faust shows. The utopian idiom of Songdo contradicts the scope, scale, and speed of the global community's unmanageable convergence of wicked problems and its departure from modernity. The discourse of Songdo's test bed urbanism, however, persists despite the global community's inability to accomplish even the modest enterprise of protecting vulnerable settlements from erratic climate phenomena.

In closing, it is timely to consider how urban models of governance are evolving and whether the alternatives are "rupture" or "splintered urbanism" (Graham and Marvin 2002; Swilling 2011). Welfare capitalism shaped the urban form through the mid-twentieth century; thereafter, neoliberalism dismantled it, engendering a splintered urbanism embodied by the distancing of social classes and hyperexploitation. In looking forward, rupture can be conceptualized as a postcritical zone, with global crises forcing a rupture in the evolution of the urban norm. This phenomenon extends the problematic conceived by Lefebvre's critical zone, the manifestation of a coherent urban model during the postindustrial epoch. As this research agenda continues to coalesce around smart city imaginaries, it is fair to ask whether the critical zone embodies "splintered urbanism," whether the smart city is merely a continuation of the current critical zone, and whether the concept of rupture is constitutive of a postcritical zone. Critical urban theory for smart city analysis can then be refashioned around this problematic while questioning the implications of governmentality for the nascent smart urban norm.

ORCID

Kris Hartley ⓘ http://orcid.org/0000-0001-5349-0427

References

Arthur, C. 2012. The thinking city. BBC Science and Technology. *Future Focus* January:55–58.

Berman, M. 1988. *All that is solid melts into air: The experience of modernity.* New York: Viking Penguin.

Brady, M., and R. K. Lippert, eds. 2016. *Governing practices: Neoliberalism, governmentality, and the ethnographic imaginary.* Toronto: University of Toronto Press.

Brenner, N. 2009. What is critical urban theory? *City* 13 (2–3):198–207.

Calvillo, C., O. Halpern, J. LeCavalier, and W. Pietsch. 2016. Test bed as urban epistemology. In *Smart urbanism: Utopian vision or false dawn?* ed. S. Marvin, A. Luque-Ayala, and C. McFarlane, 145–67. London and New York: Routledge.

Caprotti, F., R. Cowley, A. Datta, V. Castán Broto, E. Gao, L. Georgeson, C. Herrick, N. Odendaal, and S. Joss. 2017. The new urban agenda: Key opportunities and challenges for policy and practice. *Urban Research & Practice* 10 (3):367–78.

Castells, M. 1979. *The urban question: A Marxist approach,* trans. A. Sheridan. Cambridge, MA: MIT Press.

Clancy, H. 2014. Michael Berkowitz: Community is the secret of urban resilience. *GreenBizz,* August 11. Accessed April 23, 2019. http://www.greenbiz.com/blog/2014/08/12/michael-berkowitz-community-secret-ingredient-urban-resilience

Cohen, M. 2016. From Habitat II to Pachamama: A growing agenda and diminishing expectations for Habitat III. *Environment and Urbanization* 28 (1):35–48.

D.T. 2010. Sing a song of $40 billion. *The Economist,* July 22. Accessed April 23, 2019. https://www.economist.com/banyan/2010/07/22/sing-a-song-of-40-billion.

Davidoff, P. 1965. Advocacy and pluralism in planning. *Journal of the American Institute of Planners* 31 (4):331–38.

Davis, M. 2007. *Planet of slums.* New York: Verso.

Escobar, A. 2012. *Encountering development: The making and unmaking of the third world.* Princeton, NJ: Princeton University Press.

Ferguson, J., and A. Gupta. 2002. Spatializing states: Toward an ethnography of neoliberal governmentality. *American Ethnologist* 29 (4):981–1002.

Foucault, M. [1966] 1994. *The order of things: An archaeology of the human sciences.* New York: Vintage Books.

Foucault, M. 1977. *Discipline and punish: The birth of the prison,* trans. A. Sheridan. London: Allen Lane.

Glaeser, E. 2012. *Triumph of the city: How our greatest invention makes us richer, smarter, greener, healthier, and happier.* New York: Penguin.

Gordon, C. 1991. Governmental rationality: An introduction. In *The Foucault effect: Studies in governmentality,* ed. G. Burchell, C. Gordon, and P. Miller, 1–52. Chicago: University of Chicago Press.

Gordon, C., L. Whimster, and S. Lash. 1987. The soul of the citizen: Max Weber and Michel Foucault on rationality and government. *Michel Foucault: Critical Assessments* 4 (2):427–48.

Graham, S., and S. Marvin. 2002. *Splintering urbanism: Networked infrastructures, technological mobilities and the urban condition.* London and New York: Routledge.

Halpern, O., J. LeCavalier, N. Calvillo, and W. Pietsch. 2013. Test-bed urbanism. *Public Culture* 25 (270):272–306.

Hartley, K. 2015a. Smart cities and the plight of cultural authenticity. *The Global Urbanist,* March 24. Accessed April 23, 2019. http://globalurbanist.com/2015/03/24/smart-cities-cultural-authenticity.

Hartley, K. 2015b. Smart cities: Old bytes in new cables? *The Planner* (Royal Town Planning Institute UK), November 6. Accessed April 23, 2019. http://www.the-planner.co.uk/opinion/smart-cities-old-bytes-in-new-cables.

Joseph, J. 2013. Resilience as embedded neoliberalism: A governmentality approach. *Resilience* 1 (1):38–52.

Kerr, D. 1999. Beheading the king and enthroning the market: A critique of Foucauldian governmentality. *Science & Society* 63 (2):173–202.

Kitchin, R. 2014. The real-time city? Big data and smart urbanism. *Geo Journal* 79 (1):1–14.

Klauser, F., T. Paasche, and O. Söderström. 2014. Michel Foucault and the smart city: Power dynamics inherent in contemporary governing through code. *Environment and Planning D: Society and Space* 32 (5):869–85.

Kshetri, N., L. Alcantara, and Y. Park. 2014. Development of a smart city and its adoption and acceptance: The case of New Songdo. *Communications & Strategies* 96 (4):113–28.

Kuecker, G. 2013. Building the bridge to the future: New Songdo City from a critical urbanism perspective. SOAS-AKS Working Papers in Korean Studies II. School of Oriental and African Studies, University of London. Accessed April 23, 2019. http://www.soas.ac.

uk/koreanstudies/overseas-leading-university-programmes/soas-aks-working-papers-in-korean-studies-ii/.

Kuecker, G. 2014. The perfect storm: Catastrophic collapse in the 21st century. In *Transitions to sustainability: Theoretical debates for a changing planet*, ed. D. Humphreys and S. Stober, 89–105. Champaign, IL: Common Ground.

Kuecker, G. 2015. New Songdo City: A bridge to the future? In *The Second Assessment Report on Climate Change and Cities (ARC3-2)*. New York: Urban Climate Change Research Network. Accessed April 23, 2019. https://docs.wixstatic.com/ugd/2307bc_fa3293d124c34b17951a50cafb3b6fca.pdf

Larner, W. 2000. Neo-liberalism: Policy, ideology, governmentality. *Studies in Political Economy* 63 (1):5–25.

Lears, T. J. J. 1985. The concept of cultural hegemony: Problems and possibilities. *American Historical Review* 90 (3):567–93.

Lefebvre, H. [1970] 2003. *The urban revolution*. Minneapolis: University of Minnesota Press.

Lemke, T. 2002. Foucault, governmentality, and critique. *Rethinking Marxism* 14 (3):49–64.

Looser, T. 2012. The global university, area studies, and the world citizen: Neoliberal geography's redistribution of the world. *Cultural Anthropology* 27 (1):97–117.

Peters, B. 2017. What is so wicked about wicked problems? A conceptual analysis and a research program. *Policy and Society* 36 (3):385–96.

Roy, A. 2009. Why India cannot plan its cities: Informality, insurgence and the idiom of urbanization. *Planning Theory* 8 (1):76–87.

Satterthwaite, D. 2016. Editorial: A new urban agenda? *Environment & Urbanization* 28 (1):3–12.

Scott, J. 1998. *Seeing like a state: How certain schemes to improve the human condition have failed*. New Haven, CT: Yale University Press.

Shwayri, S. T. 2013. A model Korean ubiquitous eco-city? The politics of making Songdo. *Journal of Urban Technology* 20 (1):39–55.

Swilling, M. 2011. Reconceptualising urbanism, ecology and networked infrastructures. *Social Dynamics* 37 (1):78–95.

Townsend, A. 2013. *Smart cities: Big data, civic hackers, and the quest for a new Utopia*. New York: Norton.

United Nations. 2016. New urban agenda. Paper presented at the United Nations Conference on Housing and Sustainable Urban Development (Habitat III), Quito, Ecuador, October 17, 2016. Accessed April 23, 2019. http://habitat3.org/wp-content/uploads/NUA-English.pdf.

Vanolo, A. 2014. Smartmentality: The smart city as disciplinary strategy. *Urban Studies* 51 (5):883–98.

Vu, K., and K. Hartley. 2018. Promoting smart cities in developing countries: Policy insights from Vietnam. *Telecommunications Policy* 48 (10):845–59.

Wainwright, H. 2018. *A new politics from the left: Radical futures*. Cambridge, UK: Polity.

Wang, D. 2017. Foucault and the smart city. *The Design Journal* 20 (Suppl. 1):S4378–S4386.

Zamora, D., and M. C. Behrent, eds. 2016. *Foucault and neoliberalism*. New York: Wiley.

PART IV
Smart Sustainability and Policy

The Struggles of Smart Energy Places: Regulatory Lock-In and the Swedish Electricity Market

Darcy Parks (iD) and Anna Wallsten (iD)

Visions of smart energy systems are increasingly influencing energy systems around the world. Many visions entail ideas of more efficient versions of existing large-scale energy systems, where smart grids serve to balance energy consumption and demand over large areas. At the other end of the spectrum are visions of smart energy places that represent a challenge to dominant, large-scale energy systems, based on smart microgrids that facilitate the self-sufficiency of local, decentralized energy systems. Whereas smart energy places do not necessarily aim to create completely isolated microgrids, they generally aim to strengthen the connection between energy consumption and production within delimited spaces. The aim of this article is to better understand how visions of smart energy places are translated into sociomaterial configurations. Smart Grid Gotland and Climate-Smart Hyllie were two Swedish initiatives where notions of place were central to the attempts to reconfigure the local energy system. Several solutions proposed within these smart energy places struggled because of regulatory lock-in to the existing spatial arrangements of the electricity market. There was a mismatch between the larger spatial scales institutionalized in the Swedish electricity market and the smaller scales introduced in these smart energy places. The conflicting spatial arrangements between electricity market and these initiatives suggest that demonstrations of smart energy places require some degree of protection from market regulations. Without this protection, visions of smart energy places might instead result in incremental changes to existing large-scale energy systems. *Key Words: energy systems, place, smart cities, smart grids, visions.*

智慧能源系统的愿景，逐渐影响着世界的能源系统。诸多愿景承担较现有大规模能源系统更有效率的形式之概念，其中智慧电网用以平衡大区域中的能源消费与供给。该光谱的另一端，则是智慧能源地方的愿景，该愿景根据促进在地且去中心化能源系统的自给自足之智慧微电网，对主流的大规模能源系统提出挑战。智慧能源地方并不必然旨在创造全然孤立的微电网，但它们普遍致力于强化划定空间中能源消费和生产的连结。本文旨在更佳地理解智慧能源地方的愿景如何转译成社会物质配置。瑞典的智慧电网哥德兰岛和智慧气候海丽计画，是地方概念作为尝试重新配置在地能源系统之核心的两大计画。这些智慧能源地方所提出若干解决方案，因电力市场的既有空间安排的制度锁定而艰难地进行。在瑞典电力市场中制度化的较大空间规模，和这些智慧能源地方所引进的小规模之间并不协调。电力市场和上述倡议之间的空间安排冲突，显示出智慧能源地方的示范，必须受到若干程度的市场规范之保护。若缺乏这些保护，智慧能源地方的愿景，则可能反而对既有的大规模能源系统带来微不足道的改变。关键词：能源系统，地方，智慧城市，智慧电网，愿景。

Las visiones que se tienen de los sistemas energéticos inteligentes tienden a influir en alto grado los sistemas de energía alrededor del mundo. Muchas de estas visiones conllevan la idea de versiones más eficientes de los sistemas energéticos de gran escala existentes, donde redes eléctricas inteligentes sirven para balancear el consumo y la demanda de energía para áreas grandes. En el otro extremo del espectro están las visiones de lugares de energía inteligente que representan un reto a los sistemas energéticos de gran escala predominantes, basados en microrredes inteligentes que facilitan la autosuficiencia de sistemas energéticos locales y descentralizados. Mientras que los lugares con energía inteligente no necesariamente apuntan a crear microrredes completamente aisladas, aquellos generalmente buscan fortalecer la conexión entre consumo y producción de energía dentro de espacios delimitados. El propósito de este artículo es entender mejor cómo la visión de lugares con energía inteligente se traduce en configuraciones sociomateriales. La Smart Grid Gotland y la Climate-Smart Hyllie fueron dos iniciativas suecas en las que las nociones de lugar fueron centrales en los intentos de reconfigurar los sistemas energéticos locales. En varias de las soluciones propuestas dentro de estos lugares de energía inteligente se debió luchar contra el monopolio regulador impuesto en los arreglos espaciales existentes del mercado de la electricidad. Se presentaba una discordancia entre las escalas espaciales más grandes institucionalizadas en el mercado sueco de la electricidad y las escalas más pequeñas introducidas en estos lugares de energía inteligente. Los esquemas espaciales en conflicto entre el mercado de la electricidad y estas iniciativas sugieren que las demostraciones de los lugares de energía inteligente requieren algún grado de protección en los mecanismos reguladores del mercado. Sin esta protección, las visiones de lugares de energía inteligente podrían resultar inesperadamente en cambios incrementales de los sistemas existentes de energía a gran escala. *Palabras clave: ciudades inteligentes, lugar, redes eléctricas inteligentes, sistemas energéticos, visiones.*

Visions of smart energy systems are increasingly influencing energy systems around the world. What many of these visions have in common is the application of information and communication technology (ICT) to make energy systems more efficient, combined with the assumption that energy users are economically rational agents who react to price changes (Wissner 2011; Strengers 2013). These visions have been translated into policy for entire energy systems on the subnational scale (Jegen and Philion 2018; Winfield and Weiler 2018), for entire countries (Ngar-yin Mah et al. 2012; Gangale, Mengolini, and Onyeji 2013), and across multiple countries (IqtiyaniIlham, Hasanuzzaman, and Hosenuzzaman 2017). Many of these policies are based on a category of smart energy visions that imagine more efficient versions of existing large-scale energy systems, where smart grids serve to balance energy consumption and demand over large areas, in particular by adapting energy use to the variability of renewable energy production (Verbong, Beemsterboer, and Sengers 2013; Ballo 2015; Schick and Gad 2015).

At the other end of the spectrum is another category of smart energy visions that we refer to as *smart energy places*. Smart energy places represent a challenge to dominant, large-scale energy systems, based on smart microgrids that facilitate the self-sufficiency of local, decentralized energy systems (Engels and Münch 2015). Many initiatives that draw on this category of visions are explicitly framed in terms of particular types of places, such as single-city districts (McLean, Bulkeley, and Crang 2016), entire cities and urban regions (Bulkeley, McGuirk, and Dowling 2016; Levenda 2019), and islands (Pallesen and Jenle 2018). Although these initiatives do not necessarily aim to create completely isolated microgrids, they generally aim to strengthen the connection between energy consumption and production within delimited spaces. The translation of smart energy visions into sociomaterial configurations can take many different forms depending on local geography and actor coalitions (Skjølsvold and Ryghaug 2015). In smart energy places, though, this process of translation also involves decisions about choosing or defining a place in which to make a stronger connection between energy production and consumption. These decisions can be influenced by a variety of factors: for example, physical geography, as in the case of island energy systems (Pallesen and Jenle 2018); administrative boundaries, when it comes to political support from local government (Bulkeley, McGuirk, and Dowling 2016); or a high concentration of environmentally conscious residents (McLean, Bulkeley, and Crang 2016).

The aim of this article is to better understand how visions of smart energy places are translated into sociomaterial configurations. One way of understanding this translation process, based on literature on sustainability transitions, would be to conceptualize a smart energy place as a niche, where new technologies can be tested in a protected space, isolated from the selection pressures of existing sociotechnical regimes (Smith and Raven 2012). Although the spatial dimensions of sustainability transitions have traditionally been underdeveloped, more recent research has identified place-specific contributions to the formation of niches and protective space (Hansen and Coenen 2015; Valderrama Pineda and Jørgensen 2016; Torrens, Johnstone, and Schot 2018) and the political economy of urban experimentation (Bulkeley and Castán Broto 2013; Evans and Karvonen 2014). Protective space is of particular importance given the mismatch between urban responses to climate change and the ideals that guide national infrastructure regulation (Bolton and Foxon 2013; Jensen, Fratini, and Cashmore 2016; Rocholl and Bolton 2016). There is also transformative potential, though, in the frictions and tensions that arise from arenas that lack protection from established regimes (Späth and Rohracher 2012; Torrens, Johnstone, and Schot 2018).

This article analyzes the cases of two smart energy places in Sweden: Smart Grid Gotland on the island of Gotland and Climate-Smart Hyllie in the city of Malmö. Both cases were home to demonstration projects that aimed to translate smart energy visions into sociomaterial configurations. Although neither case involved attempts to establish completely isolated energy systems, they both sought to make the local energy system more self-sufficient within a defined place. Climate-Smart Hyllie began as a vision established in 2011 by the city government and an energy company, who then together received funding for a demonstration project that ran from 2011 to 2016. Smart Grid Gotland was a demonstration project that ran from 2012 to 2016. Both demonstration projects received funding from the Swedish Energy Agency. Instead of evaluating all activities that took place in relation to these initiatives, this analysis focuses on aspects of particular activities where notions of place were central to the translation of smart energy visions.

The analysis draws on in-depth studies of these two cases carried out by each of the two authors. We studied the cases using a combination of methods consisting of participant observation, interviews, and document analysis. In practice this means that we followed these initiatives for several years, attended internal meetings and public events, analyzed project descriptions and brochures used to recruit participants, and interviewed people working in the projects. During 2013 and 2014, the second author conducted nineteen interviews with fourteen people working with or in close relation to Smart Grid Gotland. During the same time period, the second author also performed participant observation of two full-day fairs and eight information meetings that aimed to recruit households to the Smart Grid Gotland project. Between 2014 and 2016, the first author conducted nineteen interviews with twenty people working with either the demonstration project in Climate-Smart Hyllie, the sustainability vision for Climate-Smart Hyllie, or properties constructed in the Hyllie city district. During 2015 and 2016, the first author also performed participant observation of thirteen meetings about the implementation of the Climate-Smart Hyllie vision, where the demonstration project was a recurring issue on the agenda.

The next section of the article explains the policy drivers of smart energy visions in Sweden, along with background about the Swedish energy market. We then analyze how the Gotland case involved attempts to adapt electricity prices to local production of electricity from wind turbines. After that we analyze the Malmö case's attempts to provide the Hyllie city district with renewable energy from the region surrounding the city. Finally, we provide conclusions about the struggles to demonstrate smart energy places within the context of liberalized energy markets.

Gotland, Hyllie, and the Support of Smart Energy Visions in Sweden

Gotland and Hyllie were home to two smart grid demonstration projects in Sweden. Both were conducted by powerful actors in the Swedish energy system, and each one was led by a large international energy company, of which only a few operate in the Swedish market. These demonstration projects were two of the three smart grid demonstration projects that received funding from the Swedish Energy Agency, the third being Stockholm Royal Seaport in Stockholm, which took place on a city district scale similar to the demonstration project in Hyllie. These demonstration projects were one part of Swedish authorities' support of smart energy visions.

Realizing smart energy visions has been high on Swedish political agendas. This relatively small country has a long history of positioning itself as a role model on how to transition to a more sustainable society (Lidskog and Elander 2012). This self-imposed role is also highly prevalent in arguments put forward by Swedish authorities in support of smart energy futures. They suggest that Sweden has the potential to become a pioneer and a global source of inspiration for the implementation of smart energy visions (e.g., Government of Sweden 2012; Swedish Coordination Council for Smart Grids 2014). According to an agreement between a majority of the political parties in the Swedish parliament, Sweden should have no net emissions of greenhouse gases by the year 2045, and 100 percent of Sweden's electricity production should be based on renewable energy resources by 2040. This agreement describes a future energy system that will increasingly rely on small-scale and distributed electricity production, in combination with the current dominance of large and centralized production units. Currently, 80 percent of Swedish electricity is produced by hydropower and nuclear power, and the remaining amounts come primarily from wind power, biofuels, and organic waste (Statistics Sweden 2017). Furthermore, the agreement calls for a higher proportion of intermittent electricity resources and active electricity users (Government Offices of Sweden 2016). The demonstration projects cofunded by the Swedish Energy Agency served to test the potential of smart grids and to bring together actors with different backgrounds, interests, and perspectives.

The Swedish energy system was liberalized in 1996 and allows electricity users to choose between competing electricity suppliers, whereas the transmission and distribution grids are operated as natural monopolies. Sweden is part of the Nord Pool electricity market that spans the Nordic and Baltic countries. Spot market prices for electricity are set by auction in predefined bidding areas, of which there are four in Sweden called SE1, SE2, SE3, and SE4. The purpose of dividing Sweden into four bidding areas was to stimulate the establishment of new power production units in areas with electricity deficits and to signal to the transmission system operator where grid enforcements are needed. Most electricity is produced in the north of Sweden, whereas most of the electricity consumption occurs in the south. Because of this imbalance, users in southern Sweden sometimes pay a higher price for their electricity than users in northern Sweden (Swedish Energy Markets Inspectorate 2012).

Sweden and Norway have a joint market for renewable electricity certificates that provides incentives to increase the amount of electricity production from renewable sources. Sweden introduced the market in 2003 and the market merged with Norway's in 2012. For each unit of renewable electricity produced, the producer receives a certificate that can be sold separately from the unit of electricity, providing additional revenue for renewable electricity production. The market is supported by quotas in each country that require electricity suppliers to maintain minimum levels of renewable electricity (NVE and Swedish Energy Agency 2018).

Making Electricity Prices Local on the Island of Gotland

Smart Grid Gotland was the name of a four-year demonstration project that took place on the

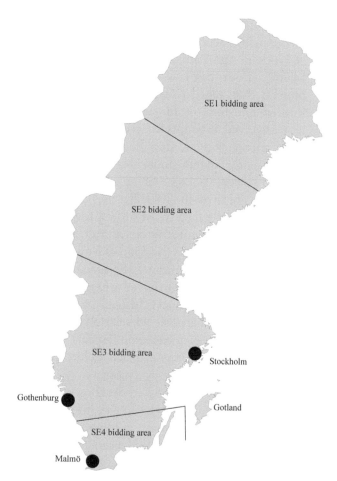

Figure 1. Map of Sweden showing the island of Gotland, the country's three largest cities, and the four electricity bidding areas (based on a map by Svenska kraftnät 2010).

Swedish island of Gotland, located in the Baltic Sea (see Figure 1). The project was conducted by a consortium of organizations from different sectors: an international energy company, the local distribution system operator and electric sales company, two international technology developers, the Swedish transmission system operator, a university, and the Swedish Energy Agency (which acted as a cofunder). The overall ambition with Smart Grid Gotland was to enhance the capacity of the island's distribution grid to handle increasing amounts of wind power production. Gotland is home to several wind farms, and many more are planned in the near future. These plans cause problems, however. The electric grid on Gotland is connected to the mainland through a single high-voltage direct current cable. The cable limits the amount of electricity transferred to and from the island, making Gotland's grid especially sensitive to mismatches in electricity supply and demand. When the project started, the grid owner had estimated that the limit of what the island's electricity grid tolerated would be reached within a few years if the planned power stations were constructed. With this backdrop, Smart Grid Gotland set out to explore whether vulnerable periods could be shortened by upgrading the existing distribution grid to a smart grid, which entailed improved control and monitoring systems as well as enhanced possibilities for flexible consumption (GEAB et al. 2011).

A subproject called Smart Customer Gotland focused on how to engage electricity users to address the imbalance in Gotland's electricity system. The subproject's objective was to examine the potential of flexible consumption, with the intention of using price signals as a motivator for such an engagement. It explored three components of the electricity price that could establish price signals: the grid tariff, the electricity tariff, and an occasional price reduction called the wind compensation. It was in exploring these different components that the subproject ran into challenges caused by the regulations of the electricity market.

The electricity tariff and wind compensation contributed to making the price of electricity more sensitive to the variability of electricity production. The electricity price was based on spot market prices; however, the daytime price peaks were enhanced by making the fluctuations between the low and high electricity prices even larger, even though the median was sustained. The initial idea behind the wind compensation was that households would be notified if the following day was expected to be windy on Gotland, in which case they would receive a price reduction. One way to interpret the intention with these fluctuating prices was simply that it established a connection between electricity production in Gotland and electricity prices:

> They see when the wind turbines spin faster and [can draw the conclusion], well, now it's cheap electricity, and then when they are standing still [they can draw the conclusion], well, now electricity is expensive. (Project Employee 1, Smart Customer Gotland)

Such a close connection between electricity prices and local production does not really exist, though. Rather, electricity prices are based on much more complicated calculations based on longer periods of access to electricity, so "when the wind blows per hour is not connected to the price per hour" (Project Employee 2, Smart Customer Gotland).

Even though the intention with Sweden's four bidding areas was to create a correlation between electricity production and electricity prices, the bidding areas are of much larger scale than the island of Gotland. Gotland belongs to the SE3 bidding area (see Figure 1) that stretches over almost a third of the Swedish mainland and includes Sweden's two largest cities, Stockholm and Gothenburg, "so there is not a local spot price on Gotland (Project Employee 2, Smart Customer Gotland). There was a mismatch between the place-specific concerns of Gotland's electricity system and the spatial organization of the Swedish electricity market.

The wind compensation component was an alternative way of making the electricity price reflect conditions of Gotland's electricity system. It was described as a price reduction offered on days that were windy on Gotland. When put into practice, though, this idea was difficult to make functional. To stick with the budget, the project had to choose a maximum number of days when participating households could obtain the wind compensation, limiting the scope of the wind compensation. Within the financial limits of the project, it was difficult to assemble a sociomaterial configuration that gave justice to local conditions.

The third component of the price signal was the grid tariff. The households recruited to the project had a time-dependent grid tariff, with higher prices from morning to late evening during the winter months. The tariff reflects the seasonal variations in the load of Gotland's electricity system. The project was not allowed to test a grid tariff with even more temporal variability, however, because Swedish regulations did not allow grid owners to offer certain households special grid tariffs, even if they were part of a demonstration project. Without a unique grid tariff, the price signal could not reflect the local wind production as accurately as would have been optimal for the project's purpose.

Smart Grid Gotland attempted to translate visions of smart energy places into sociomaterial configurations that addressed the imbalance in the island's electricity system: an increasing amount of wind power production and a limited cable that connects the island's electricity system to the mainland. The Smart Customer Gotland subproject attempted to strengthen the connection between the island's electricity consumption and production. Its attempts to establish price signals ran into two challenges due to the regulations of the Swedish electricity market, however: The spot market operated at too large a spatial scale, and it was not allowed to change grid tariffs for project participants. The wind compensation avoided these challenges but instead faced financial limitations; the cost of the wind compensation was too much to offer it to participants on an unlimited number of days. These challenges and limitations prevented Smart Grid Gotland from addressing the island's electricity imbalance to the extent envisioned in the demonstration project.

Making Renewable Energy Local in Climate-Smart Hyllie

Climate-Smart Hyllie is the second case of creating a smart energy place in Sweden. Hyllie is a new city district under development in the Swedish city of Malmö (see Figure 1). In 2011, the city government established a vision called the Climate Contract together with the international energy company that owns and operates the energy infrastructure in Malmö (City of Malmö, Eon, and VA Syd 2011). This vision presented Climate-Smart Hyllie as a sustainable city district supported by smart energy technologies. One part of the vision was the demonstration of smart grid technologies in the electricity and district heating networks, which was accomplished through a demonstration project funded by the Swedish Energy Agency. Climate-Smart Hyllie, though, was also very much a matter of establishing Hyllie as a place to increase the self-sufficiency of energy production. The Climate Contract set the goal of providing Hyllie with local, renewable energy from the Öresund region, which it defined as southern Sweden and eastern Denmark. Furthermore, the goal was defined in terms of all energy consumption in Hyllie, not just the city administration's and energy company's own consumption, in line with the city government's own 2030 goal for the entire city.

The initial plan for providing Hyllie with renewable energy had mixed success. Concerning district heating, the energy company took matters into its own hands and committed to refurbish an existing heating plant to run on biomass. The plant's annual production would be greater than the predicted 2020 heating consumption of all buildings in Hyllie. Concerning renewable electricity, the energy company proposed a division of responsibility. It committed to building new wind turbines if the city administration

would provide suitable plots of land within city limits. The city's comprehensive plan already pointed out suitable locations for wind turbines, but conflicts within the city administration meant that the energy company only eventually acquired land suitable for a single wind turbine. The company ran into further problems when it applied for building permits, as nearby residents launched legal challenges and regional authorities sided with the residents, preventing construction of the turbine at this site.

In 2015, four years after signing the Climate-Smart Hyllie vision, representatives from the energy company and the city administration found themselves looking for alternative ways to provide Hyllie with renewable electricity. They revisited the vision, which stipulated two conditions: electricity production should be located in the Öresund region and it should be constructed after the vision was signed. The energy company's representatives proposed a solution. The company had built two small wind farms within southern Sweden since 2011. These wind farms were not built with Hyllie in mind, but they did meet the conditions of the vision.

This suggestion led to debates about the principles for allocating renewable electricity to Hyllie. Representatives from the energy company and the city administration came up with two options. The first option was the most ambitious: a model where renewable electricity would only be allocated to Hyllie if it was backed by renewable electricity certificates. This model had several advantages, as laid out in slides at a presentation in 2016 (presentation to the Climate Contract steering group, May 2016):

- The model can be used in parallel in other city districts.
- The model can be scaled up and used to encompass larger areas (e.g., the City of Malmö goals for 2030) without risk for double counting.
- Hyllie would have high credibility with third parties (e.g., [potential] funders, actors, and partners in this city district; comparable initiatives in other parts of the city and outside of Sweden; journalists; researchers; etc.).

This option, however, would require tracking renewable electricity certificates for all the energy consumers in Hyllie. The second option was less complicated: The energy company would declare that these two wind farms were allocated to Hyllie, and the signatories of the vision would consider that Hyllie had received renewable electricity equivalent to the annual electricity production of these two wind farms. Such a declaration has no formal meaning, however, and this option lacked clear principles to avoid double counting. It did not provide a model that the Malmö city government could scale up to the entire city.

The energy company and the city administration debated these two options for months. The first option had the advantage of being possible to scale up to the entire city, and part of the vision for Climate-Smart Hyllie was to lead the way for the rest of the city in terms of sustainability. This option would be costly and difficult to implement, however. Because the Swedish electricity market provides electricity consumers with the freedom to choose their own suppliers, neither the energy company nor the city administration had the authority to track which consumers paid for renewable electricity or where the renewable electricity came from. The only ways to implement this option would be for the energy company to stop selling renewable electricity certificates from these turbines or for the city government to buy all the certificates from these turbines. Neither organization had planned to make such a significant financial contribution in support of Climate-Smart Hyllie. In autumn 2016, they decided in favor of the second option.

Climate-Smart Hyllie's renewable energy goal had the aim of establishing a connection between energy consumption and renewable energy production. Although the vision for Hyllie was not to become a self-sufficient electricity system, the goal called for renewable electricity to be produced locally, which the vision defined as the Öresund region. The goal was part of the framing for the smart grid demonstration project that took place in Hyllie. Hyllie was supposed to be an example for the rest of Malmö. The Swedish market for renewable electricity certificates provided a way to prove that Hyllie was provided with renewable electricity from local sources, but the market for these certificates made them far too costly for the organizations behind Climate-Smart Hyllie.

Conclusions

The aim of this article has been to better understand how visions of smart energy places are translated into sociomaterial configurations. Although these places are not necessarily intended to create completely isolated microgrids, they generally aim to strengthen the connection between energy consumption and production within a bounded space. Translating visions of smart energy places into sociomaterial configurations involves more than adapting a vision to local geography and actor coalitions. This process of translation involves decisions about choosing or defining a place in which the connection between energy production and consumption is to be strengthened. Smart Grid Gotland and Climate-Smart Hyllie were two attempts to create smart energy places that struggled in their translation processes.

Notions of place were central to the attempts of both cases to reconfigure the local energy system while drawing on visions of smart energy systems. The Smart Grid Gotland project was framed in terms of the island's energy geography: the single electricity cable connecting the island to the mainland electricity system, combined with an increasing amount of wind power production that is inherently variable. Climate-Smart Hyllie involved the definition of two places: Hyllie, a growing city district that was the site of a growing energy demand, and its energy production hinterland, defined as the Öresund region. The unique and place-specific problems addressed in these cases provide additional support to Skjølsvold and Ryghaug's (2015) claim that there is great generative capacity in the coproduction of local actor constellations and visions of smart energy systems. The cases also illustrate the challenges that the material politics of the everyday introduce in even the most well-funded smart energy projects (Bulkeley, McGuirk, and Dowling 2016). Even greater challenges in these cases were the strict regulations of the Swedish electricity market, which prohibited some proposed solutions and made others prohibitively expensive in the context of these cases.

Several solutions proposed within these smart energy places clashed with the regulations of the Swedish electricity market. In their attempts to address place-specific challenges, both cases involved struggles to deal with market principles of nondiscriminatory access and competition (Rocholl and Bolton 2016). In Gotland it was the idea of giving project participants a special grid tariff that failed. In Hyllie it was the suggestion of tracking which energy consumers purchased renewable electricity certificates from local producers; neither the city nor the energy company had access to information about electricity contracts and renewable energy certificates for electricity customers in Hyllie.

Another problem was regulatory lock-in to the existing spatial arrangements of the electricity market (Bolton and Foxon 2013). There was a mismatch between the larger spatial scales institutionalized in the Swedish electricity market and the smaller scales introduced in these smart energy places. The market for renewable electricity certificates is an example of a governance arrangement based on an "abstract territorial perspective" that conflicts with place-specific infrastructural concerns (Jensen, Fratini, and Cashmore 2016, 250). The high price of renewable electricity certificates that caused problems for Climate-Smart Hyllie was due not to demand for certificates from just those wind turbines but rather to the Swedish quota regulation that supports the demand for renewable electricity certificates across the Sweden and Norway. The market is agnostic to the distance between the purchaser of a certificate and the wind turbine that produces it. The certificates market lacks mechanisms to encourage renewable electricity production in specific electricity bidding areas. In contrast, the electricity bidding areas that caused problems for Smart Grid Gotland were not based on an abstract territorial perspective. The bidding areas were created to establish a connection between electricity production and consumption, but their scale of spatial organization was too large to accommodate the idea of a local electricity price in Gotland.

The conflicting spatial arrangements between the electricity market and these cases highlight the challenges for smart energy places in sustainability transitions. Supported by city governments and energy companies, these cases were "vested with particular interests and strategic purpose in governing" the respective energy systems (Bulkeley and Castán Broto 2013, 373). These cases were also free from political controversies that can arise when sociotechnical experiments become battlegrounds (Torrens, Johnstone, and Schot 2018). Even with the support of these organizations and a lack of political controversies, the new solutions proposed in these smart energy places were not matched by the creation of appropriate protective space (Valderrama Pineda and Jørgensen 2016). Furthermore, the regulations of the electricity market were too strongly institutionalized to allow these cases to stretch and transform the existing electricity regime (Smith and Raven 2012). To further explore the transformative potential of

smart energy places would require some degree of protection from these regulations. Exemption from market regulations would be the simplest form of protection, but additional financial support might allow projects to deal with the costs incurred by existing regulations as they test new solutions.

Without protection from market regulations, the potential of smart energy places is at risk. Although visions of smart energy places might persist despite a lack of protection, attempts to realize these visions might rather result in incremental changes to the existing large-scale energy system. We do not mean to advocate for smart energy places, or even smart energy systems at all, but the multitude of demonstration projects suggests that visions of smart energy places have a particular appeal. Supporters of such visions, and especially supporters who see a critical potential in decentralized, small-scale energy systems with alternative actor constellations, should be wary of putting their support behind small-scale visions whose adoption might lead to nominally smarter energy systems that otherwise remain large-scale and where business as usual persists.

ORCID

Darcy Parks ⓘ http://orcid.org/0000-0002-8388-7633
Anna Wallsten ⓘ http://orcid.org/0000-0003-1631-1519

References

Ballo, I. F. 2015. Imagining energy futures: Sociotechnical imaginaries of the future Smart Grid in Norway. *Energy Research & Social Science* 9:9–20. doi: 10.1016/j.erss.2015.08.015.

Bolton, R., and T. J. Foxon. 2013. Urban infrastructure dynamics: Market regulation and the shaping of district energy in UK cities. *Environment and Planning A: Economy and Space* 45 (9):2194–2211. doi: 10.1068/a45575.

Bulkeley, H., and V. Castán Broto. 2013. Government by experiment? Global cities and the governing of climate change. *Transactions of the Institute of British Geographers* 38 (3):361–75. doi: 10.1111/j.1475-5661.2012.00535.x.

Bulkeley, H., P. M. McGuirk, and R. Dowling. 2016. Making a smart city for the smart grid? The urban material politics of actualising smart electricity networks. *Environment and Planning A: Economy and Space* 48 (9):1709–26. doi: 10.1177/0308518X16648152.

City of Malmö, Eon, and VA Syd. 2011. Klimatkontrakt för Hyllie [Climate contract for Hyllie]. Accessed

May 11, 2017. http://www.hyllie.com/images/Klimatkontrakt_broschyr_SV_ENG.pdf.

Engels, F., and A. V. Münch. 2015. The micro smart grid as a materialised imaginary within the German energy transition. *Energy Research & Social Science* 9:35–42. doi: 10.1016/j.erss.2015.08.024.

Evans, J., and A. Karvonen. 2014. "Give me a laboratory and I will lower your carbon footprint!"—Urban laboratories and the governance of low-carbon futures. *International Journal of Urban and Regional Research* 38 (2):413–30. doi: 10.1111/1468-2427.12077.

Gangale, F., A. Mengolini, and I. Onyeji. 2013. Consumer engagement: An insight from smart grid projects in Europe. *Energy Policy* 60:621–28. doi: 10.1016/j.enpol.2013.05.031.

GEAB, Vattenfall, ABB, KTH. 2011. *Pre-study Smart Grid Gotland.* Accessed April 8, 2016. http://www.smart-gridgotland.se/Rapporter.pab.

Government of Sweden. 2012. Coordination council and knowledge platform for smart grids. Ministry of the Environment and Energy, Stockholm.

Government Offices of Sweden 2016. Framework agreement between the Social Democratic Party, the Moderate Party, the Green Party, the Centre Party and the Christian Democrats. Accessed November 30, 2018. https://www.regeringen.se/49cc5b/contentassets/b88f0d28eb0e48e39eb4411de2aabe76/energioverenskommelse-20160610.pdf

Hansen, T., and L. Coenen. 2015. The geography of sustainability transitions: Review, synthesis and reflections on an emergent research field. *Environmental Innovation and Societal Transitions* 17:92–109. doi: 10.1016/j.eist.2014.11.001.

IqtiyaniIlham, N., M. Hasanuzzaman, and M. Hosenuzzaman. 2017. European smart grid prospects, policies, and challenges. *Renewable and Sustainable Energy Reviews* 67:776–90. doi: 10.1016/j.rser.2016.09.014.

Jegen, M., and X. D. Philion. 2018. Smart grid development in Quebec: A review and policy approach. *Renewable and Sustainable Energy Reviews* 82:1922–30. doi: 10.1016/j.rser.2017.06.019.

Jensen, J. S., C. F. Fratini, and M. A. Cashmore. 2016. Socio-technical systems as place-specific matters of concern: The role of urban governance in the transition of the wastewater system in Denmark. *Journal of Environmental Policy & Planning* 18 (2):234–52. doi: 10.1080/1523908X.2015.1074062.

Levenda, A. M. 2019. Mobilizing smart grid experiments: Policy mobilities and urban energy governance. *Environment and Planning C: Politics and Space* 37 (4):634–51. doi: 10.1177/2399654418797127.

Lidskog, R., and I. Elander. 2012. Ecological modernization in practice? The case of sustainable development in Sweden. *Journal of Environmental Policy & Planning* 14 (4):411–27. doi: 10.1080/1523908X.2012.737234.

McLean, A., H. Bulkeley, and M. Crang. 2016. Negotiating the urban smart grid: Socio-technical experimentation in the city of Austin. *Urban Studies* 53 (15):3246–63. doi: 10.1177/0042098015612984.

Ngar-Yin Mah, D., J. M. van der Vleuten, J. Chi-Man Ip, and P. R. Hills. 2012. Governing the transition of socio-technical systems: A case study of the

development of smart grids in Korea. *Energy Policy* 45:133–41. doi: 10.1016/j.enpol.2012.02.005.

NVE and Swedish Energy Agency. 2018. A Swedish–Norwegian electricity certificate market: Annual report for 2017. Norwegian Water Resources and Energy Directorate & Swedish Energy Agency, Oslo and Stockholm.

Pallesen, T., and R. P. Jenle. 2018. Organizing consumers for a decarbonized electricity system: Calculative agencies and user scripts in a Danish demonstration project. *Energy Research & Social Science* 38:102–9.

Rocholl, N., and R. Bolton. 2016. Berlin's electricity distribution grid: An urban energy transition in a national regulatory context. *Technology Analysis & Strategic Management* 28 (10):1182–94.

Schick, L., and C. Gad. 2015. Flexible and inflexible energy engagements—A study of the Danish smart grid strategy. *Energy Research & Social Science* 9:51–59.

Skjølsvold, T. M., and M. Ryghaug. 2015. Embedding smart energy technology in built environments: A comparative study of four smart grid demonstration projects. *Indoor and Built Environment* 24 (7):878–90. doi: 10.1177/1420326X15596210.

Smith, A., and R. Raven. 2012. What is protective space? Reconsidering niches in transitions to sustainability. *Research Policy* 41 (6):1025–36. doi: 10.1016/j.respol.2011.12.012.

Späth, P., and H. Rohracher. 2012. Local demonstrations for global transitions—Dynamics across governance levels fostering socio-technical regime change towards sustainability. *European Planning Studies* 20 (3):461–79. doi: 10.1080/09654313.2012.651800.

Statistics Sweden. 2017. Electricity supply, district heating and supply of natural gas. Statistics Sweden, Stockholm.

Strengers, Y. 2013. *Smart energy technologies in everyday life: Smart utopia?* New York: Palgrave Macmillan.

Svenska kraftnät. 2010. Electricity grid areas. Accessed March 7, 2019. http://natomraden.se/.

Swedish Coordination Council for Smart Grids. 2014. Plan for effect! Final report from the coordination council for smart grids. Ministry of the Environment and Energy, Stockholm.

Swedish Energy Markets Inspectorate. 2012. Electricity bidding areas in Sweden: Analysis of the development and consequences for the market. Swedish Energy Markets Inspectorate, Stockholm.

Torrens, J., P. Johnstone, and J. Schot. 2018. Unpacking the formation of favourable environments for urban experimentation: The case of the Bristol energy scene. *Sustainability* 10 (3):879. https://doi.org/10.3390/su10030879

Torrens, J., J. Schot, R. Raven, and P. Johnstone. 2018. Seedbeds, harbours, and battlegrounds: On the origins of favourable environments for urban experimentation with sustainability. *Environmental Innovation and Societal Transitions.* Advance online publication. Accessed February 19, 2019. http://www.sciencedirect.com/science/article/pii/S2210422418301060.

Valderrama Pineda, A. F., and U. Jørgensen. 2016. Creating Copenhagen's Metro: On the role of protected spaces in arenas of development. *Environmental Innovation and Societal Transitions* 18:201–14.

Verbong, G. P. J., S. Beemsterboer, and F. Sengers. 2013. Smart grids or smart users? Involving users in developing a low carbon electricity economy. *Energy Policy* 52:117–25.

Winfield, M., and S. Weiler. 2018. Institutional diversity, policy niches, and smart grids: A review of the evolution of smart grid policy and practice in Ontario, Canada. *Renewable and Sustainable Energy Reviews* 82:1931–38.

Wissner, M. 2011. The smart grid: A saucerful of secrets? *Applied Energy* 88 (7):2509–18.

Toward Smart Foodsheds: Using Stakeholder Engagement to Improve Informatics Frameworks for Regional Food Systems

Allan D. Hollander, ⓘ Casey Hoy, Patrick R. Huber, Ayaz Hyder, Matthew C. Lange, Angela Blatt, James F. Quinn, Courtney M. Riggle, and Thomas P. Tomich

A foodshed is a concept analogous to a watershed, describing the catchment of the sources of food for a region. As such, it portrays linkages ranging from local communities out to the global food system. Inefficiencies exist at all stages of the food supply chain, resulting in the challenges of inequitable access to healthy and safe food. Many of these inefficiencies are informational; for instance, food being wasted that could be donated to food banks were there communication of the need. These informational inefficiencies can be ameliorated by a stronger semantic characterization of the links between actors and resources in the food system, allowing for the development of smarter software technologies to facilitate interconnections. We discuss an iterative process to improve informatics frameworks for the foodshed by engaging with regional stakeholders to identify important issues and information needs. *Key Words: food systems, ontologies, semantic web, smart foodsheds, stakeholder engagement.*

食物域的概念与流域相近, 即一个地区食物来源的总和, 体现了小到地方社区、大到整个全球食物系统之间千丝万缕的关联。在食物供应链的每个阶段都存在效率低下的问题, 进而引发一个重大挑战: 很多人无法公平获得健康安全的食物, 而问题的根源往往都源于信息不畅。例如, 被浪费掉的食物本可以捐赠给有需要的食物银行 (贫穷群体或无家可归者可领取捐赠食物的地点) 。有一种方法可以改善这种信息效率低下的现状: 使用更强的语义描述食品系统中参与者和资源之间的联系, 开发更智能的软件技术促进相互关联性。在本文中, 我们探讨了一个致力于改善食物域信息框架的迭代过程, 通过区域利益相关者的参与确定重要事项和信息需求。关键词: 食物系统、本体论、语义网、智能食物域、利益相关者参与。

Una cuenca alimentaria es un concepto análogo al de cuenca fluvial, que, para este artículo, describe la zona en donde están las fuentes alimentarias de una región. Como tal, ese tipo de cuenca retrata vínculos que van desde las comunidades locales hasta la totalidad del sistema global alimentario. Las deficiencias se dan en todas las etapas de la cadena de suministro de alimentos, lo cual desemboca en los retos sobre la desigualdad de acceso a alimentos saludables y seguros. Muchas de estas deficiencias son de carácter informativo; por ejemplo, el desperdicio de alimento que podría donarse a bancos alimentarios si hubiese existido comunicación sobre la necesidad. Estas deficiencias informativas pueden reducirse mediante una caracterización semántica más fuerte de los vínculos entre actores y recursos en el sistema alimentario, facilitando el desarrollo de tecnologías de software más inteligentes que faciliten las interconexiones. Nosotros discutimos un proceso iterativo designado para mejorar los marcos informáticos de la cuenca alimentaria involucrando a los interesados con el fin de identificar cuestiones importantes y necesidades de información. *Palabras clave: compromiso de interesados, cuencas alimentarias inteligentes, ontologías, red semántica, sistemas alimentarios.*

The term *foodshed* has emerged as an analogy to a watershed in describing the food system. It was likely first coined in 1929 when Hedden used it to describe the catchment of food production surrounding New York City (Hedden 1929; Peters et al. 2009). More recently the term has entered into the discourse around localizing the food system (Feenstra 1997; Butler 2013). Horst and Gaolach (2015) highlighted how the term is seen in three different ways: first, spatially, imagining a relocalized food system; second, analytically, examining the data on a region's food production and consumption patterns; and, finally, as a starting point for action in organizing local food policy and cultural shifts.

Considerable discussion exists about whether localizing the food system generally contributes to food system sustainability (e.g., Weber and Matthews 2008; Butler 2013). Taking a foodshed perspective, though, leads to interesting new analyses concerning the range of interactions in the food system, as illustrated in a review of foodshed analyses in North America (Horst and Gaolach 2015). Such analytical approaches will lead to better characterization of flows across scales and improve our ability to address some of the inefficiencies in the food system. The foodshed concept provides a good framing because it is both local (a specific region) and global (linked to the global food system). Improved information technologies have a substantial role to play in increasing sustainability in food systems and agriculture (e.g., Gebbers and Adamchuk 2010; El Bilali and Allahyari 2018), so improved characterizations of foodsheds should lead to better utilization of information technology in the food system.

In this article we discuss the prospects for creating "smart foodsheds"; using innovations in informatics to enhance the local food system. A smart foodshed would be one that could take advantage of a surfeit of data such as sensor and other Web-enabled data sources throughout the supply chain and analyze food system inefficiencies, having some of the following characteristics:

- *Traceability:* The smart foodshed would collect and synthesize data from sensors in the food supply chain, enabling tracking of food flows and mitigating food waste.
- *Transparency:* The smart foodshed would improve information flows linking producers with distributors and increase consumers' knowledge of food sources and nutrition.
- *Trust:* The smart foodshed would protect the needs for privacy and confidentiality of all actors.

As Maye (2018) elaborated, however, there is a tension between the notion of "smart cities," from which our concept of a smart foodshed derives, and the aims of the urban food movement. The smart city concept pulls together the trends of instrumentation becoming ever more pervasive in cities and the belief that a knowledge economy fosters socio-economic progress. For Maye, the urban food movement, by contrast, broadly connects issues of food security, social justice, well-being, and environmental impacts, arguing that the work of creating a "smart food city" calls for attention to cultural and social innovations and practices as much as technological solutions (Kirwan et al. 2013). Next we describe how in creating a smart foodshed we aim to merge both technical and social innovations.

Semantic Technologies for Smart Foodsheds

Our approach for developing a smart foodshed begins with the project of developing a common semantic framework for information exchange. This approach builds on a tradition of studying food systems as networked entities, examples being Chiffoleau (2009), who examined social relationships in farmers' markets in Languedoc-Roussillion; Colloredo-Mansfeld et al. (2014), who used sketch mapping to study consumer knowledge of food stores; and, more recently, Trivette (2019), who used social network analysis to examine ties between local food producers and consumers in New England. A smart foodshed is by its nature transdisciplinary, creating a need for a shared semantic characterization across diverse stakeholders. As the notion of a shared semantic web (an extension of the web where the meaning of information is formalized, enabling machine interoperability) has grown from conceptualization (e.g., Berners-Lee, Hendler, and Lassila 2001) toward maturity (e.g., Zeng and Mayr 2018), several themes have emerged. These include formalization of ontologies as shared knowledge schemas, the use of graph data to portray information networks, and prospects for linking open data.

Ontologies

In computer science terms, an *ontology* is a knowledge organization system that formally describes the types of things that exist and the relationships that connect them in a particular domain; for example, foodsheds (Allemang and Hendler 2011). Ontological technologies provide a shared semantic framework for linking disparate pieces of information together and have become important in creating a coherent knowledge network to integrate data, ensure interoperability across systems, and enable inference across the relationships that connect these data.

Ontologies build on previous knowledge organization systems such as controlled vocabularies. In the domain of food and agriculture, much of the initial work here came from the Food and Agricultural Organization (FAO) of the United Nations, beginning with the development of the controlled vocabulary AGROVOC (Caracciolo et al. 2013) in the 1980s. More recently, a number of efforts seek to aggregate these information resources, such as the AgroPortal Map of Standards (see http://vest.agrisemantics.org/), which describes existing standards for the exchange of food and agricultural data, including classification schemes and ontologies.

Representing Information in a Graph Data Structure

Because a major way we now think about relationships between actors is through social networks, it follows that graph data structures composed of nodes and linkages have become important technologies. The semantic web formulation of graph data uses a model called the resource description framework (RDF; Schreiber and Raimond 2014). This model decomposes a graph into primitive data elements called *triples*, with each triple representing a single logical statement combining a subject, predicate, and object, an example being the statement "California [subject] produces [predicate] tomatoes [object]." These triples are linked into an ontological framework using standardized nomenclature for these resources (usually termed URIs, for uniform resource identifier; Schreiber and Raimond 2014). Vocabularies such as AGROVOC have published URIs for terms, enabling shared references to these terms across a global namespace.

Linking Open Data

The principles of linked open data (Berners-Lee 2006; Heath and Bizer 2011) are summarized under Berners-Lee's (2006) five-star open data scheme: (1) Value is added to information by making it available with an open data license; (2) data should be in a machine-readable format; (3) data should be available in a nonproprietary format; (4) data should use open nomenclature to identify things; and (5) data should be linked to other data sets to provide context. Data interlinking flourishes when knowledge organization systems such as vocabularies and ontologies become more interoperable (Zeng and Mayr 2018). In the context of a smart foodshed, data sets will span the range from being fully open (e.g., governmental agricultural production data) to necessarily proprietary (e.g., private financial data). Even in the case of proprietary data, though, structuring the data using consistent globally referenced nomenclature can bring advantages in terms of data set queries.

One major aim of developing knowledge organization systems is providing a shared information space to describe resources. Some vocabularies have been set up by standards organizations, but others have emerged more locally. These ontologies are composable: resources can be described by selecting elements from different ontologies. An individual resource can be catalogued in multiple classes and linked to other resources using properties coming from several ontologies. For instance, one might describe a person's relationships to organizations using properties from the VIVO ontology (intended to enable discovery of persons and resources in academia; Mitchell 2018) and describe a person's contact information using properties from the FOAF schema, an ontology describing persons, their activities, and relationships to other people (Graves, Constabaris, and Brickley 2007).

We would like to ensure that the ontologies we develop to characterize foodsheds mesh with existing ontologies, forming what we term a multiontology framework. In so doing, we aim to fill in gaps in ontology space, because there are domains of interest that are not well covered by ontologies. For instance, conservation planning is a domain with well-developed schemas for management (e.g., the Open Standards for the Practice of Conservation; Schwartz et al. 2012), but the domain has not developed an ontology.

Developing the Smart Foodshed Iteratively

Food systems can be viewed from at least two different perspectives. One focuses on supply chain activities spanning from agricultural input supply and

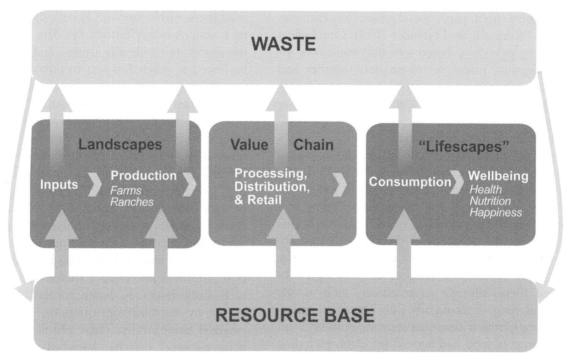

Figure 1. Food systems framing of food supply chains. From Tomich et al. (2018); adapted from a figure by Michele Grant, World Food Systems Center, ETH Zurich. Used with permission.

resource use for production through processing and marketing (Figure 1). On the supply side, these resource-intensive production activities are shaped strongly by their landscape context. Fundamental changes over the past generation (Reardon et al. 2018) require broadening our framing from fields and farms to the entire food system, extending to the health, nutrition, and overall well-being of people. Sociocultural lifescapes on the demand side of the food chain (Figure 1) are also grounded in place, an important component of local food movements described earlier. A second perspective focuses on social relationships within the food system. Even a "simple" typology of these food system stakeholders can be extremely complex (Figure 2), with most individuals holding multiple "stakeholder" identity types. Increasingly, relationships are approached in international practice as inclusive, multistakeholder processes spanning multiple spatial scales (Tomich et al. 2019). If a systems view is inescapable, the challenge becomes how to conceptualize, benchmark, and model (at least qualitatively) the myriad dimensions of economic, environmental, and social issues within these food systems (Huber et al. 2015; Springer et al. 2015). We aspire to create a food systems framework to link data sets ranging across many different data types, measurement units, and spatial scales, enabling

development of more sustainable land use practices, improved human nutrition and health outcomes, and more resilient food systems.

We envision an iterative process to improve the informatics frameworks portraying the global food system via constructing knowledge schemas at the local foodshed scale through participatory work. We assert that constructing an overall model of the food system is a very difficult task, likely without a direct path forward to generate a formal ontology. Rather, we believe that better understanding of the global food system will emerge through improved interoperability of data, especially through linked open data protocols. This interoperability will be eased by adopting common vocabularies with referenceable URIs. Initiatives such as the Global Agricultural Concept Scheme (GACS; Baker et al. 2016) are already generating a pool of identifiers for data and metadata creators to draw on.

It is widely recognized that ontologies are efficiently created by iteratively working with subject matter experts (e.g., De Nicola, Missikoff, and Navigli 2009), where a knowledge engineer refines the ontology through engagement with subject-matter experts. We expand this methodology into a participatory design process involving local communities. This notion is that communities can

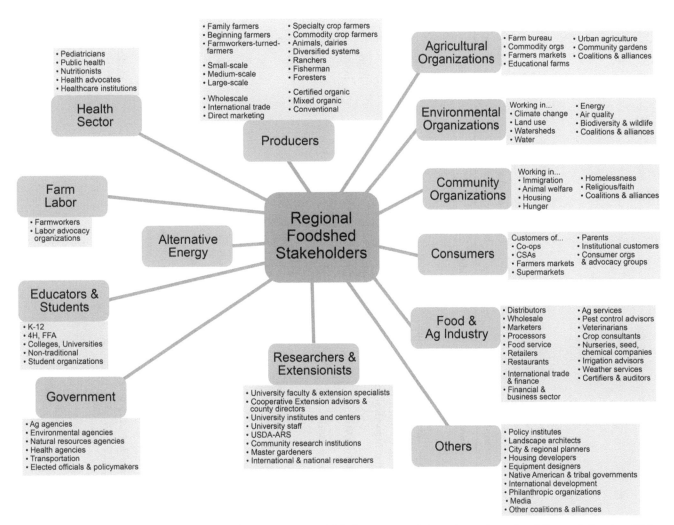

Figure 2. Stakeholders in the regional foodshed. USDA-ARS = U.S. Department of Agriculture, Agricultural Research Service; CSA = community-supported agriculture.

work with knowledge engineers to develop data schemas representing informatics concerns of local interest. An example of this process is as follows: A research group carries out interviews with stakeholders and devises a data template to code the interviews. Fields in this template are drawn, if possible, from properties in existing ontologies, but several new properties are needed to describe the content in the interviews. These properties are added into the data template, and the research group shares the template with other researchers carrying out similar projects in different cities. Later on, a knowledge engineer takes the additional properties defined in the template and extends existing ontologies with these. Data from these research projects can then be transformed into linked open data using common standards and made available as a global resource.

With these newly mobilized data and the additions to knowledge schemas describing the food system, we are able to refine our understanding of the food system. The linked open data graphs we create can be analyzed for connections through social network analysis, suggesting clusters of nodes describing resources. This in turn allows us to create better ontologies in the multiontology framework, thus continuing the iterative loop to improve the informatics framework around the food system, as well as allowing visualizations for new conceptualizations of the food system.

Approaches for Creating a Smart Foodshed

Data Sources for the Smart Foodshed

The smart foodshed can draw on a large variety of data sources, but these differ widely in their

Table 1. Examples of data sources for the smart foodshed

Sector	Data source	Accessibility to public	Degree to which data already exist	Ease of mobilization
Natural systems	Federal agency land, soil biodiversity data sets (USGS, NASA)	High (public domain)	High	High
	State agency data sets; e.g., for biodiversity	Moderately high, although confidentiality for sensitive species	High	High
Agricultural production	USDA Census of Agriculture	Moderately high, confidentiality maintained at individual scale	High	High
	County crop statistics	High	High	High
	Sensor data tracking food flow through supply chain	Low (proprietary data)	Moderate (technology still developing)	Low
	Private-sector data on precision agriculture	Low (proprietary data)	High	Low
Physical infrastructure	Government agency transportation data	High	High	High
	Private-sector "Internet of Things" data on vehicles and rail transport	Low (proprietary)	Moderate	Low
Financial	Retail sales of goods and services	Low (proprietary)	High	Low
	Banking/loans for food system investment	Moderate (proprietary)	High	Moderate
Social and political	University/other research institution data on social and religious networks	High	Moderate (research generally limited in scope)	Moderate
Human health	Public health records	Moderate (confidentiality issues)	High	Moderate

Notes: USGS = U.S. Geological Survey; NASA = National Aeronautics and Space Administration; USDA = U.S. Department of Agriculture.

characteristics for use. Table 1 shows examples of sources, arranged by sector. The first column describes accessibility to the public. Data produced by the federal government are generally in the public domain, except for cases where confidentiality needs to be protected, an example being the Census of Agriculture, which is collected at the individual farm scale and then aggregated to counties to preserve confidentiality. At the other extreme are the proprietary data collected by many precision agriculture systems. Here we distinguish between proprietary data and confidential data, the former being related to commercial concerns and the latter to privacy issues.

The second column describes the degree to which data for each example already exist. In many cases data are being collected as part of long-standing surveys (e.g., agricultural censuses) and can be readily queried. Similarly, in the private sector, financial records (e.g., loan information) are maintained as a function of business operations and are potentially available for analysis. On the other end of this scale, some data are not presently generated. For instance, the flow of food from production through the supply chain creates data, but these data are currently too fragmented to track individual food items from farm to fork.

The final column summarizes the degree to which each source of data can be mobilized for use. Rankings in these columns are as follows. High are data sets that are available online or through straightforward enquiries. Moderate are data sets that are obtainable through some investment, such as negotiated agreements to maintain confidentiality. Low are data sets that are available only with high investment in negotiations or technical development. A couple of points emerge here. First, the scale of the data set relates to the degree to which it can be mobilized. Large compilations produced by state or federal governments are readily used but are generally only available at a coarse scale (e.g., county-level statistics). Conversely, universities have the capacity to generate fine-scale studies of the networks of actors in the food system but do not typically have the resources to expand these to a wide scope. A second point is that much data in the foodshed will remain private, whether due to confidentiality concerns or due to their proprietary nature. An informatics system that successfully links across a foodshed must address the boundaries between public and private data as part of its very design.

Efforts toward Stakeholder Engagement

A wide range of stakeholder individuals and organizations can be identified in association with a regional foodshed, as illustrated in Figure 2. This network of stakeholders in a foodshed is quite complex, and it is a significant challenge to engage with a representative subset of actors. An initial step here is to solicit information about data availability and needs, particularly focusing on themes of food system coordination, development of food system infrastructure, and elimination of food system waste. These stakeholders can be engaged through focus groups and by developing use cases.

Developing Use Cases for Interaction

The purpose of a use case is to explicitly capture how an end user will use or interact with a product. The concept of use cases comes from software development and product design (e.g., Adolph, Cockburn, and Bramble 2002), and we use it to illustrate how potential stakeholders in the foodshed can use developed tools and applications. The approach for a creating a use case involves the following steps: (1) develop a use case template, usually a set of open-ended questions; (2) conduct interviews of potential stakeholders; that is, the data providers and users in the foodshed; and (3) use the answers in the use case template to design tools for specific end users. From this gathered information we will be able to refine our existing knowledge schemas for the food system.

Creating an Ontology for Stakeholder Relationships

One element of creating a smart foodshed is to be able to describe in a computable manner the networks of relationships among stakeholders. This will enable development of software to connect actors and entities for applications that are suggested by the use case surveys. Our approach here is to develop an ontology for the network of relationships. We have named this ontology PPOD, for persons–projects–organizations–data sets (Hollander 2019). Most of the classes and properties in the PPOD ontology have been extracted from existing ontologies such as VIVO. These four classes have emerged as core groupings in our experience in working with the information-sharing activities of research and

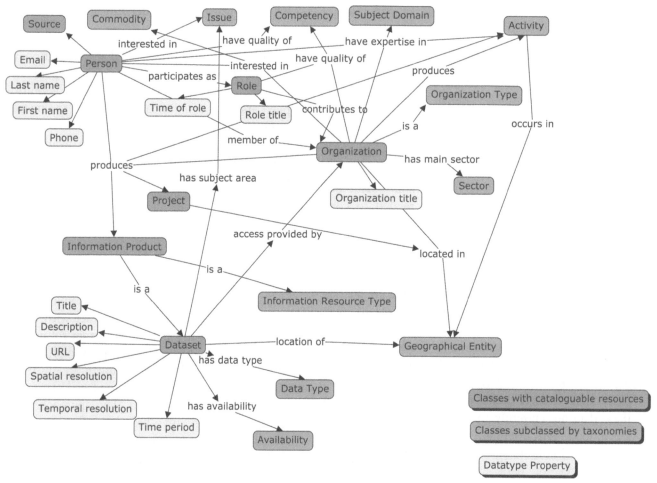

Figure 3. Concept map of the PPOD (persons–projects–organizations–data sets) ontology.

management communities. Figure 3 is a concept map illustrating the main structure of the PPOD ontology. Nodes in aqua are classes that we envision cataloguing, and nodes in green represent taxonomies. The nodes in gray are datatype properties (e.g., strings). Finally, labels annotating the links between nodes represent object properties. For instance, "located in" is an object property that links a project (a catalogable resource) with a geographical entity (a taxonomic resource coming from a gazetteer).

Participatory Design

It is important that technology for the smart foodshed has the support and interest of the stakeholders who would engage with it. Participatory design is a tradition dating back to the 1970s that could supply strategies for developing appropriate technologies (Bannon, Bardzell, and Bødker 2018). Gooch et al. (2018) discussed challenges and

opportunities for applying participatory design to the smart city. Some of the difficulties here include that of scale: The original participatory design approach featured a limited set of people (workers in a single workplace). How does one extend the approach to include all the people in a city? The approaches that have been tried to date in the smart city movement have their challenges. For instance, the strategy of releasing open data has the difficulty that often these data lack sufficient context and description. Additionally, the smart city agenda is often disconnected from the concerns of socially disadvantaged communities. Gooch et al. (2018) developed a four-stage model for citizen participation in smart city development. The first is engaging with the community to identify its problems. The second and third steps are facilitating the collection of citizens' ideas and encouraging turning these into projects. The final step is ensuring enough investment in projects such that their success is sustainable.

Discussion and Conclusions

Via our steps of appropriate engagement with stakeholders, collection of use cases, development of suitable ontologies to organize information, and attention to participatory design principles, we anticipate developing a set of informatics products in creating a smart foodshed. The following is a possible prioritization for these products.

- Tier 1: Activities to be carried out with readily available data sets and initial convenings of stakeholder groups:
  - Creating maps of food system resources, including cropping patterns, distribution points, processing centers, and preparation facilities enabling spatial identification of gaps in resources.
  - Carrying out participatory data analysis, mapping (geographic and conceptual), and modeling workshops for stakeholders on linkages between food supply, consumption, social determinants of health, and health outcomes, thus furthering information exchange and ties between stakeholders.

- Tier 2: Activities to be performed following the initial phase of stakeholder engagement and data gathering:
  - Creating a database using social network analysis exploring the network of relationships between the actors in the food system, enabling identification of key actors such as particular food distributors.
  - Preparing detailed maps of food system access, opportunities, and health outcomes to identify disparities and inequities to policymakers.

- Tier 3: Activities to be performed later on depending on availability of resources and interest of the community:
  - Developing formal ontologies about food, energy, and water to support smart foodshed applications to navigate the complexities of food system policy, management, and improving transparency of the system, an example being describing food assistance and access programs.
  - Creating estimates of current levels of traceability in sources of key food products identified by stakeholders, such as tracking the flows of specific products through the system across various spatial scales.

Opening up the data ecosystem for food and agriculture will likely become an important avenue to address the challenges of the food system in this century (Allemang and Teegarden 2016; Musker et al. 2018). Yet this is a difficult task for a variety of reasons (Allemang and Teegarden 2016). As shown in Figure 2, the set of stakeholders in the food system is broad, all of whom are potentially producers as well as consumers of data. Commercial enterprises are often concerned that opening their data will lead to competitive disadvantage. Data sharing involves strong relationships of trust between data consumers and data producers, which in turn is dependent in part on maintaining documentation on the data and ensuring its accessibility and availability. In short, trust relationships in data sharing are aided by transparency and traceability of the data. Given the wide variety of stakeholders in the system with greatly differing interests, it is expected that the degree to which data can be shared will range from complete inability to full openness. Even in cases where data cannot be shared, however, actors might be willing to share metadata about data they administer, and sharing metadata still contributes to opening up the data ecosystem.

Initiatives such as the Global Open Data for Agriculture (Musker et al. 2018) and GACS illustrate the momentum building toward open data in the global food system. These initiatives in turn rest on a backbone of common vocabularies such as GACS. From the global scale of these initiatives, though, let us return to the image of the foodshed, with food flowing mostly from rural landscapes to urban centers. Development of the knowledge organization systems backing these initiatives requires that the knowledge engineers creating these systems work closely with stakeholders in all points in the food system, whether producers, participants in the supply chain, or consumers. To generalize, most food consumers live in urban centers, so these knowledge engineers must by necessity engage with stakeholders in urban regions to further development. In short, the smart foodshed perspective embraces both local engagement and striving toward global information standards, and we believe that both are necessary for innovation in the food system moving forward.

Acknowledgments

We thank Jill K. Clark and Brian Estabrook for useful conversations about the direction of this article. Suzanna Lewis provided sage input during the beginning phase of this project. We also thank two anonymous reviewers for constructive criticism and comments on the article.

Funding

This research has been supported by the National Science Foundation (Grant No. 1737573) and the Environmental Protection Agency (Grant No. RD836938).

ORCID

Allan D. Hollander ⓘ http://orcid.org/0000-0002-2647-8235

References

Adolph, S., A. Cockburn, and P. Bramble. 2002. *Patterns for effective use cases.* Boston: Addison-Wesley Longman.

Allemang, D., and J. A. Hendler. 2011. *Semantic web for the working ontologist: Effective modeling in RDFS and OWL.* 2nd ed. Burlington, MA: Elsevier.

Allemang, D., and B. Teegarden. 2016. *A global data ecosystem for agriculture and food.* Global Open Data for Agriculture and Nutrition. Accessed November 28, 2018. https://doi.org/10.7490/f1000research.1114971.1.

Baker, T., C. Caracciolo, A. Doroszenko, and O. Suominen. 2016. GACS core: Creation of a global agricultural concept scheme. In *Metadata and semantics research,* ed. E. Geroufallou, I. S. Coll, A. Stellato, and J. Greenberg, 311–16. Cham, Switzerland: Springer International. https://link.springer.com/chapter/10.1007/978-3-319-49157-8_27.

Bannon, L., J. Bardzell, and S. Bødker. 2018. Introduction: Reimagining participatory design—Emerging voices. *ACM Transactions on Computer–Human Interaction* 25 (1):1–8. doi: 10.1145/3177794.

Berners-Lee, T. 2006. Linked data—Design issues. W3C. Accessed November 26, 2018. https://www.w3.org/DesignIssues/LinkedData.html.

Berners-Lee, T., J. A. Hendler, and O. Lassila. 2001. The semantic web. *Scientific American* 284 (5):34–43. doi: 10.1038/scientificamerican0501-34.

Butler, M. 2013. Analyzing the foodshed: Toward a more comprehensive foodshed analysis. Accessed September 24, 2019. http://pdxscholar.library.pdx.edu/geog_masterpapers/5.

Caracciolo, C., A. Stellato, A. Morshed, G. Johannsen, S. Rajbhandari, Y. Jaques, and J. Keizer. 2013. The AGROVOC linked dataset. *Semantic Web* 4 (3):341–48. https://doi.org/10.3233/SW-130106.

Chiffoleau, Y. 2009. From politics to co-operation: The dynamics of embeddedness in alternative food supply chains. *Sociologia Ruralis* 49 (3):218–35. doi: 10.1111/j.1467-9523.2009.00491.x.

Colloredo-Mansfeld, R., M. Tewari, J. Williams, D. C. Holland, A. Steen, and A.-B. Wilson. 2014. Communities, supermarkets, and local food: Mapping connections and obstacles in food system work in North Carolina. *Human Organization* 73 (3):247–57. doi: 10.17730/humo.73.3.d2n4042613u08581.

De Nicola, A., M. Missikoff, and R. Navigli. 2009. A software engineering approach to ontology building. *Information Systems* 34 (2):258–75. doi: 10.1016/j.is.2008.07.002.

El Bilali, H., and M. S. Allahyari. 2018. Transition towards sustainability in agriculture and food systems: Role of information and communication technologies. *Information Processing in Agriculture* 5 (4):456–64. doi: 10.1016/j.inpa.2018.06.006.

Feenstra, G. W. 1997. Local food systems and sustainable communities. *American Journal of Alternative Agriculture* 12 (1):28–36. doi: 10.1017/S0889189300007165.

Gebbers, R., and V. I. Adamchuk. 2010. Precision agriculture and food security. *Science* 327 (5967):828–31. doi: 10.1126/science.1183899.

Gooch, D., H. Forbes, J. Mackinnon, R. Macpherson, C. Walton, M. Barker, L. Hudson, R. Kelly, G. Kortuem, J. van der Linden et al. 2018. Amplifying quiet voices: Challenges and opportunities for participatory design at an urban scale. *ACM Transactions on Computer–Human Interaction* 25:1–34. doi: 10.1145/3139398.

Graves, M., A. Constabaris, and D. Brickley. 2007. FOAF: Connecting people on the semantic web. *Cataloging & Classification Quarterly* 43 (3–4):191–202. doi: 10.1300/J104v43n03_10.

Heath, T., and C. Bizer. 2011. Linked data: Evolving the web into a global data space. *Synthesis Lectures on the Semantic Web: Theory and Technology* 1 (1):1–136. doi: 10.2200/S00334ED1V01Y201102WBE001.

Hedden, W. P. 1929. *How great cities are fed.* Boston: D. C. Heath.

Hollander, A. D. 2019. PPOD ontology. Accessed September 24, 2019. https://github.com/adhollander/ppod.

Horst, M., and B. Gaolach. 2015. The potential of local food systems in North America: A review of foodshed analyses. *Renewable Agriculture and Food Systems* 30 (5):399–407. doi: 10.1017/S1742170514000271.

Huber, P. R., N. P. Springer, A. D. Hollander, V. R. Haden, S. Brodt, T. P. Tomich, and J. F. Quinn. 2015. Indicators of global sustainable sourcing as a set covering problem: An integrated approach to sustainability. *Ecosystem Health and Sustainability* 1 (2):1–8. doi: 10.1890/EHS14-0008.1.

Kirwan, J., B. Ilbery, D. Maye, and J. Carey. 2013. Grassroots social innovations and food localisation: An investigation of the local food programme in England. *Global Environmental Change* 23 (5):830–37. doi: 10.1016/j.gloenvcha.2012.12.004.

Maye, D. 2018. "Smart food city": Conceptual relations between smart city planning, urban food systems and innovation theory. *City, Culture and Society* 16:18–24. doi: 10.1016/j.ccs.2017.12.001.

Mitchell, S. 2018. VIVO ontology for researcher discovery. Accessed July 9, 2019. https://bioportal.bioontology. org/ontologies/VIVO.

Musker, R., J. Tumeo, B. Schaap, and M. Parr. 2018. *GODAN's impact 2014–2018—Improving agriculture, food and nutrition with open data.* Oxfordshire, UK: Global Open Data for Agriculture and Nutrition. https://doi.org/10.7490/f1000research.1115970.1.

Peters, C. J., N. L. Bills, J. L. Wilkins, and G. W. Fick. 2009. Foodshed analysis and its relevance to sustainability. *Renewable Agriculture and Food Systems* 24 (1):1–7. doi: 10.1017/S1742170508002433.

Reardon, T., R. Echeverria, J. Berdegué, and B. Minten. 2018. Rapid transformation of food systems in developing regions: Highlighting the role of agricultural research & innovations. *Agricultural Systems* 172:47–59. doi: 10.1016/j.agsy.2018.01.022.

Schreiber, G., and Y. Raimond. 2014. RDF 1.1 primer. W3C Working Group Note 24 June. Accessed November 26, 2018. https://www.w3.org/TR/rdf11-primer.

Schwartz, M. W., K. Deiner, T. Forrester, P. Grof-Tisza, M. J. Muir, M. J. Santos, L. E. Souza, M. L. Wilkerson, and M. Zylberberg. 2012. Perspectives on the open standards for the practice of conservation. *Biological Conservation* 155:169–77. doi: 10.1016/j.biocon.2012.06.014.

Springer, N. P., K. Garbach, K. Guillozet, V. R. Haden, P. Hedao, A. D. Hollander, P. R. Huber, C. Ingersoll, M. Langner, G. Lipari et al. 2015. Sustainable sourcing of global agricultural raw materials: Assessing gaps in key impact and vulnerability issues and indicators. *PLoS ONE* 10 (6):e0128752. doi: 10.1371/journal. pone.0128752.

Tomich, T. P., P. Lidder, M. Coley, D. Gollin, R. Meinzen-Dick, P. Webb, and P. Carberry. 2018. Food and agricultural innovation pathways for prosperity. *Agricultural Systems* 172:1–15. doi: 10.1016/j.agsy.2018.01.002.

Tomich, T. P., P. Lidder, J. Dijkman, M. Coley, P. Webb, and M. Gill. 2019. Agri-food systems in international research for development: Ten theses regarding impact pathways, partnerships, program design, and priority-setting for rural prosperity. *Agricultural Systems* 172.101–9. doi: 10.1016/j.agsy.2018.12.004.

Trivette, S. A. 2019. The importance of food retailers: Applying network analysis techniques to the study of local food systems. *Agriculture and Human Values* 36 (1):77–90. doi: 10.1007/s10460-018-9885-1.

Weber, C. L., and H. S. Matthews. 2008. Food-miles and the relative climate impacts of food choices in the United States. *Environmental Science & Technology* 42 (10):3508–13. doi: 10.1021/es702969f.

Zeng, M. L., and P. Mayr. 2018. Knowledge organization systems (KOS) in the semantic web: A multi-dimensional review. *International Journal on Digital Libraries* 20:209–30. doi: 10.1007/s00799-018-0241-2.

Smart Transportation for All? A Typology of Recent U.S. Smart Transportation Projects in Midsized Cities

Scott B. Kelley, Bradley W. Lane, Benjamin W. Stanley, Kevin Kane, Eric Nielsen, and Scotty Strachan

Greater integration of advanced vehicle technologies is commonly discussed as a component of developing smart cities, potentially leading to a host of benefits. Final impacts of such benefits are uncertain, though, given research that illustrates induced travel by initial adopters of emerging vehicle technologies and services and mixed effects in transit use and active transportation. The locations within cities where interventions of advanced vehicle technologies are envisioned, geographic scope and extent of integration, and the characteristics of these areas are all likely to influence these effects, and these relationships have received limited investigative attention. To address this, we conducted a comprehensive review of proposals submitted by 78 midsized cities in the United States to create a typology that considers (1) the geographic scope of intervention and (2) the degree of integration of connected and automated vehicles, generating five distinct types of projects. Characteristics of the areas within cities identified for intervention are compared to those of their U.S. Metropolitan Statistical Area (MSA). We identified indicators of comprehensive planning efforts as they relate to sustainability and resilience outcomes in each city. Results show that areas identified by cities for advanced vehicle technology interventions differ in important ways from each city's broader population that warrant attention relative to known demographic characteristics and behavior of early adopters of transportation technologies. There is also variation in project motivation and municipal planning indicators across typology classifications. These are essential considerations as smart city–aligned transportation interventions continue to develop. *Key Words: automated vehicle, connected vehicle, smart city, typology.*

先进车辆技术的近一步整合，被普遍认为是发展智慧城市的构成要素，并具有潜力产生大量益处。但有鉴于描绘早期采用新兴车辆技术与服务者所引发的旅次之研究，以及在运输使用和主动运输中的混合效应，此般益处的最终影响却是不确定的。城市中先进车辆技术预期介入的地点、整合的地理范围与程度，以及这些区域的特徵，皆有可能影响上述效应，但这些关系却仅获得有限的研究关注。为了应对此一问题，我们对美国七十八座中型城市所提出的计画进行综合性的回顾，以创造能考量以下面向的类型学：(1) 介入的地理范畴，以及 (2) 连结和自动化车辆的整合程度，并生产五大区别的计画类型。我们将城市中指认进行介入的区域特徵和其于美国大都会统计区 (MSA) 中的特徵进行比较。我们指认综合规划的指标，它们关乎每个城市的可持续性和回复力结果。研究结果显示，城市为先进车辆技术介入指认的地区，以重要的方式不同于各自城市的广泛人口，因而需要关注已知人口特徵和运输科技早期采用者的行为。该计画动机和市政规划指标，亦在类型学的区分上有所差别。随着与智慧城市紧密合作的运输介入持续发展，这些皆为关键的考量。关键词: 自动化车辆，连结车辆，智慧城市，类型学。

Una integración más amplia de tecnologías avanzadas para vehículos comúnmente se discute como un componente para el desarrollo de ciudades inteligentes, que potencialmente conducen a una multitud de beneficios. Sin embargo, son inciertos los impactos finales de tales ventajas, dada la investigación que ilustra el viaje inducido por los adoptantes iniciales de tecnologías emergentes y servicios para vehículos, y los efectos mixtos por el uso del tránsito y el transporte activo. Los sitios urbanos donde se prevén intervenciones de tecnologías vehiculares avanzadas, el alcance geográfico y extensión de la integración, y las características de estas áreas, probablemente influirán estos efectos; y estas relaciones han recibido atención investigativa limitada. Para abocar esta situación, adelantamos una exhaustiva revisión de las propuestas presentadas por 78 ciudades de mediano tamaño en los Estados Unidos, con el propósito de crear una tipología que considere (1) el alcance geográfico de la intervención, y (2) el grado de integración de vehículos conectados y automatizados, generando cinco tipos distintos de proyectos. Luego se compararon las características de las áreas dentro de las ciudades seleccionadas para intervención con las de su Área Estadística Metropolitana americana (MSA). Se identificaron indicadores de los esfuerzos de planificación comprensivos en cuanto se relacionan con sustentabilidad y resultados resilientes en cada ciudad. Los descubrimientos muestran que las áreas identificadas por las ciudades para intervenciones avanzadas de tecnología vehicular difieren de modo importante de la población más grande de cada ciudad que justifique una relativa atención a las características demográficas y comportamentales conocidas de adoptantes pioneros de las tecnologías de transporte. Hay también variación en la motivación del proyecto y los indicadores de planificación municipal a través de las clasificaciones de la tipología. Estas son las consideraciones esenciales en la medida en que las intervenciones al transporte de la ciudad inteligente alineada sigan su proceso de desarrollo. *Palabras clave: ciudad inteligente, tipología, vehículo automatizado, vehículo conectado.*

Smart city initiatives are continuing to develop around the world. Generally, these plans aim to develop sociotechnical systems that rely on investments in information and communications technology (ICT), data sharing strategies, and strong collaborations between agencies and companies that aim to improve the quality of life for urban residents (Hollands 2008; Caragliu, Del Bo, and Nijkamp 2011; Damiani, Kowalczyk, and Parr 2017). Beyond investments in technology and infrastructure, the importance of including human and social capital in smart city development is considered essential to effective creation of sustainable and equitable urban environments (Paskaleva 2011; Schuurman et al. 2012; Angelidou 2014). These efforts are evolving in tandem with those of municipal urban planning agencies. In the past twenty years, comprehensive planning has been profoundly influenced by sustainability and resilience philosophies that emphasize a holistic balance among economic, environmental, and social equity concerns. Many municipalities have created city-wide sustainability or resilience plans intended to provide an overarching framework to guide other planning efforts. Because more traditional comprehensive plans already tend to include economic and environmental components, this sea change in municipal planning has explicitly focused on social equity outcomes when devising planning goals—an essential consideration in smart city planning as well. Meanwhile, transportation systems, particularly those framed as "smart mobility," are receiving more attention as a means to address sustainability and resilience-oriented objectives (Ben Letaifa 2015; Benevolo, Dameri, and D'Auria 2016).

This is an important moment, then, to critically examine how cities are planning to deploy and integrate emerging vehicle technologies as part of smart city planning efforts, given the recent developments in connected and/or automated vehicle (C/AV) technologies. At present, personal vehicle travel and its supporting ecosystem have contributed to well-known issues that stand in opposition to sustainable urban development, leading some to call for a dramatic shift in how transportation systems operate, primarily through greater C/AV integration (e.g., Sperling 2018). It is uncertain, though, that such a shift and its associated projected benefits will be realized, in part due to uncertainties in public use and adoption of these technologies. Indeed, some initial studies of early users of ride-hailing services that closely mirror how people would travel via popularly discussed smart mobility vehicle-based modes suggest that users travel more often and substitute trips previously made by transit, bicycling, or walking (Rayle et al. 2016; Clewlow and Mishra 2017).

As cities strive to meet sustainability and resilience goals in this changing landscape that might include the introduction of C/AVs in some way, understanding the spatial characteristics of projects, potential affected populations, and degree of reliance on C/AVs in future plans is important to consider. Our research, then, uses a recent nationwide set of proposals developed by seventy-eight midsized cities that participated in the U.S. Department of Transportation's (USDOT) Smart City Challenge (USDOT 2016). We primarily consider where within cities projects are proposed and how they propose to integrate C/AVs to construct a typology that explicitly considers the following dimensions: (1) the geographic scope and extent of the intervention proposed and (2) the degree of proposed integration of C/AVs. After classifying projects along these two dimensions, we then consider how the following vary across project type: (1) similarity or dissimilarity of neighborhood characteristics relative to the rest of the U.S. Metropolitan Statistical Area (MSA) that relate to known early transportation technology adopter profiles in the literature, (2) evidence of coordinated municipal sustainability and resilience planning in proposals put forward by the seventy-eight midsized cities, and (3) project equity motivations.

This effort helps to provide a comprehensive insight into how midsized cities across the country are considering introducing new vehicle technologies and services that align with smart city objectives, equity concerns, and sustainability and resilience outcomes.

Background

There is growing optimism that integration of disruptive transportation technologies and services will help address a number of issues in the present-day transportation sector, leading to more sustainable and resilient urban futures, particularly in the vision for a future where shared, automated, electric vehicles are central to urban transportation systems (Fulton, Mason, and Meroux 2017; Sperling 2018). Some expect that such a future could enhance mobility and accessibility for many and reduce total urban vehicle travel, reallocate urban space occupied by parking garages and streets, and lower total emissions from the transportation sector (Greenblatt and Shaheen 2015; Zhang et al. 2015). There is uncertainty, however, about the comprehensive nature of projected benefits (Fagnant and Kockelman 2014; Thomopoulos and Givoni 2015; Wadud, MacKenzie, and Leiby 2016), which are contingent on public consideration, adoption, and wide use of these technologies across diverse populations and urban areas (Cohen, Jones, and Cavoli 2017).

There is also extensive research demonstrating that early adopters of new transportation technologies and services are distinct from the general population in ways that are worth examining in this changing landscape. Early adopters of alternative fuel vehicles (AFVs) tend to be predominantly wealthier, well-educated people living in multicar households with longer commutes, more

proenvironmental ideals, and an awareness of how their lifestyle is compatible with changes necessary to adopt an AFV (Sangkapichai and Saphores 2009; Ziegler 2012; Lane et al. 2018). There is more limited understanding of early adopters of C/AVs, because these vehicles and their supporting infrastructure have only begun to be integrated into existing transportation systems. Recent studies, however, have found that stated willingness to adopt and use them tends to be concentrated in the higher income, technologically savvy, male, younger segments of the population who live in denser urban areas (Bansal, Kockelman, and Singh 2016; Hulse, Xie, and Galea 2018). Additionally, initial users of on-demand ride-hailing services—which closely resemble the ways in which people would access future C/AVs on a shared, as-needed basis—are predominantly younger, male, higher educated people living in dense urban areas and, important for longer term sustainability goals, many users substitute trips previously made by transit, biking, or walking while inducing travel (Rayle et al. 2016; Clewlow and Mishra 2017). Taken together, these findings suggest that there are common sociodemographic and socioeconomic characteristics among transportation technology adopters that are essential to consider when recommending new vehicle technologies as a component of smart cities.

At present, though, ride-hailing use nationwide is far from ubiquitous, with much of it concentrated in a few select major urban areas and, important for this research, within certain parts of those cities, leading some to note that introducing new transportation technologies such as C/AVs will face spatial barriers to widespread adoption (Celsor and Millard-Ball 2007; Clewlow 2016; Dias et al. 2017; Litman 2017). This suggests that a geographical assessment of proposed smart city–related transportation interventions is a priority consideration when examining proposed benefits to local populations and municipal sustainability and resilience goals and that attention should be devoted to the kinds of areas being recommended for C/AV integration. Given the rapid market growth in emerging transportation technologies and services, the wide range of uncertain outcomes regarding the introduction of C/AVs and their role in sustainable transportation futures, and the central role that such technologies might play in a number of proposed smart city efforts, this is a crucial area of research focus.

Further confounding the issue is that proposed smart city–related transportation interventions vary in scale, scope, and mode, and there have been only limited efforts to classify them in a generalizable way (Haynes 2018). Typologies have been developed to help clarify municipal operations, governance, citizen services, and scenario planning for developing smart cities (e.g., Batty et al. 2012; Anthopoulos and Fitsilis 2013; Lee and Lee 2014), although these are not focused on the geographic nature of proposed transportation interventions. Nam and Pardo (2011) developed a typology of smart cities along the dimensions of technology, people, and institutions, although transportation is only one component of smart cities considered. These typologies provide a useful framework for constructing one focused on advanced vehicle technologies in smart cities.

Proposal Data

We reviewed all seventy-eight proposals submitted to the USDOT Smart City Challenge in 2015, which included submissions from cities with populations between 200,000 and 850,000 as of the 2010 U.S. Census (USDOT 2016). These data have been made available to the public and represent a comprehensive and recent assessment of how midsized cities across the country are considering transportation-related technological interventions that align with smart city efforts. Each proposal was crafted to align with solicitation requirements that included twelve so-called vision elements, including "connected vehicles" and "urban automation," and cities could elect to address some or all of these elements in their proposals. Although the two aforementioned vision elements certainly helped shape the nature of the submissions as they relate to C/AV integration, cities were free to identify local issues that could be addressed with an intervention, coordinate with municipal planning actors and efforts as they saw fit, and identify locations within their municipal boundaries that warranted an intervention.

Typology Framework

Our typology first considers two primary dimensions: geographic extent and scope of the proposed intervention and degree of integration of C/AVs. This identification of primary themes as the core of the typology is similar in structure to those applied by previous efforts that focused on emerging topics (e.g., Nam and Pardo 2011; Malek, Maine, and McCarthy 2014). The key dimensions of the typology reflect how cities chose to address the following: (1) where within cities transportation-related interventions were planned and (2) how cities considered introducing C/AVs in their proposals. Each of the seventy-eight proposals was then categorized based on its location along these dimensions (Figure 1). For the geographic dimension, we identify the degree to which cities propose to concentrate their interventions in a specific area or disperse them throughout the city. For the vehicle technology dimension, we consider whether or not cities propose to introduce C/AVs into their transportation systems. Using this framework, five distinct classifications of projects emerged: concentrated C/AVs, zones of C/AV integration, mobility hubs, infrastructure first, and transit. The first three include clear plans for C/AV deployment and have a distinct geographic extent and scope of integration, whereas the other two lack one of these criteria.

First, transit projects are those that do not center on the introduction of C/AVs but do identify a clear geographic location for the project. Infrastructure first projects are the reverse case: They are proposed by cities that do not identify a targeted geographic area within the city to include new transportation technologies but instead focus on investment in a distributed, connected, city-wide network of sensors, smart signals, fiber, or other cyberinfrastructure that can eventually support C/AV travel. Generally, these projects are framed as investments in technological capability, and intervention area designations will follow at a later time. We focus next on the three classifications that include C/AV integration and clear geographic specificity.

Concentrated C/AV projects are those that recommend new vehicle technology integration along fixed routes and corridors. Examples included automating existing or planned shuttle or circulator routes, transitioning existing vehicles on dedicated routes into C/AVs, or developing "smart" corridors along clearly identified arterials or highways. Projects with zones of C/AV integration clearly identify an area on a map for C/AV deployment. Within such zones, vehicles would not be restricted only to fixed routes but to flexible travel in the identified area. Proposals in this category often label these zones as demonstration areas or a similar term and favor on-

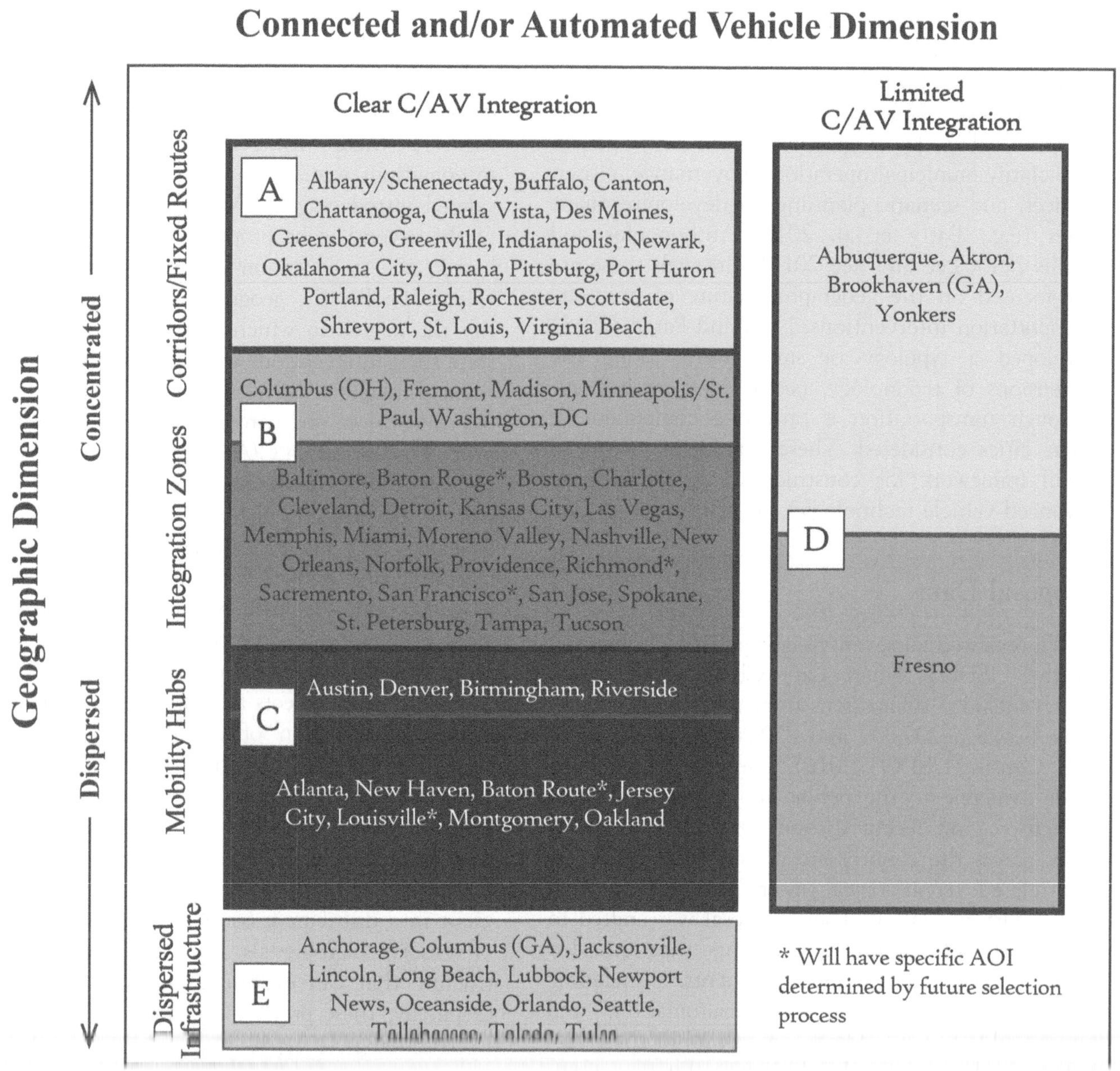

Connected and/or Automated Vehicle Dimension

Figure 1. Typology schematic, including list of participating cities. Five classifications on graphic are labeled as (A) concentrated C/AVs, (B) zones of C/AV integration, (C) mobility hubs, (D) transit, and (E) infrastructure first. AV = automated vehicle; C/AV = connected and/or automated vehicles; AOI = area of intervention.

demand C/AV models. Mobility hub projects, whose name comes from a term referenced in Haynes (2018), feature strategic mobility aggregation points located throughout a city. Once at the strategic points, users can access mobility options that could include immediate or future C/AV integration, sometimes on an on-demand, as-needed basis, offering access to locations throughout the city.

In some cases, there were multiple geographic integration types in a city's proposal. These are identified in the transition zones between classifications in Figure 1, although we group projects according to

their most geographically dispersed project for the purposes of this study.

Alignment with Local Characteristics

Each proposal included an annotated map that identified specific planned project locations within cities, which we use to classify projects on the geographic dimension of our typology, if applicable. We use the term *area of intervention* (AOI) to describe these locations, for the four project classifications that have them. All AOIs were digitized and stored in a geographic information systems environment. Next, we intersected AOIs with a national-level U.S. Census block group polygon spatial data set to measure the socioeconomic and demographic composition of these areas using American Community Survey (ACS) 2012–2016 five-year estimates (U.S. Census Bureau 2017). An example of an AOI is shown in Figure 2. We then compiled job availability data within each AOI using the U.S. Census's Longitudinal Employer–Household Dynamics (LEHD) data for 2015 (U.S. Census Bureau 2018) and collected daytime populations relative to residential population for each AOI. These are all factors identified in previous literature that are known to relate to transportation technology adoption.

All metrics were summarized for each AOI (Table 1). AOIs generally included multiple block groups, with an average of twenty per AOI. All metrics were then compared to those of the MSAs in which the project was located. We independently compared AOI metrics to the outlying remainder of the MSA and then conducted difference-of-proportions tests for each characteristic in Table 1 to determine whether the observed percentage in the AOI significantly differed from that of the MSA.

Alignment with Comprehensive Municipal Planning

We then reviewed publicly available planning documents from the same seventy-eight cities that submitted proposals to the USDOT Smart City Challenge, including comprehensive plans, sustainability and resilience plans, or others with a similar city-wide scope and emphasis on sustainability. We also reviewed language within the institutional Web sites hosting these public documents, usually those of municipal planning departments, mayor's offices, or dedicated sustainability offices. Four key planning elements relevant to the aims of smart city efforts as well as sustainability and resilience planning were considered for each city: (1) the existence of a sustainability or resilience plan, (2) the existence of a municipal office of sustainability, (3) the existence of thematic coordination between comprehensive and sustainability plans, and (4) the ongoing use of data indicators in planning, including baselines, targets, and regular public dissemination. In

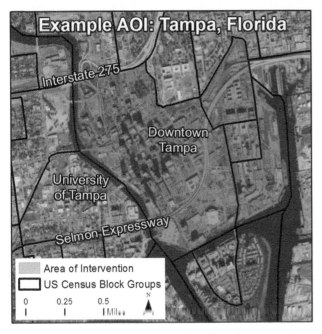

Figure 2. Example AOI designation based on the proposal from Tampa, Florida. AOI = area of intervention.

Table 1. Area of intervention characteristics collected at the U.S. Census block group level

Demographics	Housing	Commuting and vehicle ownership	Employment
% Male	% Owner-occupied units	% Drive alone	% Household income <$50,000/year
% White (non-Hispanic)	% Renter-occupied units	% Carpool	% Household income >$200,000/year
% Age 20–39	% Vacant units	% Transit	% Jobs, workers under age 30
% Age 55+	% Single-family units	% Bike or walk	% Jobs, workers over age 55
% Bachelor's degree or higher	% Multifamily units	% Households with no vehicles	% Jobs with annualized wage <$15,000/year
		% Households with two or more vehicles	% Jobs with annualized wage >$40,000/year

conjunction, we studied the proposals to assess the degree to which city or municipal planning departments were identified as key actors or stakeholders in the submitted proposal, in addition to involvement from metropolitan planning organizations (MPOs) and state transportation agencies. Finally, we identified whether or not a proposal explicitly mentioned providing transportation to a disadvantaged population as a motivating factor for the intervention to assess the relative inclusion of equity. We also considered the extent to which public–private partnerships were involved in the proposal.

Typology Classification Comparisons

Table 2 summarizes proposal-level and AOI characteristics across all five classifications in the typology. Public–private partnerships are most commonly identified in the three project classifications with geographically defined C/AV integration, and these three also have comparatively high rates of MPO involvement in the proposal. Interestingly, as dispersion of project increases, we note a higher percentage of cities that had sustainability and resilience plans at the time of the study. Mobility hub projects most frequently identify providing mobility to a disadvantaged population as a motivating factor, whereas less than half of all other project types do.

Figure 3 independently compares the differences in rates of AOI characteristics relative to that of the MSA by the four typology classifications with a clearly identified geographic intervention area. Generally, residents in AOIs live in areas with comparatively lower rates of driving alone to work and higher rates of

Table 2. Project, planning, and AOI characteristics by typology classification

Factor	Concentrated C/AVs[a]	Zones of C/AV integration[b]	Mobility hubs[c]	Transit[d]	Infrastructure first[e]
Project characteristics					
Public–private	95%	100%	82%	40%	62%
App	100%	100%	100%	100%	85%
Equity motivation	50%	48%	72%	0%	38%
Planning characteristics					
Sustainability or resilience plan	59%	67%	82%	60%	69%
Office of Sustainability or Resilience	68%	89%	64%	20%	77%
Planning indicators	50%	41%	45%	20%	54%
Thematic coordination	27%	48%	36%	40%	15%
Lists MPO	72%	78%	82%	20%	31%
Lists city planning department	18%	30%	18%	40%	15%
AOI metrics					
AOIs	37	55	25	17	—
Block groups	686	1,370	408	207	—
AOI population	954,548	1,744,198	666,231	293,294	—
Downtown AOI	86%	70%	72%	100%	—
University AOI	27%	38%	55%	60%	—

AOI comparison with MSA (%)[f]	<MSA	>MSA	<MSA	>MSA	<MSA	>MSA	<MSA	>MSA	—
Drive alone	70	16	80	11	76	16	71	12	—
Walk or bicycle	8	73	20	67	24	68	24	65	—
Transit use	27	59	18	80	12	76	18	76	—
Two or more vehicles	86	11	87	7	76	12	94	6	—
Age 20–39	22	76	18	69	24	64	6	88	—
Degree (any)	49	46	51	45	44	48	71	29	—
White (non-Hispanic)	78	22	76	22	80	12	65	29	—
Income <$50,000	19	78	15	75	20	72	0	100	—
Jobs >$3,333/month	27	68	29	64	40	48	41	41	—

Notes: C/AVs = connected and/or automated vehicles; MPO = metropolitan planning office; AOI = area of intervention; MSA = Metropolitan Statistical Area.
[a]$n = 22$.
[b]$n = 27$.
[c]$n = 11$.
[d]$n = 5$.
[e]$n = 13$.
[f]Remaining percentages within each classification comprised of AOIs that do not significantly differ from MSA metric.

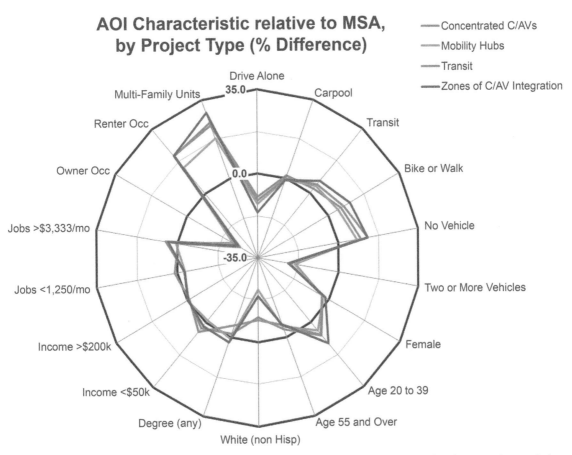

Figure 3. Comparison of AOI characteristic differences relative to MSA value, by project classifications that include geographic specificity. AOI = area of intervention; MSA = Metropolitan Statistical Area; C/AVs = connected and/or automated vehicles.

transit use, bicycling and walking to work, and those who do not own a vehicle. They also are comparatively younger, have higher minority populations, and have higher rates of renter-occupied, multifamily housing units, although these differences are commonly more pronounced in zones of C/AV integration.

Table 2 demonstrates that the majority of AOIs have significantly higher percentages of transit use, walking, and bicycling and lower percentages of driving alone compared to the rest of the MSA, which is essential for considering how C/AVs are deployed in these areas. Concentrated C/AV projects and zones of C/AV integration tended to have AOIs with higher percentages of higher paying jobs compared to the rest of the MSA, which signals an important difference and variation in potential job access compared to mobility hub and transit projects. Meanwhile, relatively concentrated C/AV AOIs tended to have significantly higher proportions of residents making less than $50,000 per year compared to the rest of the MSA, which is notable for considering the mismatch between the employment of local residents and these relatively well-paying jobs. Transit projects had a

noticeably different distribution of degree-holders in their populations compared to the other three groups: The majority of these AOIs had significantly lower proportions of their populations without degrees, whereas this was relatively evenly split across AOIs for the other classifications.

Finally, we note that the relative prevalence of AOIs with daytime ratios higher than 2.0 declines as C/AV interventions become more dispersed in the typology (Figure 4). This suggests that more dispersed projects offer potential access to a greater variety of neighborhoods. The relatively high number of AOIs with high daytime population ratios in more concentrated projects indicates that a greater share of those who could interact with the C/AV intervention includes more than just residents, which is a consideration for how benefits accrue to users of these interventions.

Discussion and Conclusions

The typology constructed indicates that the majority of proposals submitted by these seventy-

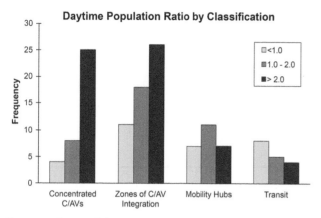

Figure 4. Ratio of daytime area of intervention populations to area of intervention populations, by project classifications that include geographic specificity. C/AVs = connected and/or automated vehicles.

eight midsized cities do recommend introducing C/AVs, although there are differences in geographic concentration. Although we observe variation across typology classifications, AOIs are generally found to be comparatively younger and less wealthy, with a higher share of minority populations, whose residents live in multifamily, renter-occupied units. Residents in these AOIs drive alone to work less often, use transit and active transportation at higher rates than other parts of the city, and own multiple vehicles at lower rates.

These findings carry a number of implications for future smart transportation planning efforts. First, the relative similarity of AOI neighborhood characteristics across proposals is an important finding, because that suggests that these seventy-eight midsized cities are proposing to implement projects in neighborhoods that differ from the rest of the MSA in a consistent manner. Second, given the high focus on C/AV integration, it is worth noting that enhancing access to vehicle travel in these areas that are already less automobile dependent than other areas of the city is unlikely to bring about short-term reductions in vehicle travel in these cities, and therefore in emissions from the transportation sector, without addressing vehicle travel elsewhere in the city. Such counterbalancing efforts do not appear to be directly considered in these projects.

On the other hand, these interventions could allow AOI residents to reach services or higher paying jobs elsewhere through enhanced mobility, and that such access varies by geographic dispersion across the typology warrants attention. Residents in areas that are more reliant on transit and active transportation might be unable to afford personal vehicles at present

and would prefer to travel more by automobile. Therefore, to what extent adoption and use of C/AVs by AOI residents occurs should be closely watched in the short-term future as proposals move to reality.

The observed characteristics of residents in AOIs do align with some known characteristics of adopters of new transportation technologies and services in a manner that does not support near-term optimism for sustainable outcomes but also diverges in important ways. Early adopters are known to be wealthier, less racially diverse, and more educated than the profile of residents targeted by these smart city projects. The income component is particularly logical, because personally owned AFVs and C/AVs require a significant personal investment to obtain. On the other hand, there might be opportunities for AOI residents to access these vehicles in the widely proposed shared vehicle ownership model, although it remains uncertain how pervasive such travel would be. It is also unclear from these proposals to what extent subsidies or assistance would be provided to those who might need it to access C/AVs in this manner.

The predominant integration of C/AVs among proposals signals a possible parallel to the reintegration of rail transit in North American urban areas in the 1980s through the 2000s, when initial lines were built in targeted and visible areas, often for economic redevelopment purposes (Lane 2008). The importance of successful demonstration is also seen from early adopters of AFVs, where prior experience is consistently a primary predictive adoption factor (Lane et al. 2018). Targeting initial deployment to areas of a city where early adopters are most likely to use it aligns with diffusion theory's predictions regarding the dissemination of disruptive technologies (Norton and Bass 1987). It is unclear how this will proceed with C/AV introduction, although visible areas already seem to be popular areas for interventions in midsized cities.

Taken together, this could represent an uncertain inflection point in early transportation technology adopter profiles as proposals move to interventions: Different populations than those in the past might comprehensively adopt these emerging technologies and services, or the residents in these areas might not use them to the degree hoped for a variety of reasons. Cities and the academic research community generally cite more long-term societal benefits as motivations for these efforts, but implementation at present appears to depend on a convergence of societal classes toward common technological interactions and capability, currently built on mobile applications. There might be mixed personal technology access or engagement in the AOIs, though, therefore limiting the use and potential benefits to residents. Purposeful development of institutional intelligence and infrastructure efficiency at scale, with perhaps longer term expectations of individual engagement and evolution of networked inhabitants, will be important for future planning efforts.

Given the relatively high daytime resident ratios in more concentrated projects that recommend introducing C/AVs, there is a risk that benefits might accrue to those who work in or travel to these areas instead of residents. Indeed, we observed that public–private partnerships are more commonly observed in these proposals than the explicit inclusion of equity as a motivating factor, which warrants attention. Municipal planning agencies might be motivated by system optimization, which could be improved by inclusion of the private sector and its innovative environment, but there is a risk of friction between longer term or societal benefits and shorter term data monetization and profit-driven motives when private actors are involved, especially if efforts are not tightly aligned with publicly devised comprehensive planning and sustainability goals.

Finally, there are notable differences in percentages of sustainability plans and offices in cities across typology classifications, along with observed alignment of planning elements by regional and municipal agencies. Smart city proposals infrequently mentioned municipal planning agencies as key actors despite the existence of many preexisting municipal plans establishing separate "smart" planning goals. This indicates that there is an opportunity for comprehensive planning deeply informed by sustainability and resilience principles, typically generated by a representational swath of the local community in a transparent public process, to become more tightly coordinated with future smart transportation plans. This disconnect could limit the future ability of a city to align its suite of efforts with longer term sustainability and resilience planning, should such proposals develop in the future.

There are limitations to this study that should be highlighted. The majority of these areas did not translate their reviewed proposal into construction, at least in the state proposed. Second, there was a limitation placed on city sizes for the competition, notably leaving out very large cities, which might

have introduced C/AVs in notably different ways. Other characteristics of the AOI could also be considered, along with known interactive effects of the independent metrics identified in this study, which should be considered in future modeling studies. Although the intent was to provide an initial indication of project alignment with key residential characteristics, more detailed information on the types of residents and visitors in these areas would be helpful to consider in certain cities. Further, there might be historical and political reasons why certain cities elected certain intervention types and geographic areas that are not stated in the proposal. These factors are not captured in this analysis and would be an important avenue for future research.

References

Angelidou, M. 2014. Smart city policies: A spatial approach. *Cities* 41:S3–S11. doi: 10.1016/j.cities.2014.06.007.

Anthopoulos, L., and P. Fitsilis. 2013. Using classification and roadmapping techniques for smart city viability's realization. *Electronic Journal of e-Government* 11 (2):326–36.

Bansal, P., K. Kockelman, and A. Singh. 2016. Assessing public opinions of and interest in new vehicle technologies: An Austin perspective. *Transportation Research Part C: Emerging Technologies* 67:1–14. doi: 10.1016/j.trc.2016.01.019.

Batty, M., K. W. Axhausen, F. Giannotti, A. Pozdnoukhov, A. Bazzani, M. Wachowicz, G. Ouzounis, and Y. Portugali. 2012. Smart cities of the future. *The European Physical Journal Special Topics* 214 (1):481–518. doi: 10.1140/epjst/e2012-01703-3.

Benevolo, C., R. P. Dameri, and B. D'Auria. 2016. Smart mobility in smart city. In *Empowering organizations*, ed. T. Torre, A. M. Braccini, and R. Spinelli, 13–28. Cham, Switzerland: Springer International.

Ben Letaifa, S. 2015. How to strategize smart cities: Revealing the SMART model. *Journal of Business Research* 68 (7):1414–19. doi: 10.1016/j.jbusres.2015.01.024.

Caragliu, A., C. Del Bo, and P. Nijkamp. 2011. Smart cities in Europe. *Journal of Urban Technology* 18 (2):65–82. doi: 10.1080/10630732.2011.601117.

Celsor, C., and A. Millard-Ball. 2007. Where does carsharing work?: Using geographic information systems to assess market potential. *Transportation Research Record: Journal of the Transportation Research Board* 1992 (1):61–69. doi: 10.3141/1992-08.

Clewlow, R. 2016. Carsharing and sustainable travel behavior: Results from the San Francisco Bay area. *Transport Policy* 51:158–64. doi: 10.1016/j.tranpol.2016.01.013.

Clewlow, R., and G. S. Mishra. 2017. Disruptive transportation: The adoption, utilization, and impacts of ride-hailing in the United States. Report No. UCD-

ITS-RR-117-07, Institute of Transportation Studies, Davis, CA.

Cohen,T., P. Jones, and C. Cavoli. 2017. *Social and behavioural questions associated with automated vehicles.* London: University College London Transport Institute.

Damiani, E., R. Kowalczyk, and G. Parr. 2017. Extending the outreach: From smart cities to connected communities. ACM *Transactions on Internet Technology* 18 (1):1–7. doi: 10.1145/3140543.

Dias, F., P. S. Lavieri, V. M. Garikapati, S. Astroza, R. M. Pendyala, and C. R. Bhat. 2017. A behavioral choice model of the use of car-sharing and ride-sourcing services. *Transportation* 44 (6):1307–23. doi: 10. 1007/s11116-017-9797-8.

Fagnant, D., and K. Kockelman. 2014. The travel and environmental implications of shared autonomous vehicles, using agent-based model scenarios. *Transportation Research Part C: Emerging Technologies* 40:1–13. doi: 10.1016/j.trc.2013.12.001.

Fulton, L., J. Mason, and D. Meroux. 2017. Three revolutions in urban transportation. Accessed September 16, 2018. https://steps.ucdavis.edu/wp-content/uploads/2017/05/STEPS_ITDP-3R-Report-5-10-2017-2.pdf.

Greenblatt, J. B., and S. Shaheen. 2015. Automated vehicles, on-demand mobility, and environmental impacts. *Current Sustainable/Renewable Energy Reports* 2 (3):74–81. doi: 10.1007/s40518-015-0038-5.

Haynes, D. 2018. Shared use mobility, transportation technology, and intercity transit services. Atlanta: Federal Transit Administration, Region IV.

Hollands, R. 2008. Will the real smart city please stand up? *City* 12 (3):303–20. doi: 10.1080/13604810802479126.

Hulse, L. M., H. Xie, and R. Galea. 2018. Perceptions of autonomous vehicles: Relationships with road users, risk, gender and age. *Safety Science* 102:1–13. doi: 10.1016/j.ssci.2017.10.001.

Lane, B. W. 2008. Significant characteristics of the urban rail renaissance in the United States: A discriminant analysis. *Transportation Research Part A: Policy and Practice* 42 (2):279–95. doi: 10.1016/j.tra.2007.10.001.

Lane, B. W., S. Carley, S. Siddiki, J. Dumortier, K. Clark-Sutton, R. M. Krause, and J. D. Graham. 2018. All plug-in electric vehicles are not the same: Predictors of preference for a plug-in hybrid versus a battery-electric vehicle. *Transportation Research Part D: Transport and Environment* 65:1–13. doi: 10.1016/j.trd. 2018.07.019.

Lee, J., and H. Lee. 2014. Developing and validating a citizen-centric typology for smart city services. *Government Information Quarterly* 31:S93–S105. doi: 10.1016/j.giq.2014.01.010.

Litman, T. 2017. Autonomous vehicle implementation predictions: Implications for transport planning. Report of the Victoria Transport Policy Institute, Victoria, BC, Canada.

Malek, K., E. Maine, and I. P. McCarthy. 2014. A typology of clean technology commercialization accelerators. *Journal of Engineering and Technology Management* 32:26–39. doi: 10.1016/j.jengtecman.2013.10.006.

Nam, T., and T. A. Pardo. 2011. Conceptualizing smart city with dimensions of technology, people, and institutions.

In *Proceedings of the 12th Annual International Digital Government Research Conference: Digital Government Innovation in Challenging Times,* 282–91. New York: ACM. http://doi.acm.org/10.1145/2037556.2037602.

Norton, J. A., and F. M. Bass. 1987. A diffusion theory model of adoption and substitution for successive generations of high-technology products. *Management Science* 33 (9):1069–86. doi: 10.1287/mnsc.33.9.1069.

Paskaleva, K. 2011. The smart city: A nexus for open innovation? *Intelligent Buildings International* 3 (3):153–71. doi: 10.1080/17508975.2011.586672.

Rayle, L., D. Dai, N. Chan, R. Cervero, and S. Shaheen. 2016. Just a better taxi? A survey-based comparison of taxis, transit, and ridesourcing services in San Francisco. *Transport Policy* 45:168–78. doi: 10.1016/j. tranpol.2015.10.004.

Sangkapichai, M., and J. D. Saphores. 2009. Why are Californians interested in hybrid cars? *Journal of Environmental Planning and Management* 52 (1):79–96. doi: 10.1080/09640560802504670.

Schuurman, D., B. Baccarne, L. De Marez, and P. Mechant. 2012. Smart ideas for smart cities: Investigating crowdsourcing for generating and selecting ideas for ICT innovation in a city context. *Journal of Theoretical and Applied Electronic Commerce Research* 7 (3):11–62. doi: 10.4067/S0718-18762012000300006.

Sperling, D. 2018. *Three revolutions: Steering automated, shared, and electric vehicles to a better future.* Washington, DC: Island.

Thomopoulos, N., and M. Givoni. 2015. The autonomous car—A blessing or a curse for the future of low carbon mobility? An exploration of likely vs. desirable outcomes. *European Journal of Futures Research* 3 (1):14. doi: 10.1007/s40309-015-0071-z.

U.S. Census Bureau. 2017. 2012–2016 American Community Survey 5-year estimates. Accessed July 1, 2018. https://factfinder.census.gov/faces/nav/jsf/pages/index.xhtml.

U.S. Census Bureau. 2018. Longitudinal Employer Household Dynamics. Accessed November 15, 2018. https://lehd.ces.census.gov/.

U.S. Department of Transportation (USDOT). 2016. Smart City Challenge vision statements. Accessed July 1, 2018. https://www.transportation.gov/smartcity/visionstatements/index.

Wadud, Z., D. MacKenzie, and P. Leiby. 2016. Help or hindrance? The travel, energy and carbon impacts of highly automated vehicles. *Transportation Research Part A: Policy and Practice* 86:1–18. doi: 10.1016/j.tra. 2015.12.001.

Zhang, W., S. Guhathakurta, J. Fang, and G. Zhang. 2015. Exploring the impact of shared autonomous vehicles on urban parking demand: An agent-based simulation approach. *Sustainable Cities and Society* 19:34–45. doi: 10.1016/j.scs.2015.07.006.

Ziegler, A. 2012. Individual characteristics and stated preferences for alternative energy sources and propulsion technologies in vehicles: A discrete choice analysis for Germany. *Transportation Research Part A: Policy and Practice* 46 (8):1372–85. doi: 10.1016/j.tra.2012.05.016.

Challenges and Opportunities for Coping with the Smart Divide in Rural America

Ruopu Li, ⓘ Kang Chen, ⓘ and Di Wu

Recent success in the many applications of information and communications technologies (ICTs) in urban settings largely defines our understanding of sustainable development in the context of smart societies and communities. The footprints of such smart transformation barely imprint in the countryside, however, where rural communities face tremendous barriers, such as the absence of ICT infrastructure, geographic isolation from technological advances, and various social inequalities. The conventional wisdom of smart societies and communities built on well-established broadband access and ICT applications in urban hubs might not be suitable for decentralized rural regions. We contend that a new form of digital divide in the context of recent ICT advancement—a smart divide—is evolving into a new challenge to smart societies and communities. This study aims to characterize the smart divide, an emerging type of social inequality. To address this aim, we conduct a literature review with a focus on the technological aspects of the smart divide in rural regions and identify potential gaps in smart infrastructure and smart applications and services between urban and rural America. The article culminates in a synthesis of potential strategies for bridging the smart divide in rural America. The study is expected to benefit scientists, planners, entrepreneurs, and policymakers interested in smart rural development. *Key Words: broadband, digital divide, ICT, rural, smart divide.*

近年来，信息和通信技术（ICT）在城市环境中的广泛成功应用，在很大程度上决定了我们对智能社会和社区背景中可持续发展的认识。但这种智能化转型几乎没有在农村地区留下任何足迹，因为农村社区面临着巨大的发展障碍，比如缺少 ICT 基础设施、与技术进步的地理隔离以及各种社会不平等现象。一直以来，智能城市和社区的理念都基于完善的宽带接入以及 ICT 在都市中心的应用，这样的概念可能并不适合分散的农村地区。我们认为，近年来的 ICT 发展开始出现一种新型数字鸿沟——智能鸿沟，正在逐步发展成智能社会和社区的新挑战。本研究旨在分析智能鸿沟这种新兴社会不平等现象的特征。为此我们进行了文献回顾，重点关注农村地区智能鸿沟的技术层面，找出美国城市和乡村在智能基础设施以及智能应用和服务之间的潜在差距。本文在结尾还提出了弥合美国农村地区智能鸿沟的潜在综合性策略。该研究尤其适合关注智能农村发展的科学家、规划人士、企业家和决策者。关键词: 宽带，数字鸿沟，ICT，农村，智能鸿沟。

El reciente éxito que registran las numerosas aplicaciones de tecnologías de la información y las comunicaciones (TIC) en escenarios urbanos define en gran medida nuestro entendimiento del desarrollo sustentable en el contexto de sociedades y comunidades inteligentes. Sin embargo, las huellas de tal transformación inteligente escasamente se notan en el entorno rural, donde las comunidades enfrentan tremendas barreras, tales como la ausencia de infraestructura de las TIC, el aislamiento geográfico de los avances tecnológicos y varias desigualdades sociales. La sabiduría convencional de las sociedades y comunidades inteligentes, construida sobre un bien establecido acceso de banda ancha y aplicaciones de las TIC en núcleos urbanos, podría no ser apropiada para las regiones rurales descentralizadas. Nuestra opinión es que está evolucionando una nueva forma de divisoria digital en el contexto del reciente avance de las TIC—una divisoria inteligente—en la forma de un nuevo reto para las sociedades y comunidades inteligentes. Este estudio se propone caracterizar la divisoria inteligente, un tipo emergente de desigualdad social. Para abocar este propósito, llevamos a cabo una revisión de literatura enfocada en los aspectos tecnológicos de la divisoria inteligente en regiones rurales, e identificamos brechas potenciales entre la América urbana y la rural en términos de infraestructura inteligente y aplicaciones y servicios inteligentes. El artículo culmina con una síntesis de estrategias potenciales para tender un puente en la divisoria inteligente de la América rural. Se espera que el estudio sea benéfico para científicos, planificadores, empresarios y legisladores interesados en un desarrollo rural inteligente. *Palabras clave: banda ancha, divisoria digital, divisoria inteligente, rural, TIC.*

With the growing availability of information and communication technology (ICT) infrastructure and affordable smart devices and services, our society and communities have been progressing toward being "smart." Without consensus on the definitions of smart cities and smart communities, the meanings of the term *smart* range from ideological and strategic (e.g., place branding) to practical and instrumented (e.g., city operation) with explicit or implicit intentions to address sustainable development and other pressing problems (Cocchia 2014; Albino, Berardi, and Dangelico 2015; Visvizi and Lytras 2018; Zavratnik, Kos, and Stojmenova Duh 2018). The concepts, however, commonly include technological and social factors (Schaffers et al. 2011; Kitchin 2014; Kitchin, Cardullo, and Feliciantonio 2019). For example, Levy and Wong (2014) defined smartness as the potential for using ICT infrastructure, connected devices, and digital networks to improve people's lives. Smart innovations are expected by urban technocrats to provide convenience in daily life and deliver targeted public services, although concerns have arisen about how the public, instead of corporate interests, benefits from "actually existing smart cities" (Shelton, Zook, and Wiig 2015; Kitchin et al. 2017).

There are still communities and individuals, though, who might largely be isolated from future smart cities and communities (SCCs). For example, the imbalanced capacity and penetration of ICT infrastructure could create a prominent division among regions and communities (Grubesic 2003; Warf 2013; Riddlesden and Singleton 2014). This problem is especially acute in rural America with aging and decentralized populations, low-density user markets, and a general shortage of information and other resources (Malecki 2003). ICT infrastructure resources pivotal to the transformation of SCCs tend to focus on urban centers or are largely developed in urban contexts. Although the development of SCCs is still in its infancy, the potential divide of this smart transformation is likely to widen in the foreseeable future as SCCs mature. In this study, we investigate the dire prospect that the advancement of SCCs could lead to a new form of digital divide among geographic areas and our citizens; that is, a smart divide. A review of the literature is necessary to assess the concept and context of the smart divide in the age of SCCs. In this article, we review the conceptual differences between the smart divide and the digital divide and focus on the gaps in smart infrastructure and smart applications and services (SASs) between urban and rural America, as well as potential strategies for bridging the smart divide. We broadly review multidisciplinary literature in geography, electrical engineering, and rural studies in support of these research perspectives.

This article is structured as follows. In the next section, we introduce the redefined concept of the smart divide in a broader context of SCCs. We then review the technological origins of the smart divide and clarify how it differs from the digital divide. After that, we further elaborate on the vulnerability of rural America to the smart divide and then synthesize strategies that could mitigate the looming divide in rural regions. Finally, we summarize the research findings and envision future directions in this field.

What Is the Smart Divide?

The term *smart divide* was coined by Lee (2016), who defined the term as a qualitative divide determined by the intellectual and skill gap within groups of smart device users. In this context, varying levels of adoption of smart devices contribute to the division among people. A subsequent paper (Nam and Park 2017) used a similar concept and definition with case studies to examine adoption differences among disadvantaged individuals. The authors viewed the smart divide as the latest evolvement from the digital divide with close ties to the digital literacy of people in association with variable income, race, gender, age, and geographic locations (Nam and Park 2017). We found that this definition of the smart divide largely aligns with similar demographic and socioeconomic dimensions of the digital divide; that is, digital inequalities, which have been well elaborated in the literature (Van Deursen and Van Dijk 2014; Shelton, Zook, and Wiig 2015; Friemel 2016; Leszczynski 2016; Kitchin, Cardullo, and Feliciantonio 2019). The rise of SCCs, however, requires the deployment of more advanced, networked, and functional ICT infrastructure, as well as the associated SASs, which are beyond the traditional technological domain of the digital divide. This profound smart ICT advancement could potentially bring the smart divide to our communities. The traditional definition of the digital divide does not fully reflect the technological contexts of progressing ICT in the age of SCCs. Our discussion of the smart divide focuses on its technological dimension, which, we contend, largely defines the smart divide as a new concept emerging from the traditional digital divide.

Departing from Lee (2016), we redefine the term smart divide in the broader context of SCCs. The smart divide is the type of social inequalities that are associated with imbalanced penetration and matureness of smart infrastructure and SASs, as well as the intellectual and skill gap among communities and regions in the age of smart cities and communities. The smart divide can be regarded as a progressive evolution of the digital divide in the age of SCCs. In this article, we name the period preceding SCCs "the digital age" and the period of SCCs "the SCCs age." Such a smart disparity is not a surprise; Shelton, Zook, and Wiig (2015) contended that SCCs are "internally differentiated" and "geographically uneven at a variety of scales" (15). Hosseini et al. (2018) were concerned about the general negligence of smart rural towns in the research on smart cities and smart solutions. Despite a strong research agenda focusing on the critical social dimension of SCCs (Kitchin 2014; Shelton, Zook, and Wiig 2015; Kitchin 2019; Shelton and Lodato 2019), research on the technological foundations of the smart divide and its prospects in rural America is scant.

Technological Foundation of the Smart Divide

SCCs have evolved during the development of mutually reinforcing ICT infrastructure (Damiani, Kowalczyk, and Parr 2017) and SASs (Neirotti et al. 2014; Alaa et al. 2017). The advances in ICT infrastructure enable new SASs to be adopted and applied. Our need for SASs drives further deployment and expansion of ICT infrastructure. Figure 1 shows the overall architecture of smart ICT infrastructure and SASs in SCCs. Although both foundations (particularly the ICT infrastructure) are not brand new in comparison with their counterparts in the digital age, they present new challenges in terms of capability and penetration issues. In the following subsections, we detail the foundations and then present their link to the smart divide.

Cornerstone Smart ICT Infrastructure

The ICT infrastructure translates modern computing and communication technologies into powerful capabilities of collecting, transferring, and processing information related to our lives (termed *data capability* hereafter), which can be further classified into computing, networking, and software infrastructure. Different from the digital age, SCCs substantially raise the scale of the demands on data capability, especially with the emergence of big data (Song, Zhou, and Zhu 2013; Wolfert et al. 2017). As a result, the ICT infrastructure developed decades ago cannot offer the full capability of enabling SCCs. Compared with conventional ICT infrastructure, its SCC-oriented counterpart differs regarding not only the evolution of technologies (e.g., from 4G to 5G) but also the approaches for scaling data capability. Table 1 summarizes the major differences in ICT infrastructure in the digital and SCCs ages in terms of computing, networking, and software infrastructure.

Computing Infrastructure. The computing infrastructure directly supports the collection and storage of data and associated computing. The state-of-the-art computing infrastructure aims to provide high computing capability and ubiquitous availability. Such goals are typically achieved through three layers of infrastructure: a centralized computing infrastructure (CCI), an edge computing infrastructure (ECI), and a device-side computing infrastructure (DCI). The CCI brings a highly powerful computing capability by pooling and virtualizing a large quantity of servers and making them remotely and easily accessible through the Internet. Cloud computing (Fox et al. 2009) is a representative form of the CCI. Because physical distance and network bandwidth often constrain access to the CCI, the ECI arises to push computing close to end users (Reis et al. 2014; Taleb et al. 2017). Finally, the DCI provides on-site data collection and preprocessing capability, which mostly refers to individual computing devices, such as sensors and vehicles.

Compared with ICT infrastructures in the digital age, SCCs have three unique characteristics in the contemporary computing infrastructure. First, the capability of a single central processing unit (CPU) has increased by orders of magnitude (i.e., Moore's Law; Schaller 1997). Second, parallelism and aggregation have been widely adopted to boost data processing beyond the capacities of individual CPUs and hard disks (Kumar 2002; Dean and Ghemawat

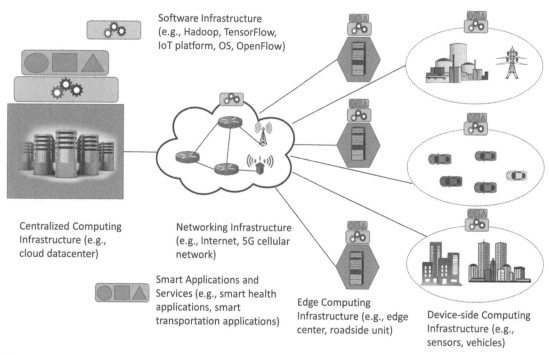

Figure 1. Information and communications technology infrastructure and supporting smart applications and services for smart cities and communities. *Note:* IoT = Internet of Things; OS = operating systems.

Table 1. Major differences in information and communication technology infrastructure between the digital and SCC ages

	Digital age	SCC age
Computing infrastructure	Mainly in the form of either personal computing devices or privately owned computing clusters	Three-layer computing resources that have ubiquitous availability and high-performance capability
Networking infrastructure	The availability mostly applies to developed areas, and the bandwidth is very limited	The availability is much more pervasive, and the bandwidth has evolved into broadband
Software infrastructure	The functioning scope of the software mostly is limited to individual computing and networking equipment	The functioning scope of the software has been substantially expanded in accordance with the development of the scale of computing and networking infrastructures

Note: SCC = smart city and community.

2008). Third, virtualization technologies and networking infrastructure make computing resources readily sharable and accessible and lower the costs of obtaining high-capability computing resources (Joseph et al. 2010).

Networking Infrastructure. The networking infrastructure interconnects end devices with servers as well as storage and computing resources by providing high-bandwidth and universally accessible networking services. This infrastructure consists of a high-throughput, globally deployed wired core network and various access networks (Kurose 2005). Access networks are deployed on the edge of the core network to Internet connectivity to end devices or

servers. The access network might be implemented by different technologies, such as Asymmetric Digital Subscriber Line (Goralski 2000), Fiber to the Home (Koonen 2006), Wi-Fi (Gast 2005), and 4G Long Term Evolution (Dahlman, Parkvall, and Skold 2013).

Core networks and access networks that support SCCs require significant upgrades compared to those established during the digital age. The advancement of technologies has greatly enhanced the bandwidth of the core network (e.g., switch/routing devices and transmitting media) and the access network (e.g., cable-based broadband and 4G Long Term Evolution wireless broadband). Meanwhile, because of continuous public and private investment, the networking infrastructure has been broadly established to provide ubiquitous network services. The bandwidth and accessibility of the networking infrastructure are expected to further coevolve with other new technologies, such as optical fiber and 5G.

Software Infrastructure. The software infrastructure serves as the interface between SASs and computing and networking infrastructures. In the lower layer of the computer system, the software infrastructure provides an interface for managing and accessing raw computing and networking resources, such as operating systems running on servers and sensors and OpenFlow for software-defined networking (Kreutz et al. 2015). In the upper layer of the computer systems, which is built on the lower layer, the software infrastructure anchors frameworks that turn the raw computing or networking resources into the powerful capabilities needed by SASs. Example frameworks include cloud resource management frameworks (e.g., OpenStack; Beloglazov and Buyya 2015), big data cloud frameworks (e.g., Hadoop and Spark; K. Li and Qin 2017), artificial intelligence frameworks (e.g., TensorFlow; Abadi et al. 2016), and Internet of Things (IoT) application development platforms (e.g., Amazon Web Service; Guth et al. 2016).

The evolution of the software infrastructure is often synchronized with that of computing and networking infrastructures. It aims to converge a large-scale computing infrastructure (i.e., CCI, ECI, and DCI) and a network infrastructure (i.e., core and access networks) into easily accessible resources for SASs. Thus, the service scope and capability of the software infrastructure have been substantially expanded in the SCC age. For example, in the lower layer, customized operating systems have been developed to support emerging computing devices, such as autonomous vehicles and specialized sensors. In the upper layer, the software frameworks and systems have been developed to aggregate computing resources and offer powerful data capabilities.

Smart Applications and Services

The high data capability demonstrated by an advanced ICT infrastructure stimulates the development of SASs. Based on the application domains, SASs can be classified into twelve categories (see Lim, Kim, and Maglio 2018) that are integral to the economy and livelihood of SCCs, such as smart transportation (Golub et al. 2018), smart agriculture (Fan 2013), and smart homes (X. Li et al. 2011). For example, smart transportation applications target transportation efficiency and safety through timely scheduling of road traffic and intelligent technologies like autonomous vehicles. Smart agriculture aims to more precisely control the inputs and environmental conditions in various phenology stages of cropping. Compared with the digital age, the SAS age presents two peculiarities. First, to achieve the desired level of smartness, SASs require a higher data capacity for timely collecting, transferring, and processing massive amounts of data on resource consumption and traffic, which requires the development of matching ICT infrastructure. Second, SASs are often achieved through a closed loop involving tremendous distributed end devices for collecting, transferring, and processing the data.

Technological Links to the Smart Divide

A smart ICT infrastructure and SASs, which are largely established in the urban context, could contribute to the smart divide. The disparity in data capability directly affects the varying abilities to develop and offer SASs, leading to the emergence of the smart divide.

ICT Infrastructure for the Smart Divide. Although the digital divide has been largely resolved through continuous investments in ICT infrastructure, the evolution of smart infrastructure and SASs creates new challenges to the accessibility and affordability of SCCs. We elaborate on the smart divide related to ICT infrastructure in terms of investment and maintenance.

First, the smart ICT infrastructure demanded by SCCs requires intensive investment in construction and technology adoption, which creates division among areas with different economic and market capacities. The economic returns for such investment are naturally higher in urban centers with dense user bases. In addition, the deployment of smart infrastructure involves not only the construction investment but also, more critically, the adoption of advanced ICT technologies. For example, 5G cellular communication systems rely on new wireless ICT, such as massive multiple-input and multiple-output and mmWave communications, to provide gigabit wireless broadband capability (Agiwal, Roy, and Saxena 2016). These technologies, such as the mmWave communication technology that limits a single 5G base station's coverage range to only hundreds of meters (Roh et al. 2014), are naturally more suited to urban scenarios.

Second, the maintenance of an advanced ICT infrastructure poses challenges to the long-term economic sustainability of SCCs in remote rural regions with limited economic capacities and a qualified technical workforce. In addition to the initial setup investment, the ICT infrastructure for SCCs involves other operational costs for long-term professional maintenance and periodic replacement (Kozma and Vota 2014). One example is the sensors deployed to support SCCs in the context of the IoT (Saxe 2019). When millions of sensors are deployed, the potential replacement cost is formidable, even for large cities. On this point, a contemporary smart ICT infrastructure is not necessarily sustainable (Lövehagen and Bondesson 2013). Without longer term maintenance, ICT infrastructure tends to be unstable or perform poorly in support of SCCs.

SASs for the Smart Divide. Based on the work by Stratigea (2011), we summarize three SAS-related barriers that potentially contribute to the smart divide. First, the smartness enabled by SASs requires the deployment of high-performance data capability. Thus, although SASs are mostly implemented as scalable software, the availability, affordability, and quality of certain SASs might vary due to regional disparities in the capacity and penetration of the ICT infrastructure. Second, the deployment of SASs, the development of their content, and maintenance of services are discouraged by the limited rural market scale. Some SASs (e.g., ridesharing apps) might be financially unsustainable in the first place from the perspectives of business models. Therefore, SAS providers are incentivized to provide services first in urban centers. Third, the adoption and effective use of certain SASs are deterred by a lack of end users, including the public, businesses, and agencies. The success of SASs relies on appropriate service consumption by public users (for smart transportation), entrepreneurial activities (for virtual meetings), and training opportunities (for smart education).

The Smart Divide for Rural America

The contemporary paradigm of SCCs was largely invented as a strategy in response to the foreseeable intensified urbanization. Smart technologies are designed to address the chronic urban symptom, the overscale problem as urban complexities agglomerate (Batty 2008; Trindade et al. 2017), which strikes a profound contrast with the rurality that commonly suffers from a lack of scale (Naldi et al. 2015) and, in many cases, remoteness (Roberts et al. 2017). The lack of scale problem in rural regions, represented by aging and decentralized populations, low-density user markets, and a general shortage of information and other resources, contradicts the essential socioeconomic contexts of deploying smart infrastructure and SASs and suggests an intrinsic tendency for smart exclusion in the SCC age. Geographic remoteness further complicates the accessibility and provision of smart technologies (Roberts et al. 2017).

Of course, the preceding characterization of rural regions exhibits geographic variability. As noted by Naldi et al. (2015) and Porter et al. (2004), the term *rural* should not be treated as a homogeneous unit; instead, suburban or peri-urban areas do not exhibit the same profiles as remote rural regions but often benefit from "the size advantage and spillover advantages of nearby cities" (Naldi et al. 2015, 95). The proximity to urban core areas, along with lower living costs, allows these areas to gain access to the same or additional technologies, markets, and qualified workforces. Incorporated as a critical metropolitan component, these suburban areas have been found to dominate the urban development in terms of investment (Short, Hanlon, and Vicino 2007). Porter et al. (2004) defined rural regions as a counterpart to metropolitan regions and recognized the internal heterogeneity of rural regions based on their proximity to metropolitan areas. This definition of rural encompasses those micropolitan areas and other remote rural towns and villages. Accordingly, our concerns about the rural smart divide focus on those remote areas, which experience a shortage of scale and geographic links to urban centers. From a technological perspective, these areas face two major challenges for potential transformation into SCCs: (1) fragile capabilities and penetration of existing infrastructure and (2) low capacities for establishing an advanced ICT infrastructure and peripheral SASs.

First, most remote rural regions have only a legacy ICT infrastructure that was developed mainly to address digital coverage. For example, despite recent development efforts to bridge the urban–rural gap in broadband coverage, many rural communities are still left with a limited number of broadband providers according to the Federal Communications Commission (FCC 2018). A similar situation applies to cellular communication networks that support mobile Internet access. More critical, the availability, affordability, and quality of the ICT infrastructure vary widely in rural regions (Warf 2013; Pant and Odame 2017). As the evolution of ICT drives SCCs at an accelerated pace, remote rural regions are expected to be further isolated from SCCs.

Second, remote rural regions generally have a limited economy, market capacity, and workforce to establish, maintain, and update the smart infrastructure, preventing sustainable SCC operations. Thus, it is likely that an advanced ICT infrastructure would be deployed in populated or affluent urban centers first (van Winden and van den Buuse 2017). Early signs have indicated that this divisive trend exists. For example, according to the 2018 Smart Cities Survey (United States Conference of Mayors 2018), 100 percent of large cities and 35 percent of midsized cities have at least one smart city project, whereas only 7.5 percent of small towns have at least one smart city project. Consequently, remote rural regions are not as ready as their urban counterparts to establish, maintain, and update smart infrastructure.

For SASs, remote rural regions are also subject to numerous inherent obstacles when shifting to SCCs. These obstacles could delay the research and development of smart services for these areas and aggravate smart division. For example, a limited rural customer base makes it not financially viable to deploy ride-sharing apps and improve availability (e.g., incentives to recruit ride-sharing drivers), affordability (e.g., attractive service rate), and quality (e.g., optimization of vehicle scheduling). Numerous studies have indicated that rural populations are not as active as urban populations in terms of adopting SASs in daily life (Gilbert, Karahalios, and Sandvig 2008; Hecht and Stephens 2014; Johnson et al. 2016). Another important obstacle lies in potential infrastructure capacity and penetration issues. Although the establishment of ICT infrastructure to solve the digital divide has been successful (e.g., the broadband coverage ratio has surpassed 95 percent; Brogan 2018), the capability level and penetration rate often vary dramatically between urban and rural regions, as witnessed by highly variable broadband providers and speeds in different regions (FCC 2018). For example, real-time video-based remote medical services are easier to deploy in cities than in rural regions due to the disparate conditions of ICT infrastructures (Khan and Driessen 2018).

Strategies for Coping with the Rural Smart Divide

Although the smart divide has not yet materialized, it is helpful to extend the discussion to potential coping strategies. As argued, ICT-based smart solutions formulated in the context of smart cities do not necessarily fit the rural reality. Thus, we discuss only potential strategies that are relevant for technology and rural regions.

Enhancing the ICT Infrastructure for Rural Contexts

Considering the critical roles of smart infrastructure in enabling SCCs, enhancing the ICT infrastructures in rural contexts can effectively address the smart divide. Specific ICT enhancement strategies can be developed in multiple ways. First, the ICT could be innovated to match particular rural contexts. Although seemingly technocentric and ideal, there have been emerging efforts in developing low-cost but efficient smart solutions, such as the IoT and cloud computing, to achieve the goal of ubiquitous accessibility of data capabilities in support of SASs. There are also countless emerging efforts to promote affordable wireless broadband service in rural regions through hot air balloons, drones, and low north orbit satellites (see del Portillo, Cameron, and Crawley 2019). The emerging network virtualization technologies are expected to reduce the operational costs of network operation by 40 percent to 50 percent (Moore 2017). Second, enhancing the data capacity of the network infrastructure in remote rural communities is worthwhile. As rural sparseness continues to haunt rural smart development, it is important to bridge the geographic remoteness to improve broadband connections (Naldi et al. 2015). The full spectrum of socioeconomic benefits from broadband access has been documented by Prieger (2013) and Pant and Odame (2017). In future SCCs, broadband service is expected to connect local needs virtually via computing and software resources accessible over the Internet, which partially alleviates the need to establish smart computing and software infrastructures in rural regions.

Third, policies and techniques could be developed to encourage better utilization of existing resources by promoting ICT resource sharing in rural regions. The FCC's recent release of the 3.65 GHz Citizens Broadband Radio Service spectrum could greatly lower the cost of providing broadband service in rural regions (see CairoNet 2017; FCC 2019). The development of a volunteer cloud (Mengistu et al. 2018) and spectrum-sharing technologies (Kang, Liang, and Yang 2018) would allow more efficient multiplexing of computing and wireless networking resources to support SCCs.

Developing SASs to Meet Rural Needs

Developing SASs suitable for remote rural regions is another important direction to constrain the smart divide, envisioning no quick solution to address a long-lasting infrastructure gap between rural and urban areas. SASs that do not require a high-performance networking and computing infrastructure can be promoted in these areas. There have been emerging efforts in developing SASs, which satisfy practical needs but are compatible with low-bandwidth wireless or wired networks and intermittent connectivity in remote rural regions. Typical applications include last-mile transportation scheduling and dispatch, intelligent farming, and smart education. Furthermore, despite the envisioned division in accessibility to SASs, SASs designed for rural consumers suggest new opportunities. For example, smart health technologies, such as telemedicine and remote monitoring, will enable patients living in rural regions to gain access to quality health care and effective treatment (Deloitte Luxembourg 2018). It is foreseeable that such SASs are more likely to be established before smart ICT infrastructure in rural areas.

Improving Smart Planning and Development

In addition to technological perspectives, solutions for the smart divide should be anchored in a holistic, integrated (instead of piecemeal) rural development plan (Stratigea 2011) that focuses on strengthening intrarural and rural–urban links for positive synergies. First, considering the imbalanced dependencies between urban and rural areas regarding access to resources (Naldi et al. 2015), it is important to enhance the virtual connectivity of rural regions to metropolitan areas (Stauber 2001;

Porter et al. 2004; Roberts et al. 2017). We envision the ideal mode of future SCCs to be an urban–rural continuum instead of a binary dichotomy. Second, rural regions with the potential for matching ICT resources and sharing markets should be connected into competitive conglomerates as cross-sectoral networks, regional clusters, or other coordinated partners in an integrated development framework. To enable internal connectivity, physical and virtual connections could be innovatively strengthened among towns and villages in the same regions (van Gevelt et al. 2018). Third, new initiatives for smart rurality could be developed and promoted as demonstrative projects to break cultural lock-in and facilitate the introduction of innovation. A plethora of literature shows that place-based, rather than top-down prescribed, innovation is vital to the development of rural smart towns (Porter et al. 2004; Naldi et al. 2015; Roberts et al. 2017, Hosseini et al. 2018; Zavratnik, Kos, and Stojmenova Duh 2018), given the heterogeneous resources and amenities of rural regions. A new initiative, Smart Villages, was recently proposed by the European Union as a catalyst for innovating rural services in support of smart rural development (European Network for Rural Development 2018; van Gevelt et al. 2018; Zavratnik, Kos, and Stojmenova Duh 2018). Such an effort seems to be gaining momentum worldwide (Holmes, Jones, and Heap 2015; Visvizi and Lytras 2018). Similar initiatives seem to be largely absent on paths to smart rural America, however.

Conclusions

This research concerns the existing digital divide and inequalities in our disadvantaged rural communities that might evolve into a new form of divide, the smart divide, as ICTs continue to progress, which will affect the sustainable and smart growth of our cities and communities. This article characterizes the smart divide, an emerging sociotechnical concept in SCCs, by focusing on its technological dimension and urban–rural differences. We discuss the implications of the ICT infrastructure and SASs for the smart divide in the context of rural America. To cope with the impending smart divide, we propose technological and planning strategies. With this study, we aim to inform government decision makers and other stakeholders to avoid pitfalls similar to the historic digital divide as we progress toward

SCCs. Although we focus on the smart divide occurring in rural America, we argue that the concept and implications of the smart divide extend beyond the urban–rural geographic divide and possibly to disadvantaged and socially segregated urban neighborhoods (which should be investigated further).

Although we focus on the technological dimension of the smart divide, its human dimension is worthy of further research. The technological dimension of the smart divide represents the imbalanced availability, affordability, and quality of smart infrastructure and SASs, and the human dimension is concerned with the unequal access to smart infrastructure and SASs partially arising from existing social inequalities. As argued by Hindman (2000), creating the ICT infrastructure does not automatically erase the economic disadvantages of certain citizens. The expansion of smart infrastructure and SASs could exacerbate the accessibility to smart technologies for certain groups of individuals, communities, and regions. Despite the seemingly evolutional successive link to the digital divide, researchers might further examine whether the smart divide has the same social characteristics as the digital divide. A plethora of recent critical studies (Kitchin et al. 2017; Kitchin 2019; Shelton and Lodato 2019) have suggested that the complexity of human dimension factors associated with the smart divide warrants more research.

Acknowledgments

We thank the editor and two anonymous reviewers for constructive comments and suggestions on the structure and various details of this article.

ORCID

Ruopu Li http://orcid.org/0000-0003-3500-0273
Kang Chen https://orcid.org/0000-0002-2376-2898

References

Abadi, M., P. Barham, J. Chen, Z. Chen, A. Davis, J. Dean, M. Devin, S. Ghemawat, G. Irving, M. Isard, et al. 2016. Tensorflow: A system for large-scale machine learning. In *12th Symposium on Operating Systems Design and Implementation (OSDI)*, 265–83. Savannah, GA: USENIX.

Agiwal, M., A. Roy, and N. Saxena. 2016. Next generation 5G wireless networks: A comprehensive survey. *IEEE Communications Surveys & Tutorials* 18 (3):1617–55.

Alaa, M., A. A. Zaidan, B. B. Zaidan, M. Talal, and M. L. M. Kiah. 2017. A review of smart home applications based on Internet of Things. *Journal of Network and Computer Applications* 97:48–65. doi: 10.1016/j.jnca.2017.08.017.

Albino, V., U. Berardi, and R. M. Dangelico. 2015. Smart cities: Definitions, dimensions, performance, and initiatives. *Journal of Urban Technology* 22 (1):3–21. doi: 10.1080/10630732.2014.942092.

Batty, M. 2008. The size, scale, and shape of cities. *Science* 319 (5864):769–71. doi: 10.1126/science.1151419.

Beloglazov, A., and R. Buyya. 2015. OpenStack Neat: A framework for dynamic and energy-efficient consolidation of virtual machines in OpenStack clouds. *Concurrency and Computation: Practice and Experience* 27 (5):1310–33. doi: 10.1002/cpe.3314.

Brogan, P. 2018. U.S. broadband availability year-end 2016. Accessed December 20, 2018. https://www.ustelecom.org/sites/default/files/USTelecom%20Research%20Brief%202.22.18.pdf.

CairoNet. 2017. Home page. Accessed September 3, 2019. http://caironet.net/.

Cocchia, A. 2014. Smart and digital city: A systematic literature review. In *Smart city*, ed. R. P. Dameri and C. Rosenthal-Sabroux, 13–43. Cham, Switzerland: Springer International. Accessed April 29, 2019. http://link.springer.com/10.1007/978-3-319-06160-3_2.

Dahlman, E., S. Parkvall, and J. Skold. 2013. *4G: LTE/LTE-Advanced for Mobile Broadband*. Cambridge, MA: Academic.

Damiani, E., R. Kowalczyk, and G. Parr. 2017. Extending the outreach: From smart cities to connected communities. *ACM Transactions on Internet Technology* 18 (1):1–7. doi: 10.1145/3140543.

Dean, J., and S. Ghemawat. 2008. MapReduce: Simplified data processing on large clusters. *Communications of the ACM* 51 (1):107–13. doi: 10.1145/1327452.1327492.

Deloitte Luxembourg. 2018. A journey towards smart health. Accessed April 28, 2019. https://www2.deloitte.com/lu/en/pages/life-sciences-and-healthcare/articles/journey-towards-smart-health.html.

del Portillo, I., B. G. Cameron, and E. F. Crawley. 2019. A technical comparison of three low earth orbit satellite constellation systems to provide global broadband. *Acta Astronautica* 159:123–35. doi: 10.1016/j.actaastro.2019.03.040.

European Network for Rural Development. 2018. Smart villages: Revitalising rural services. *EU Rural Review*. Accessed July 9, 2019. https://enrd.ec.europa.eu/sites/enrd/files/enrd_publications/publi-enrd-rr-26-2018-en.pdf.

Fan, T. 2013. Smart agriculture based on cloud computing and IOT. *Journal of Convergence Information Technology* 8 (2):210–16. doi: 10.4156/jcit.vol8.issue2.26.

Federal Communications Commission. 2018. Fixed broadband deployment. Accessed December 10, 2018. https://broadbandmap.fcc.gov/#/.

Federal Communications Commission. 2019. 3650-3700 MHz radio service. Accessed September 9, 2019. https://www.fcc.gov/wireless/bureau-divisions/broadband-division/3650-3700-mhz-radio-service.

Fox, A., R. Griffith, A. Joseph, R. Katz, A. Konwinski, G. Lee, D. Patterson, A. Rabkin, and I. Stoica. 2009. Above the clouds: A Berkeley view of cloud computing. Report No. UCB/EECS 28, No. 13, Department of Electrical Engineering and Computer Sciences, University of California, Berkeley.

Friemel, T. N. 2016. The digital divide has grown old: Determinants of a digital divide among seniors. New Media & Society 18 (2):313–31. doi: 10.1177/1461444814538648.

Gast, M. 2005. 802.11 wireless networks: The definitive guide. Newton, MA: O'Reilly.

Gilbert, E., K. Karahalios, and C. Sandvig. 2008. The network in the garden: An empirical analysis of social media in rural life. In Proceedings of the Twenty-Sixth Annual CHI Conference on Human Factors in Computing Systems—CHI '08. Florence, Italy: ACM Press. Accessed June 17, 2019. http://portal.acm.org/citation.cfm?doid=1357054.1357304.

Golub, A., M. Serritella, V. Satterfield, and J. Singh. 2018. Community-based assessment of smart transportation needs in the city of Portland. NITC-RR1163, Transportation Research and Education Center (TREC), Portland, OR.

Goralski, W. J. 2000. ADSL and DSL technologies. New York: McGraw-Hill, Inc.

Grubesic, T. H. 2003. Inequities in the broadband revolution. The Annals of Regional Science 37 (2):263–89. doi: 10.1007/s001680300123.

Guth, J., U. Breitenbücher, M. Falkenthal, F. Leymann, and L. Reinfurt. 2016. Comparison of IoT platform architectures: A field study based on a reference architecture. In 2016 Cloudification of the Internet of Things (CIoT), 1–6. Paris: IEEE.

Hecht, B., and M. Stephens. 2014. A tale of cities: Urban biases in volunteered geographic information. In Proceedings of the Eighth International AAAI Conference on Weblogs and Social Media, 197–205. Palo Alto, CA: The AAAI Press.

Hindman, D. B. 2000. The rural–urban digital divide. Journalism & Mass Communication Quarterly 77 (3):549–60. doi: 10.1177/107769900007700306.

Holmes, J., B. Jones, and D. Heap. 2015. Smart villages. Science 350 (6259):359. doi: 10.1126/science.aad6521.

Hosseini, S., L. Frank, G. Fridgen, and S. Heger. 2018. Do not forget about smart towns: How to bring customized digital innovation to rural areas. Business & Information Systems Engineering 60 (3):243–57. doi: 10.1007/s12599-018-0536-2.

Johnson, I. L., Y. Lin, T. J.-J. Li, A. Hall, A. Halfaker, J. Schöning, and B. Hecht. 2016. Not at home on the range: Peer production and the urban/rural divide. In Proceedings of the 2016 CHI Conference on Human Factors in Computing Systems, 13–25. Santa Clara, CA: ACM Press. Accessed June 17, 2019. http://dl.acm.org/citation.cfm?doid=2858036.2858123.

Joseph, A. D., R. Katz, A. Konwinski, L. E. E. Gunho, D. Patterson, and A. Rabkin. 2010. A view of cloud computing. Communications of the ACM 53 (4):50–58. doi: 10.1145/1721654.1721672.

Kang, X., Y. Liang, and J. Yang. 2018. Riding on the primary: A new spectrum sharing paradigm for wireless-powered IoT devices. IEEE Transactions on Wireless Communications 17 (9):6335–47. doi: 10.1109/TWC.2018.2859389.

Khan, F., and J. Driessen. 2018. Bridging the telemedicine infrastructure gap: Implications for long-term care in rural America. Public Policy & Aging Report 28 (3):80–84. doi: 10.1093/ppar/pry027.

Kitchin, R. 2014. The real-time city? Big data and smart urbanism. GeoJournal 79 (1):1–14. doi: 10.1007/s10708-013-9516-8.

Kitchin, R. 2019. The timescape of smart cities. Annals of the American Association of Geographers 109 (3):775–90. doi: 10.1080/24694452.2018.1497475.

Kitchin, R., P. Cardullo, and C. D. Feliciantonio. 2019. Citizenship, justice and the right to the smart city. In The right to the smart city, ed. P. Cardullo, C. D. Feliciantonio, and R. Kitchin, 1–28. Bingley, UK: Emerald.

Kitchin, R., C. Coletta, L. Evans, L. Heaphy, and D. MacDonncha. 2017. Smart cities, epistemic communities, advocacy coalitions and the "last mile" problem. IT - Information Technology 59 (6):275–84. doi: 10.1515/itit-2017-0004.

Koonen, T. 2006. Fiber to the home/fiber to the premises: What, where, and when? Proceedings of the IEEE 94 (5):911–34.

Kozma, R. B., and W. S. Vota. 2014. ICT in developing countries: Policies, implementation, and impact. In Handbook of research on educational communications and technology, ed. M. Spector, M. D. Merrill, J. Elen, and M. J. Bishop, 885–94. New York: Springer.

Kreutz, D., F. M. V. Ramos, P. Verissimo, C. E. Rothenberg, S. Azodolmolky, and S. Uhlig. 2015. Software-defined networking: A comprehensive survey. Proceedings of the IEEE 103 (1):14–76. doi: 10.1109/JPROC.2014.2371999.

Kumar, V. 2002. Introduction to parallel computing. Boston: Addison-Wesley Longman.

Kurose, J. F. 2005. Computer networking: A top-down approach featuring the Internet. 3rd ed. Delhi: Pearson Education.

Lee, S. 2016. Smart divide: Paradigm shift in digital divide in South Korea. Journal of Librarianship and Information Science 48 (3):260–68. doi: 10.1177/0961000614558079.

Leszczynski, A. 2016. Speculative futures: Cities, data, and governance beyond smart urbanism. Environment and Planning A: Economy and Space 48 (9):1691–1708. doi: 10.1177/0308518X16651445.

Levy, C., and D. Wong. 2014. How smarter healthcare and smarter cities are changing our world. Accessed December 23, 2018. https://www.techradar.com/news/world-of-tech/future-tech/how-smarter-healthcare-and-smarter-cities-are-changing-our-world-1256940.

Li, K., and X. Qin. 2017. One platform rules all: From Hadoop 1.0 to Hadoop 2.0 and Spark. In Big data management and processing, ed. K. Li, H. Jiang, and A. Y. Zomaya, 191–214. New York: Chapman and Hall/CRC.

Li, X., R. Lu, X. Liang, X. Shen, J. Chen, and X. Lin. 2011. Smart community: An Internet of things application. IEEE Communications Magazine 49 (11):68–75.

Lim, C., K. J. Kim, and P. P. Maglio. 2018. Smart cities with big data: Reference models, challenges, and considerations. *Cities* 82:86–99. doi: 10.1016/j.cities.2018.04.011.

Lövehagen, N., and A. Bondesson. 2013. Evaluating sustainability of using ICT solutions in smart cities—Methodology requirements. In *International Conference on Information and Communication Technologies for Sustainability*, 175–82. Zürich: ETH E-Collection.

Malecki, E. J. 2003. Digital development in rural areas: Potentials and pitfalls. *Journal of Rural Studies* 19 (2):201–14. doi: 10.1016/S0743-0167(02)00068-2.

Mengistu, T. M., A. M. Alahmadi, Y. Alsenani, A. Albuali, and D. Che. 2018. Cucloud: Volunteer computing as a service (VCAAS) system. In *International Conference on Cloud Computing*, ed. M. Luo and L. Zhang, 251–64. New York: Springer.

Moore, G. 2017. NFV and SDN: The future of service delivery. Accessed December 5, 2018. http://www.telco-professionals.com/blogs/35443/1187/nfv-and-sdn-the-future-of-service-delivery.

Naldi, L., P. Nilsson, H. Westlund, and S. Wixe. 2015. What is smart rural development? *Journal of Rural Studies* 40:90–101. doi: 10.1016/j.jrurstud.2015.06.006.

Nam, S. J., and E. Y. Park. 2017. The effects of the smart environment on the information divide experienced by people with disabilities. *Disability and Health Journal* 10 (2):257–63. doi: 10.1016/j.dhjo.2016.11.001.

Neirotti, P., A. De Marco, A. C. Cagliano, G. Mangano, and F. Scorrano. 2014. Current trends in smart city initiatives: Some stylised facts. *Cities* 38:25–36. doi: 10.1016/j.cities.2013.12.010.

Pant, L. P., and H. H. Odame. 2017. Broadband for a sustainable digital future of rural communities: A reflexive interactive assessment. *Journal of Rural Studies* 54:435–50. doi: 10.1016/j.jrurstud.2016.09.003.

Porter, M. E., C. H. M. Ketels, K. Miller, and R. Bryden. 2004. *Competitiveness in rural U.S. regions: Learning and research agenda.* Washington, DC: U.S. Economic Development Administration.

Prieger, J. E. 2013. The broadband digital divide and the economic benefits of mobile broadband for rural areas. *Telecommunications Policy* 37 (6–7):483–502. doi: 10.1016/j.telpol.2012.11.003.

Reis, A. B., S. Sargento, F. Neves, and O. K. Tonguz. 2014. Deploying roadside units in sparse vehicular networks: What really works and what does not. *IEEE Transactions on Vehicular Technology* 63 (6):2794–2806. doi: 10.1109/TVT.2013.2292519.

Riddlesden, D., and A. D. Singleton. 2014. Broadband speed equity: A new digital divide? *Applied Geography* 52:25–33. doi: 10.1016/j.apgeog.2014.04.008.

Roberts, E., B. A. Anderson, S. Skerratt, and J. Farrington. 2017. A review of the rural-digital policy agenda from a community resilience perspective. *Journal of Rural Studies* 54:372–85. doi: 10.1016/j.jrurstud.2016.03.001.

Roh, W., J.-Y. Seol, J. Park, B. Lee, J. Lee, Y. Kim, J. Cho, K. Cheun, and F. Aryanfar. 2014. Millimeter-wave beamforming as an enabling technology for 5G cellular communications: Theoretical feasibility and prototype results. *IEEE Communications Magazine* 52 (2):106–13. doi: 10.1109/MCOM.2014.6736750.

Saxe, S. 2019. I'm an engineer, and I'm not buying into "smart" cities. Accessed July 30, 2019. https://www.nytimes.com/2019/07/16/opinion/smart-cities.html.

Schaffers, H., N. Komninos, M. Pallot, B. Trousse, M. Nilsson, and A. Oliveira. 2011. Smart cities and the future Internet: Towards cooperation frameworks for open innovation. *Future Internet Assembly, LNCS* 6656:431–46.

Schaller, R. R. 1997. Moore's law: Past, present and future. *IEEE Spectrum* 34 (6):52–59. doi: 10.1109/6.591665.

Shelton, T., and T. Lodato. 2019. Actually existing smart citizens: Expertise and (non)participation in the making of the smart city. *City* 23 (1):35–52. doi: 10.1080/13604813.2019.1575115.

Shelton, T., M. Zook, and A. Wiig. 2015. The "actually existing smart city." *Cambridge Journal of Regions, Economy and Society* 8 (1):13–25. doi: 10.1093/cjres/rsu026.

Short, J. R., B. Hanlon, and T. J. Vicino. 2007. The decline of inner suburbs: The new suburban gothic in the United States. *Geography Compass* 1 (3):641–56. doi: 10.1111/j.1749-8198.2007.00020.x.

Song, Y., G. Zhou, and Y. Zhu. 2013. Present status and challenges of big data processing in smart grid. *Power System Technology* 37 (4):927–35.

Stauber, K. 2001. Why invest in rural America—And how? A critical public policy question for the 21st century. In *Center for the study of rural America, exploring policy options for a new rural America*, ed. M. Drabenstott and K. Sheaff, 57–87. Kansas City, MO: Federal Reserve Bank of Kansas City.

Stratigea, A. 2011. ICT for rural development: Potential applications and barriers involved. *Netcom* 25 (3–4):179–204. doi: 10.4000/netcom.144.

Taleb, T., K. Samdanis, B. Mada, H. Flinck, S. Dutta, and D. Sabella. 2017. On multi-access edge computing: A survey of the emerging 5G network edge cloud architecture and orchestration. *IEEE Communications Surveys & Tutorials* 19 (3):1657–81. doi: 10.1109/COMST.2017.2705720.

Trindade, E. P., M. P. F. Hinnig, E. M. da Costa, J. S. Marques, R. C. Bastos, and T. Yigitcanlar. 2017. Sustainable development of smart cities: A systematic review of the literature. *Journal of Open Innovation: Technology, Market, and Complexity* 3 (11):1–14. doi: 10.1186/s40852-017-0063-2.

United States Conference of Mayors. 2018. 2018 smart cities survey. Accessed December 22, 2018. http://www.usmayors.org/wp-content/uploads/2018/06/2018-Smart-Cities-Report.pdf.

Van Deursen, A. J., and J. A. Van Dijk. 2014. The digital divide shifts to differences in usage. *New Media & Society* 16 (3):507–26. doi: 10.1177/1461444813487959.

van Gevelt, T., C. Canales Holzeis, S. Fennell, B. Heap, J. Holmes, M. H. Depret, B. Jones, and M. T. Safdar. 2018. Achieving universal energy access and rural development through smart villages. *Energy for Sustainable Development* 43:139–42. doi: 10.1016/j.esd.2018.01.005.

van Winden, W., and D. van den Buuse. 2017. Smart city pilot projects: Exploring the dimensions and conditions of scaling up. *Journal of Urban Technology* 24 (4):51–72.

Visvizi, A., and M. Lytras. 2018. It's not a fad: Smart cities and smart villages research in European and global contexts. *Sustainability* 10 (8):2727. doi: 10.3390/su10082727.

Warf, B. 2013. Contemporary digital divides in the United States. *Tijdschrift voor Economische en Sociale Geografie* 104 (1):1–17. doi: 10.1111/j.1467-9663.2012.00720.x.

Wolfert, S., L. Ge, C. Verdouw, and M. Bogaardt. 2017. Big data in smart farming—A review. *Agricultural Systems* 153:69–80. doi: 10.1016/j.agsy.2017.01.023.

Zavratnik, V., A. Kos, and E. Stojmenova Duh. 2018. Smart villages: Comprehensive review of initiatives and practices. *Sustainability* 10 (7):2559. doi: 10.3390/su10072559.

Index